1st International Symposium on Electrical Arc and Thermal Plasmas in Africa

(ISAPA 2011)

IOP Conference Series: Materials Science and Engineering
Volume 29

Ouagadougou, Burkina Faso.
17 - 22 October 2011

ISBN: 978-1-63439-840-4
ISSN: 1757-8981

1st International Symposium on Electrical Arc and Thermal Plasmas in Africa

(ISAPA 2011)

IOP Conference Series: Materials Science and Engineering
Volume 29

Ouagadougou, Burkina Faso
17 - 22 October 2011

ISBN: 978-1-63439-840-4
ISSN: 1757-8981

IOP Institute of Physics

Printed from e-media with permission by:

Curran Associates, Inc.
57 Morehouse Lane
Red Hook, NY 12571

Some format issues inherent in the e-media version may also appear in this print version.

Printed by Curran Associates, Inc. (2015)

For permission requests, please contact the Institute of Physics
at the address below.

Institute of Physics
Dirac House, Temple Back
Bristol BS1 6BE UK

Phone: 44 1 17 929 7481
Fax: 44 1 17 920 0979

techtracking@iop.org

Additional copies of this publication are available from:

Curran Associates, Inc.
57 Morehouse Lane
Red Hook, NY 12571 USA
Phone: 845-758-0400
Fax: 845-758-2634
Email: curran@proceedings.com
Web: www.proceedings.com

Table of contents

Volume 29

1st International Symposium on Electrical Arc and Thermal Plasmas in Africa (ISAPA)
17–22 October 2011, Ouagadougou, Burkina Faso

Accepted papers received: 16 December 2011
Published online: 27 February 2012

Electrical Arc and Plasmas

Alternative Energies

1st International Symposium on Electrical Arc and Thermal Plasmas in Africa (ISAPA) IOP Publishing
IOP Conf. Series: Materials Science and Engineering **29** (2012) 011001 doi:10.1088/1757-899X/29/1/011001

FIRST INTERNATIONAL SYMPOSIUM ON ELECTRICAL ARC AND THERMAL PLASMAS IN AFRICA (ISAPA)

17th TO 21th OCTOBER, 2011

OUAGADOUGOU, BURKINA FASO

PROCEEDINGS PREFACE

LABORATORY OF MATERIAL AND ENVIRONMENT UNIVERSITY OF OUAGADOUGOU, OUAGADOUGOU BURKINA FASO	LABORATORY OF ELECTRIC ARC AND THERMAL PLASMAS BLAISE PASCAL UNIVERSITY, CLERMONT-FERRAND FRANCE

On behalf of **RAMSES**

(Réseau Africain pour la Mutualisation et le Soutien des pôles d'Excellence Scientifique)

And the financial support of

EU-ACP EDULINK Program

OUAGADOUGOU 2011

IOP Conf. Series: Materials Science and Engineering **29** (2012) 011001 doi:10.1088/1757-899X/29/1/011001

Africa (especially Sub-Sahara Africa) is a continent where electrification is at low level. However, the development of electrical power sector is a prerequisite for growth of the other industrial activities, that is to say for the social and economical development of African countries. Consequently, a large number of electrification projects (rural electrification, interconnection of different countries grids) takes place in many countries. These projects need of expertise and make Africa a continent of opportunity for companies in different domains for business and research on energy and energetic: production, transmission, distribution and protection of electricity, supplying of cable, construction, engineering and expertise in solar and wind power field.

The first International Symposium on electrical Arc and thermal Plasma in Africa (ISAPA) was held for the first time in Ouagadougou, Burkina Faso to impulse and develop the research of new physical developments, technical breakthroughs, and ideas in the fields of electrical production and Electrical applications. The ISAPA aimed to encourage the advancement of the science and applications of electrical power transformation in Africa by bringing together the specialists of many areas from Africa and all other the world. Such considerations have led us to define a Scientific Committee including representatives from many countries.

This first meeting has been an innovating opportunity for researchers and engineers from academic and industrial sectors to exchange views and knowledge. Both fundamental aspects as thermal plasma, electrical arc, diagnostics and applied aspects as circuit breakers, ICP analyses, photovoltaic energy conversion, alternative energies, and space applications have been covered.

The Laboratory of Material and Environment (LAME) from Ouagadougou University and the Laboratory of Electric Arc and Thermal Plasmas (LAEPT) from Blaise Pascal University have worked in close collaboration within the framework of the Organizing Committee of this new and first symposium ISAPA in Africa.

We have registered 40 participants from France, Portugal, Belgium, Mali, Niger, Togo, Tchad and Burkina Faso of course and also through collaborative works from Russia, Poland and Ukraine. 20 papers, one poster and 3 oral contributions have been presented for this first ISAPA.

The ISAPA Symposium has been held with material and financial supports of the following organizations:

- EDULINK Program of EU-ACP;
- SCAC Service of France Embassy in Burkina Faso;
- IRD (Institute of Research for the Development) Burkina Faso;
- ASDI/SAREC project in Burkina Faso;
- University of Ouagadougou, Burkina Faso;
- Blaise Pascal University of Clermont Ferrand, France.

The opening ceremony of ISAPA Symposium have been presided by two ministers: the minister in charge of secondary and higher education (MESS) and the minister in charge of scientific research and innovation (MRSI). Thus, they have marked the interest given by the government of Burkina Faso for RAMSES Scientific Meetings such as ISAPA.

Zacharie Koalaga (LAME, University of Ouagadougou, Burkina Faso)
Pascal André (LAEPT, Blaise Pascal University, France)
ISAPA 2011
Co-Chairmen of the ISAPA International Organizing Committee and editors

IOP Conf. Series: Materials Science and Engineering **29** (2012) 011003 doi:10.1088/1757-899X/29/1/011003

ISAPA INTERNATIONAL SCIENTIFIC COMMITTEE

Pascal André	France	LAEPT, Blaise Pascal University
Yacine Babou	Belgium	Von Karman Institute
Mario Da Silva	Portugal	IST University of Lisbonne
Michel Dudeck	France	IJLRA, University of Paris VI
Zacharie Koalaga	Burkina Faso	LAME University of Ouagadougou
Tony Murphy	Australia	CSIRO Materials Science and Engineering Lindfield
Sergey Shkol'nik	Russia	AF Ioffe Physico-Tech. Institute, St Petersburg
Francois Zougmore	Burkina Faso	LAME University of Ouagadougou

ISAPA INTERNATIONAL ORGANIZING COMMITTEE

Zacharie Koalaga	LAME, University of Ouagadougou, Burkina Faso
Pascal André	LAEPT, Blaise Pascal University, France

ISAPA LOCAL ORGANIZING COMMITTEE

Francois Zougmore	LAME, University of Ouagadougou
Zacharie Koalaga	LAME, University of Ouagadougou
Sosthène Tassembedo	LAME, University of Ouagadougou
Stéphane Ouédraogo	Pôle GM/RAMSES, University of Ouagadougou
Bernard Zouma	LAME, University of Ouagadougou
Martial Zoungrana	LAME, University of Ouagadougou
Abdoul K. Kagoné	LAME, University of Ouagadougou
François Ouédraogo	LAME, University of Ouagadougou

PARTNERSHIPS AND SPONSORS

Local Administration

University of Ouagadougou
http://www.univ-ouaga.bf

SCAC of France Embassy in Burkina Faso
http://www.ambafrance-bf.org/

IRD, Burkina Faso
http://burkina-faso.ird.fr/

ASDI/SAREC Project, Burkina Faso

Associations and Learned Society

Association Arc Electrique
http://www.association-arc-electrique.org/

SFP Auvergne
http://sfp.in2p3.fr/auvergne/

Centre National des Risques Industriels (CNRI bourges)
http://www.cnri-bourges.org

ISAPA Symposium Web site:
http://www.univ-bpclermont.fr/CONGRES/ramses2011/isapa.html

Symposium Web site Editor:
http://www.elforia-design.fr

IOP Conf. Series: Materials Science and Engineering **29** (2012) 012001 doi:10.1088/1757-899X/29/1/012001

Electric fuses operation, a review: 1. Pre-arcing period

W Bussière

Clermont Université, Université d'Auvergne, LAEPT, BP 10448, F-63000 Clermont-Ferrand, France

E-mail: william.bussiere@univ-bpclermont.fr

Abstract. Electrical needs are continously growing because of many various factors linked to increasing consumption and a green perception. This development – geographical, increasing production, growth of transport network, interconnection of continental networks, diversity of the transport technologies … – can not be dissociated from electrical safety considerations whatever the voltage level. For the three main levels of electric network – High Voltage, Middle Voltage and Low Voltage – one has to provide efficient electrical safety techniques or schemas which integrate different electrical safety apparatus. Among well-known apparatus we can cite SF_6 breakers, HV and MV switchgears (such as MV cells, MV and LV high current vacuum switchgears), and fuses. Electrical fuses are especially used in the MV and LV domains, sometimes as an additional safety device and sometimes as the main electrical safety component which is linked to the electrical current breaking function of electric fuse. In the paper, we will quickly depict the various kinds of electric fuses. We will especially focuss on the physical mechanisms – whatever the type of work, experimental, theoretical, modelling or empirical – prevailing during the pre-arcing period of the electric fuse operation.

1. Introduction

Electric fuses are widely used for electrical safety purposes from many years in the different levels of the electric network: in High Voltage (HV) level corresponding to power houses, in Middle Voltage (MV) level corresponding to electric network and specific applications (such as motors of trains or subways), and Low Voltage (LV) level for electric supply both in industry and individual use. Within the LV domain we can add particular fuses such as miniature fuses for electronics safety designs. The considerations detailed in the paper are typically those of High Breaking Capacity fuses but most of them can be extended to other types of fuses.

Due to the wide domain of applications, electric fuses are the result of various technical solutions which have been constinuously designed from the 19th century [1]. These improvements are linked to the increase of the current and voltage values of the network or device to protect. Up to now many studies have been carried out both in industrial research and development departments and in academic research laboratories. These studies can be of theoretical or experimental type, but due to the hardness for a whole understanding of the physical mechanisms, many empirical or semi-empirical results have been published to describe by means of numerical laws some fundamental quantities. At the end the whole of this work is done in order to build a physical modelling of the fuse operation able to predict the fuse operation and the fuse ability to ensure the breaking of a *fault* current. Such a modelling has to be built on reliable experimental assessment of fundamental physical and electrical quantities. Moreover such a physical modelling could significantly help to reduce the number, the duration and the cost spent to test prototypesin industry. These tests performed by means of power stations up to tens or hundreds of MVA remain necessary in the industrial environment in order to test the expected performance in specific conditions imposed by international standards and especially by the IEC [2].

At the present time a typical electric fuse is made of four main components: one or several fuse elements (typically a fuse strip), generally in silver (sometimes in copper), equipped with one or many reduced sections (also called constrictions) of various shapes (half circular, rectangular, triangular for the mains), the fuse element being winded on an insulating core (often in ceramics with a helicoidal section for HBC fuse) ; the two metal electrodes at the ends of the fuse element ; the outer cartridge whose role is to ensure a good mechanical and thermal protection against outer environment ; the arc quenching material, usually silica sand (SiO_2 or quartz sand) of high purity and controlled granulometry and packing density, which fills the void space between the core and the cartridge. In the whole silver (for the fuse element material) and silica (for the arc quenching material) are widely used in the LV and MV ranges. Different fillers have been studied for example in [9] but silica sand (granular filler) appears to be the best compromise to this day because of various benefits (facilities for production and supply, morphometric and physico-chemical properties, energetic efficiency to dissipate the energy brought by the electric *fault*). Main researches and developments are currently dedicated to additives and binders for example in the case of specific LV fuses (filler under the form of agglomerates) and to compaction methods in the case of MV fuses (dispersed granular filler). Other materials or geometries or designs are used for example in the case of new fault-tolerant inverter and interleaved chopper and are currently under investigation [70]. From the academic research point of view main of the current works are focussed around the building of consistent modellings. In the last fourty years major published contributions have been produced by North-American, English and Australian co-workers whose theoretical investigations have been led in close collaboration with industry and also by Polish co-workers. In the more recent past significant contributions was published by European (France, Greece, Italy, Slovenia, Spain) and Chinese co-workers also within the framework of a close collaboration with industry. These works are partly illustrated by the quoted references. At the current time French co-workers make a special effort to assess fundamental processes linked to industrial considerations (GEMH, LAEPT, LAGEP, SIMAP INPG, SPIN ENSM).

Typically the fuse operation is as follows. When a *fault* current situation occurs, the high level current implies the increase of the temperature inside the fuse element due to Joule effect. Thus the electric resistance of the fuse element increases. The temperature increase is especially high in the constrictions where ohmic resistance is higher than in other parts of the fuse element. Due to the energy increase by Joule effect the fuse element constriction is melted and hence vaporized, thus leading to the ignition of an electric arc between the two parts of the fuse element. This time range is called the *pre-arcing* period which ends in the sharp increase of the voltage across the fuse. It is followed by the *arcing* period. The electric arc quickly interacts with the surrounding materials, namely the air in the interstices of the silica sand and the silica sand grains. This interaction leads to the ignition of the fuse arc plasma. As a consequence this plasma is composed of various chemical species resulting from the vaporization of Ag and SiO_2 mainly. In the same time the erosion of the fuse element goes on in the two opposite directions and the plasma column takes place in the inter-electrode space. The fuse arc plasma expands in the gap between the two parts of the fuse element up to the end of the fuse operation. As a result of the material vaporization – both silver and silica – and of the temperature increase, the pressure in the plasma column increases strongly and especially in the direction normal to the fuse strip. This implies a strong widening of the thickness of the plasma column which can reach many tens times the initial thickness of the fuse strip. As soon as the voltage across the fuse can not be maintained any more by the supply voltage, the current decreases down to zero. This defines the end of the *arcing* period. Thus the contribution of Joule effect decreases quickly and doesn't exist any more at the end. The energy of the liquid or vaporized material is gradually removed mainly by thermal conduction and re-condensation of the vaporized species. The plasma column cools down and finally solidifies under the shape of a dense aggregate called *fulgurite*. The total duration of the fuse operation is the sum of the *pre-arcing* and *arcing* times.

Consequently it is very complex to describe the fuse operation in the whole by means of a single tool. In fact one has to deal with the *pre-arcing* period and the *arcing* period separately. Moreover for each of the operation times one has to take into account the dual role played by the fuse, from the physical and the electrical points of view. And whatever the operation time considered the physical depiction has to be done for the three phases of the two main materials (Ag and SiO_2) namely solid, liquid and vapour.

Academic research and research/development activities are currently concerned by some unknowns or physical processes or key technological features of very high interest. Some of these problematics are pointed out in either *Part 1 (Pre-arcing period)* or *Part 2 (Arcing period)* of the review. Among the whole of these key points one may quote a modelling tool both for the *pre-arcing* and the *arcing* periods built on physical considerations only, the prediction of the electrical conductivity on the whole fuse operation taking into account the time dependent relative ratio of the solid, liquid and vapour phases, the growth of the interaction volume and the resulting fulgurite volume and the transition between the *pre-arcing* and the *arcing* periods. As a result the increase of the knowledge focussed around the whole of these problematics is the condition necessary to improve the control of the electric fuse operation. In most cases from the *fault* current features point of view (especially the peak value and the appearance time of the electric *fault*) the *pre-arcing* period duration is significantly higher than the *arcing* period duration (see table 2 for the definition of the *pre-arcing* period according to its duration, and consider a value ~ 10 ms for the *arcing* period duration). The shortening of the fuse operation duration depends at the same time on the ability to decrease the *pre-arcing* time (mainly by fitting of the fuse elements features inside the fuse box, especially in the case of low overload) and to improve the conversion of the energy (electrical to thermal) together with the increase of the full voltage drop linked to the fuse element disruption (transition between the *pre-arcing* and the *arcing* periods) and to the time-dependent electric field of the plasma column (during the *arcing* period). The control of these understandings from the theoretical point of view is a necessary condition to provide new effective solutions in electrical safety industry. As a result a perfect electric fuse should be able to break the electric *fault* whatever the overload level in the shorter time possible with an efficient ability to withdraw the energy brought by the *fault*.

The current paper (*Part 1. Pre-arcing period*) is divided into two main sections respectively focussed on general considerations dedicated to the whole of the fuse operation and to the *pre-arcing* period only. Section 2 is mainly concerned by general considerations about studies (section 2.1) and features (section 2.2) dedicated to electric fuses. The fuse operation in its whole – *pre-arcing* period and *arcing* period – is depicted in section 2.3. General considerations linked to electrical and thermal points of view are given in sections 2.4 and 2.5. Section 3 is dedicated to the *pre-arcing* period only with a depiction of the physical processes governing this step: main results are highlighted both for experimental and theoretical/modelling works.

Part 2 of the current review (*Part 2. Arcing period*) is concerned by the *arcing* period which is more complex to depict as many different physical processes occur in the same time in the whole of the interaction volume.

2. General considerations

2.1. About electric fuse studies

The most comprehensive survey work about research and applications dedicated to electric fuse is published in [3] with some of the most significant works, especially those related to the North American types of fuses. Among the key phenomena which are depicted, we can cite the *pre-arcing* behaviour and the *arcing* behaviour of cartridge fuses mainly of Low Voltage HBC type. Since the *pre-arcing* period ends before the appearance of the electric arc, the *pre-arcing* period study is done up to the time for which the energy for melting is obtained, that is to say that the heat of melting is

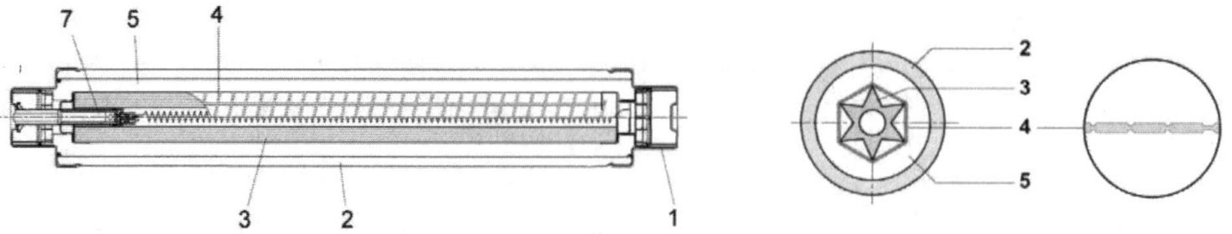

Figure 1. Sketch section of a MV fuse (from [5]). 1: electrodes or end contacts – 2: cartridge – 3: core – 4: fuse element – 5: arc quenching material – 6: striker.

fully provided once the melting temperature has been reached. The *pre-arcing* period duration is directly linked to the level of *fault* current. Especially at low current level near the *rated current* – for this current, the equilibrium between input power and the power dissipated by the fuselink is obtained, resulting in a nearly equilibrium situation – the *pre-arcing* period can last more than one hour. All intermediate levels of current are considered between the two extreme situations: very high currents, high currents, intermediate current levels and currents near the minimum fusing level (see [2] for I_1, I_2 and I_3 test series). Many mathematical and experimental studies are reported for the whole current level range. Once the *pre-arcing* period is ended, it is immediately followed by the *arcing* period which is considered by underlining on the main physical processes controlling the electric arc growth and extinction down to the cancellation of the *fault* current. Among them *Wright* and *Newbery* [3] underline the cathode-fall region, the anode-fall region and the arc plasma column. For this latter region the length and the cross-section area are especially studied because they are essential for the calculation of the electrical conductivity which is one of the most significant physical quantities when dealing with electric fuse.

Another significant contribution is those published by *Daalder* in [4] which deals with the *arcing* period mainly. The whole study is divided into five parts: model of fuse arc, *burn-back* rate of fuse elements, arc channel expansion, course of the probe voltage and the influence of the notch disintegration on the arc voltage. Main of the results is built on experiments leading to semi-empirical or empirical modellings, these latters having to be fitted to the characteristics of the electric fuse under consideration.

Many other studies have been published dealing either with the *pre-arcing* or the *arcing* phenomena. As the *pre-arcing* period is usually defined as the time necessary to reach the melting point of the fuse element metal, many modellings can be found including comparisons with measurements. Among the whole of these studies one may distinguish *pre-arcing* period considered under adiabatic assumptions [15] (very short *pre-arcing* times or high *fault* current level), *pre-arcing* period considered by taking into account the whole of the physical processes responsible for the energy withdrawal [13, 15] and all the intermediated cases where various simplifications are made depending on the operating conditions [52]. Phenomenological modellings have been also published [25]: the electric fuse is described by means of an equivalent electrical circuit built on resistive and capacitive contributions, according to parallel and series schemas. Because of its high complexity the *arcing* period has not been fully described by a physical modelling whose simulated results could be compared with experimental results. As a consequence main of the published studies can be found under the form of semi-empirical or empirical modellings. Most of these latters deal with phenomena or quantities at the macroscopic scale occuring or observed during the *arcing* period such as the fuse element erosion [4, 62, 63] (also known as *burn-back*), the arc column [37] or the arc channel [65, 69] (also known as lumen), the electrical conductivity [4]. Only few studies deal with the *arcing* period as a whole or deal with the phenomena or quantities at the microscopic scale in relation to the fuse arc plasma properties [59]. This is mainly due to the numerous phenomena occuring in the same time, the various states of the present materials – solid, liquid and gas for silver and silica – and the simultaneity of each of the phenomena.

IOP Conf. Series: Materials Science and Engineering **29** (2012) 012001 doi:10.1088/1757-899X/29/1/012001

2.2. Brief review of existing electric fuses

Many types of electric fuses can be found depending in the same time on the construction and the application field. We refer to [3] to summarize briefly the main features.

First one can classify electric fuses according to the voltage range of application: high-voltage (HV), low-voltage (LV) and miniature, this classification being the reference defined by the International Electrotechnical Commisssion [6]. As quoted in [3] the division between HV and LV occurs at 1000 V AC.

Table 1. Various types of electric fuses defined according to the type of applications, from [3].

Type	Application (characteristic)	Breaking range
gG	General application, mainly for conductor protection	Full range
gM	Motor circuit protection	Full range
gN	North American general application for conductor protection	Full range
gD	North American general application time delay	Full range
gR, gS	Semiconductor protection	Full range
gU	Wedge tightening fuse for utilities	Full range
gL, gF, gI, gII	Former type of fuse for conductor protection (replaced by gG type)	Full range
aM	Short-circuit protection of motor circuit	Partial range (back-up)
aR	Semiconductor protection	Partial range (back-up)

Secondly one can refer to an electric fuse by taking into account the constructions. Among basic constructions one can cite enclosed fuses (such as in figure 1) sometimes equipped with a fuse older especially for the LV range, mainly of cylindrical form with silica sand as quenching material (other materials can be found especially in the case of liquid fuse in the HV range). Vacuum fuses have been designed for medium voltage applications but they are limited and their development can not be compared with those linked to switching in vacuum. Some semi-enclosed fuses can also be found but their inferior performances lead to a decreasing demand. Other types of electric fuses can be found – especially "*fault* limiters" including "resettable fuses", "permanent fuses" and "superconducting fuses" – but their operation deviates from previous ones which work as typical fuses (after melting of the current carrying element, the electric circuit is permanently opened).

Thirdly electric fuses can be defined according to operating specifications quoted in applications guides. To illustrate the diversity of applications we reproduce in table 1 an example from [3]. Full range fuse refers to "current-limiting fuse capable of breaking under specified conditions of use and behaviour, all currents that cause melting of the fuse element(s), up to its maximum breaking point", and back-up fuses refer to "current-limiting fuse capable of breaking, under specified conditions of use and behaviour, all currents from the rated maximum breaking current down to the rated minimum breaking current".

Fourthly electric fuses can be named according to the specific application they are designed for. One may especially cite fuses for the protection of semiconductors for which there exist many theoretical and experimenal studies [45].

Finally practical information about use, normalized test and specifications can be found in the standards from IEC: the 60269 IEC series for LV, the 60282 IEC series for HV and the 60127 series for miniature fuses [6].

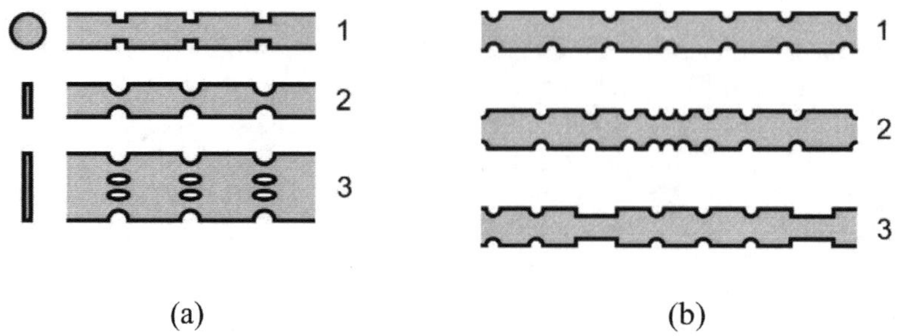

Figure 2. (a) Examples of fuse element with various sections. 1: wire – 2: strip – 3: thin plate. (b) Examples of various reduced section or notches. 1: regular – 2: gradual – 3: intermediate case [5].

2.3. General considerations about the electric fuse operation

The depiction of the main components of an electric fuse is shown in figure 1 in the case of a MV fuse. Typically an electric fuse comprises four main components: the outer cartridge mechanically and thermally resistant, the two electrodes, the fuse element(s) (put in parallel if more than one fuse element, each of them being equipped with reduced sections in series) round wound the helical core in insulating material (often ceramics), and the arc quenching material (usually silica sand of high purity checked during the industrial process).

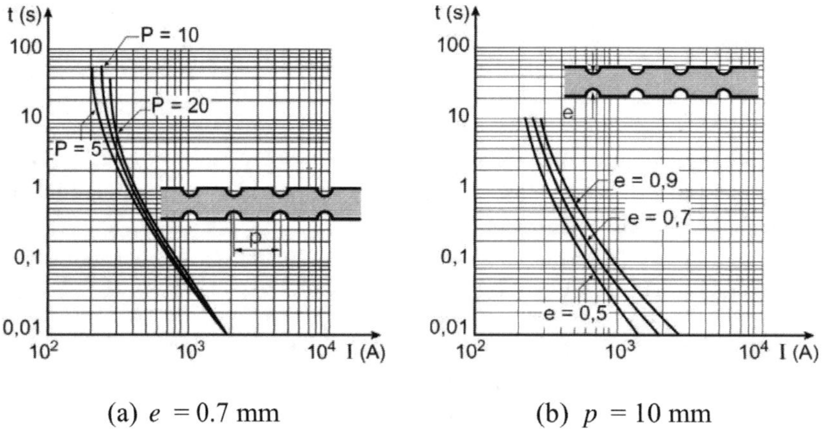

(a) e = 0.7 mm (b) p = 10 mm

Figure 3. Time-current curves showing the influence of: (a) the step p of the fuse element reduced sections, and (b) the width of the fuse element reduced section in the case of circular section [5]. The fuse element is a strip with many reduced sections in series.

The fuse element is certainly the most important component with the arc quenching material. In figure 2 we show various types of fuse elements: variable section (figure 2(a)) and various shapes for the reduced sections (figure 2(b)). Different fuse element constitutive materials have been tested in the past: tin, zinc, aluminium, copper, silver … [26]. Among these specifications the number, the distribution and the shape of the reduced sections play a great role during the *pre-arcing* period because they influence directly the time (or the energy) necessary to reach the melting point. To illustrate this influence we show in figure 3 two time-current curves where the variable parameter is the step between two consecutive reduced sections in figure 3(a) and the width of the reduced section in figure 3(b). These diagrams are established for a notched fuse strip with silver as assumed material.

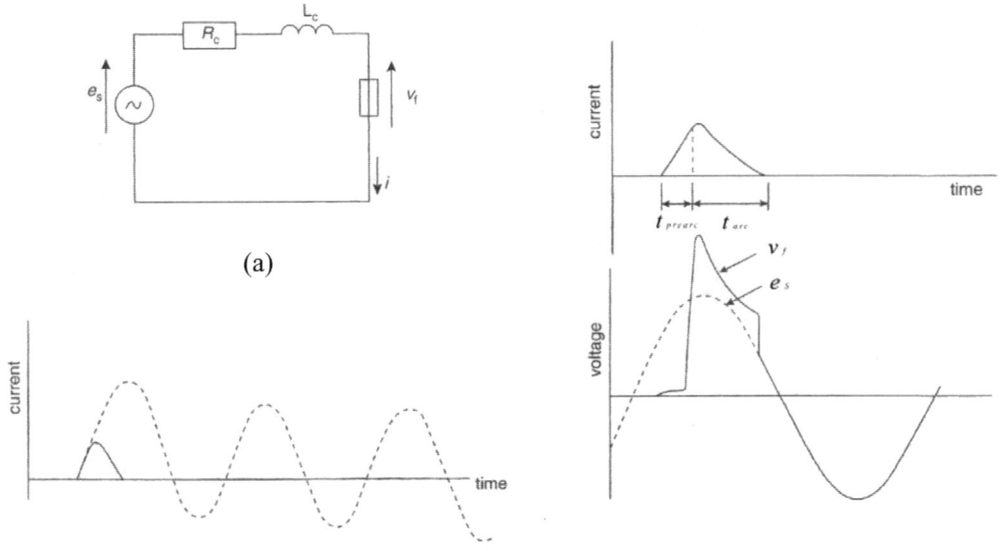

(b) (c)

Figure 4. (a) Typical equivalent circuit: e_S, source *emf*; R_c, resistance of the circuit (the fuse resistance being disregarded) ; L_c : inductance of the circuit ; v_f : voltage across the fuse ; i : current of the circuit. (b) Illustration of the prospective current (dashed line) and the actual circuit current flowing through the fuse. (c) Current in and voltage across the fuse during the fuse operation: t_{prearc}, *pre-arcing* period time ; t_{arc}, *arcing* period time. From [3].

In industry the length of the reduced section is fitted too (not shown in figure 3). This very well-known curve in the fuse community gives the time necessary to reach the fuse element melting point as a function of the current AC value. From figure 3(a) where e = 0.7 mm we see that the time to reach the melting point depends weakly on the step p at high current level for $I > 10^3$ A, and is three times higher for p = 20 mm than for p = 5 mm at current level around 3×10^2 A. From figure 3(b) where the step between two consecutive reduced sections is kept constant we see that at $I \sim 10^3$ A, for e = 0.5 and 0.9 mm we have respectively $t \sim 25$ ms and $t \sim 85$ ms. Thus even for high current level range the time to reach the melting point depends strongly on the width of the reduced section, while for low current level range this time also depends on the step of the reduced sections. Finally the fuse element is the result of the best compromise between these parameters.

The time necessary to reach the melting point is one of the key properties of the electric fuse as it has to break the *fault* current after the shorter operation time possible but it has to ensure the highest dielectric rigidity once the current falls down to zero to avoid a possible re-arcing situation [27].
The typical fuse operation is shown in figure 4. The typical electric circuit given figure 4(a) in the case of a sinusoidal source is defined by an impedance ($\sqrt{R_c^2 + L_c^2\omega^2}$, ω being the pulsation of the circuit source) resulting from resistive (R_c) and inductive (L_c) contributions, from whose is calculated the *power factor* of the circuit by: $\cos\varphi = R_c \big/ \sqrt{R_c^2 + L_c^2\omega^2}$. Due to the resistive feature of the fuse link the circuit current generates a voltage across the fuse depending on various parameters, such as the geometrical features of the reduced sections of the fuse element discussed above. The current limiting role played by the fuse is shown in figure 4(b). The dashed current curve is the current produced in the electric circuit without the fuse also called *prospective* current. In *fault* conditions this *prospective*

current is the *fault* current which has to be cut by the fuse. The *limiting* role of the fuse results in the solid line current curve in figure 4(b). We notice that the fuse current peak is significantly smaller than the *fault* current peak with a shorter duration thus justifying the *limiting* role of the fuse. The current in the fuse is detailed in figure 4(c) together with the voltage across the fuse. At the appearance of the fuse current – beginning of the *pre-arcing* period – and due to the initial electrical resistivity of the fuse strip we observe a sharp but restricted increase of the fuse voltage. It is followed by a roughly steady increase itself followed by a significant and abrupt increase: it defines the end of the *pre-arcing* period of duration t_{prearc} for which the reduced section has been melted and vaporized thus opening a gap between the two parts of the fuse element. An electric arc is ignited between these two electrodes. A plasma mainly composed of silver vapours appears. From this step starts the *arcing* period. The energy brought by the *fault* current is given by $\int_{t_{prearc}}^{t_{arc}} r(t)\cdot i^2(t)\cdot dt$ up to the current interruption at the end of the *arcing* period of duration t_{arc} (the inductive contribution $1/2\,L_c I_{peak}^2$ being more or less significant). Due to this energy the two parts of the fuse element are eroded (melting and vaporization) in two opposite directions thus lengthening the inter-electrode distance. This erosion process is called *burn-back*. During this lengthening the silver plasma quickly interacts with the granular silica sand of given porosity (with air in the interstices) with SiO_2 as the main chemical species. The plasma composition (initially of silver type) quickly evolves to a plasma resulting both from the vaporization of the silver fuse element and from the dissociation and hence from the excitation and the ionization of silicon (Si) and oxygen (O) atoms. From spectroscopic measurements no evidence of emitting species resulting from the dissociation of air molecules has been pointed out without any doubt [66,67,68]. It suggests that the air surrounding the fuse element reduced section is pushed away through the interstices of the silica sand during the vaporization of the fuse element metal. In the plasma column one can observe in the same time radiation from monoatomic species and ions (mainly first ions such as Si^+) which are issued from different plasma subvolumes from the center to the edges of the plasma column. In the plasma column there is a strong increase of the pressure due to the vaporization of materials – metal and silica – and due to the high temperature. Tens of atmospheres have been measured [19] for maximum mean temperatures ranging approximately from 16 kK to 22 kK [22] depending on the morphometric properties of the silica sand, especially mean granulometry and packing density. More details are given in Part 2 of the review.

During the *arcing* period the lengthening of the plasma column implies the increase of the voltage across the fuse. This increase can be linked to the equation describing the circuit given in figure 4(a): $e_S = i\cdot R_c + d/dt\,(L_c i) + v_f \approx i\cdot R_c + L_c\, di/dt + v_f$. If we consider the situation described in figure 4(c) the time derivative of current di/dt is positive at the beginning of the arcing period. The breaking of the *fault* current can be only obtained in the case where di/dt becomes negative. To satisfy this latter condition in a given electrical circuit – the value of R_c being set – one has to increase v_f in $e_S - i\cdot R_c - v_f$. In other words one has to increase the fuse resistance ($v_f = r_f \cdot i$) which justifies the well-known resistive role played by the electric fuse. As a consequence one of the obvious solutions is to increase the number of reduced sections as it is shown in figure 2(b). The benefit lies in the voltage fall obtained by the ignition of many arcs in series. If we consider a simple situation where n arcs in series are ignited we roughly obtain $n\times(V_{af}+V_{cf})$ with V_{af} the anode fall and V_{cf} the cathode fall. A comprehensive experimental database for V_{cf} can be found in [7]. For silver V_{cf} ranges from 8.0 V to 16.5 V. V_{af} numericals values can be taken from [8]: for silver V_{af} = 9.5 V to 10 V and they also give V_{cf} = 9 V to 9.5 V. Thus for one arc we obtain $V_{cf}+V_{af}$ ~ 17.5 V to 26.5 V. For example in the case of an industrial fuse for the MV range and one fuse element equipped with one hundred reduced sections the fuse voltage at the beginning of the *arcing* period can be estimated in the range from 1750 to 2650 V. During the *arcing* period the fuse voltage will be raised according to the fuse resistance which partly depends on the morphomotric properties of the porous quenching material which change the plasma composition.

At the end of the *arcing* period the temperature of the plasma column decreases. This plasma column surrounded by more or less melted silica sand grains solidifies and the temperature decreases down to the ambiant temperature with a time scale in the order of some minutes. The solid aggregate obtained at the end is called *fulgurite*. We show in figure 5(a) the picture of a *fulgurite* obtained in an industrial fuse and in figure 5(b) a cross-section view of a *fulgurite* obtained in laboratory. In the case of the industrial fuse many bulges are visible at the initial location of the reduced sections. They are the result of the very high pressure increase in the initial reduced section area. In the case of the *fulgurite* obtained in laboratory a cross-section is shown to illustrate the significant increase of the channel: in the case of figure 5(b), the fuse element thickness is initially around 0.1 mm ; after fuse operation and dismantling, this thickness where the arc plasma took place is multiplied up to between 20 to 50 times the initial value, this factor depending on the morphometric properties and the peak current.

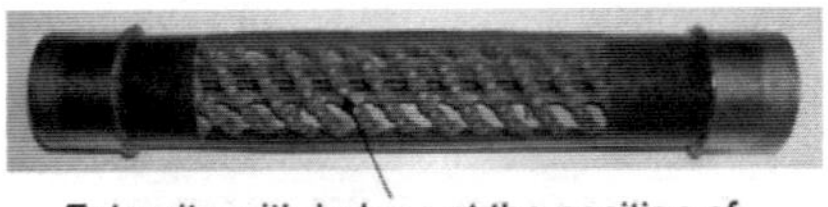

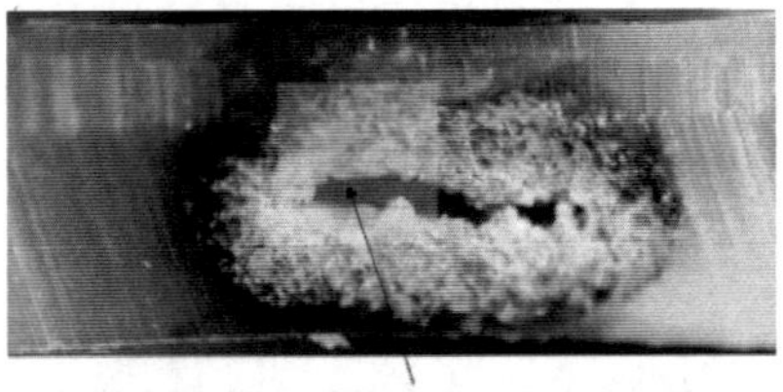

(a) (b)

Figure 5. Fulgurites observed after the fuse operation and dismantling: (a) for an industrial fuse and (b) for a single reduced section in the case of a fuse test performed in laboratory.

2.4. Electrical and thermal considerations

As it can be logically understood the electric curves observed during the fuse operation are the results of the physical mechanisms occuring both during the *pre-arcing* and the *arcing* periods and taking place both in the plasma column and in the near vicinity. At the end the aim of industrialists and academic researchers is to formulate a modelling of the whole of the physical mechansims in order to calculate the electrical outputs of a fuse taking into account industrial features: the fuse element geometry, the number of fuse elements in parallel and the number of reduced sections in series, the shape of the reduced sections which may vary on the whole length of a fuse element, the morphometric properties of the arc quenching material – usually silica sand but other material have been tested [9] – especially the mean granulometry and the packing density, the distribution of the interstices within the arc quenching material.

Up to now no there exists no fully physical modelling of both the *pre-arcing* and the *arcing* periods. Among all the works published to this day one may highlight some contributions such as those detailed in section 3 for the *pre-arcing* period and Part 2 which have to be considered together with some experimental studies or empirical modellings dealing with a particular step during the fuse operation or a particular mechanism. Whatever the approach which is considered one of the main restrictions lies in the knowledge of electrical and thermal basic data. In figure 6 we give for silver the temperature evolution of the enthalpy, the electrical conductivity and the thermal conductivity. The solid (s) to liquid (l) transition appears at 1235 K and the liquid to vapour (b) transition appears at 2433 K which correspond respectively to the latent heats $L_m = 1.05\times10^5$ J·kg^{-1} and $L_b = 2.43\times10^6$

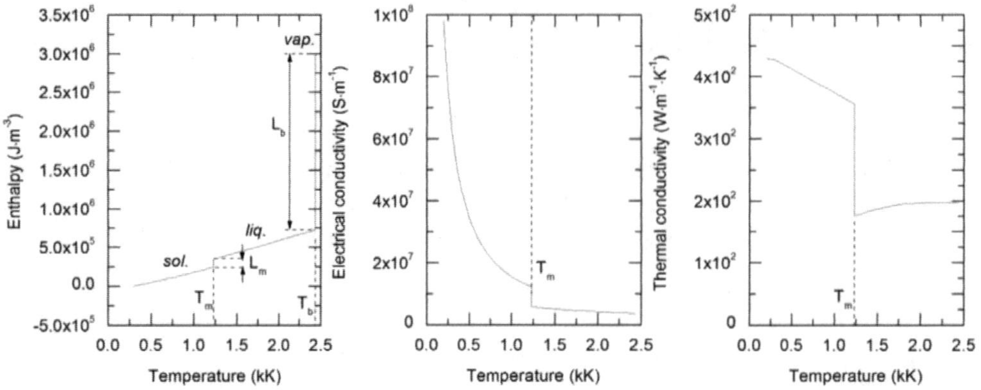

Figure 6. For the temperature range corresponding to the *pre-arcing* period, evolutions (at atmospheric pressure) of the enthalpy [10], the electrical conductivity [12] (extrapolation from 1670 K [13]) and the thermal conductivity [11] in the case of silver. T_m, melting temperature ; T_b, boiling temperature ; L_m, latent heat for melting (sol. $\rightarrow$ liq.) ; L_b, latent heat for boiling (liq. $\rightarrow$ vap.) .

J·kg^{-1}. The heat capacities are c_s = 261.36 J·kg^{-1}·K^{-1} and c_l = 310.4 J·kg^{-1}·K^{-1} (the weak dependence with the temperature can be obtained from [10]).

At the beginning of the *pre-arcing* period the fuse element has a typical electrical resistance of a few mΩ or tens of mΩ. The heating of the fuse element in the solid state is quite low and becomes faster and more efficient as fast as the current increases and the electrical resistance becomes higher, especially once the fuse element is in the liquid state. Considering the very low value of the reduced section to fuse element section ratio the most significant heating takes place in the center of the reduced section where the width is the smallest. In spite of the fact that $L_b >> L_m$ the vaporization of the reduced section arrives very quickly once the reduced section is in the liquid state especially as the current reaches its peak value at the end of the *pre-arcing* period. The Joule heating which rules the *pre-arcing* period is especially influenced by the current increase and the change observed for electrical conductivity in the solid state and during the solid to liquid transition: $\sigma_{sol,300K} / \sigma_{liq,1235K}$ ~ 10. On the contrary the same ratio for the thermal conductivity is around 2.5. The influence of the thermal conductivity will be weak or even insignificant with high current *fault* corresponding to short *pre-arcing* time – typically around the milliseconde – and more significant for lower *fault* current: in this latter case especially if the reduced section to fuse element section ratio is very small the heat transfer towards the two parts on each side of the fuse element around the reduced section will be more significant thus decreasing slightly the temperature of the reduced section.

The influence of radiation during the *pre-arcing* period in now considered. In figure 7(a) we give the luminance corresponding to a black body emitter for the melting temperature and the boiling temperature in the spectral domain up to 10 µm. In figure 7(b) we consider the silver fuse element as a grey emitter taking into account emissivity value in the liquid state. Values are deduced from measurements published in [14]: for 865 K < T_r < 1750 K, $\varepsilon(T_r) = -0.0444 + 1.5775\times10^{-4} \times T_r$ where $\varepsilon(T_r)$ is the normal spectral emissivity at 684.5 nm and T_r is the radiance temperature. For 1750 K < T_r < 2433 K values are extrapolated considering the latter formulation valid because there exist no data at our knowledge. First we see that the gap is rather weak between the actual case – silver fuse element considered as a grey emitter – and the assumption of maximal thermal radiation – silver fuse element considered as a black body emitter. Secondly from the maximal values of the luminance at T_m and T_b we obtain respectively 0.12 mJ and 3.4 mJ: these energies are calculated for $\Delta\lambda$ = 9.75 µm, a radiating area 1 mm × 1 mm (which is overestimated considering typical reduced sections used

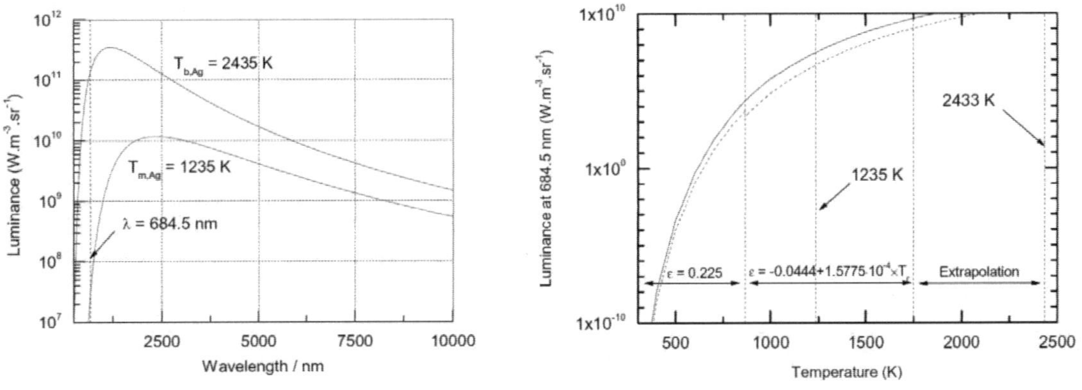

(a) (b)

Figure 7. (a) Luminance calculated from the Planck's law at melting temperature (T_m) and boiling temperature (T_b) for the spectral domain up to 10 μm. (b) Luminance at 684.5 nm for silver considered as a grey emitter taking into account emissivities from [14].

in industry) and a time operation of 1 ms which is typical of short *pre-arcing* period. These contributions are negligible in comparison with the heat transfer by conduction in the fuse element and the energy necessary for the heating and the phase changes.

For longer durations of the *pre-arcing* period – known as medium and long *pre-arcing* times – the role played by the silica sand grains becomes more significant especially in the heat transfer by conduction. The thermal conductivity of silica is low in comparison with silver, $k_{SiO_2} = 8$ W·m^{-1}·K^{-1} << $k_{Ag} = 429$ W·m^{-1}·K^{-1} [15] (at 293 K) but >> $k_{Air} = 0.026$ W·m^{-1}·K^{-1} [16] (at 300 K) for the air in the interstices of the silica sand. As a result the silica sand grains in contact with the fuse element can play a great role especially those in contact with the reduced section where the temperature gradient is high.

The *arcing* period is more complex to describe because many different physical mechanisms play a great role in the same time, the intensity of each mechanism depending on the current value and the location *i.e.* in the plasma core or in the plasma column surroundings. Especially the *arcing* period is the fuse operation step governed by the arc quenching material properties. The composition of the plasma column initially dominated by silver vapours is changed due to the injection of chemical species resulting from the dissociation and hence the ionisation of silicon and oxygen. Thermodynamic properties and transport coefficients are the results of these new chemical species. In figure 8 we give the evolution versus temperature of basic thermodynamic properties for the temperature range from 5000 K to 25 000 K: the pressure (figure 8(a)), the electronic concentration (figure 8(b)), the internal energy (figure 8(c)), the heat capacity at constant volume (figure 8(d)). The plasma composition is also shown in figure 8(e). These calculations made at constant volume are based on the Helmoltz free energy minimization [18] and thermodynamic equilibrium is assumed. This last assumption is questionable in regard to the time of observation during *arcing* period and the closeness with the core of the plasma column. Two explanations among several can be highlighted (discussed in detail in *Part 2. Arcing period*). First, at the beginning of the *arcing* period immediately after the fuse element disruption, the silver plasma interacts with the silica sand grains leading to the silver-silica plasma ; cold material is injected into the hot metallic plasma ; the existence of a thermal disequilibrium should not be disregarded. Secondly, from the hot plasma core to the colder plasma edges, there are strong gradients (temperature, density and pressure) ; consequently these strong inhomogeneities imply material flow (different temperatures and densities) and strong temperature gradients and a possible thermal disequilibrium. On the contrary the inner part of the plasma core can

IOP Conf. Series: Materials Science and Engineering **29** (2012) 012001 doi:10.1088/1757-899X/29/1/012001

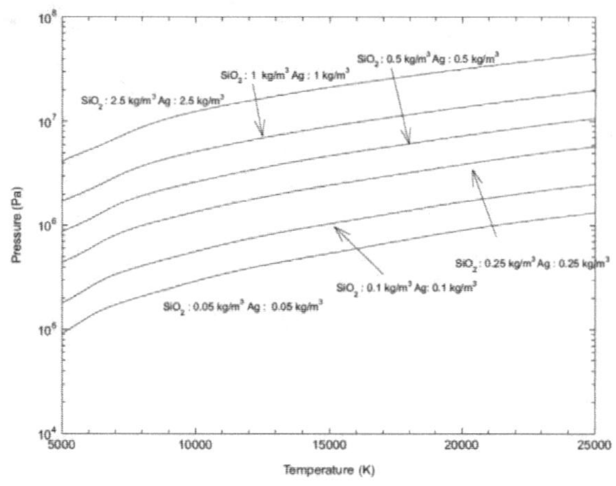

(a)

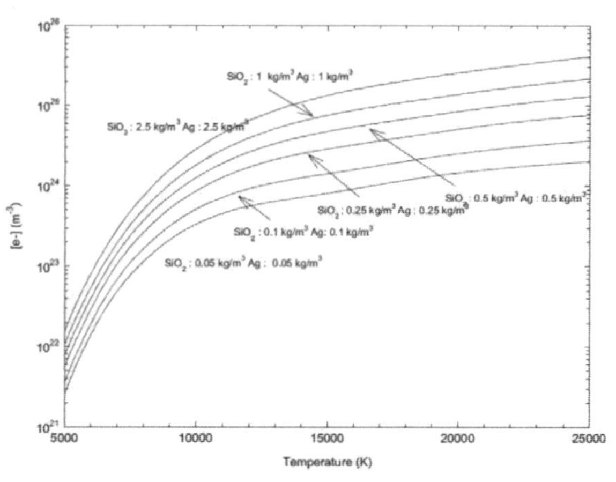

(b)

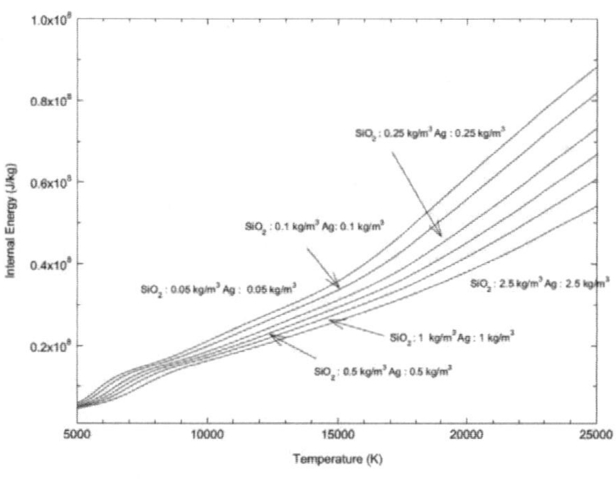

(c)

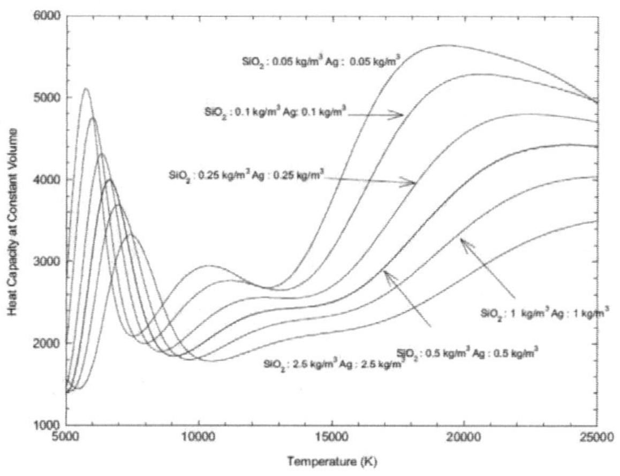

(d)

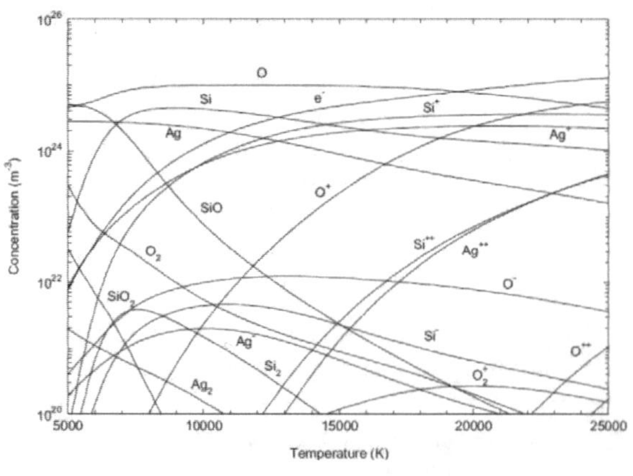

(e)

Figure 8. For various mixtures of Ag and SiO_2 (proportion is given in weight percentage) evolution versus temperature of [17]: (a) the pressure, (b) the electronic concentration, (c) the internal energy, (d) the heat capacity at constant volume, (e) the plasma composition. Initial densities (the same for SiO_2 and Ag): 2.5 kg/m^3, 1 kg/m^3, 0.5 kg/m^3, 0.25 kg/m^3, 0.1 kg/m^3 and 0.05 kg/m^3.

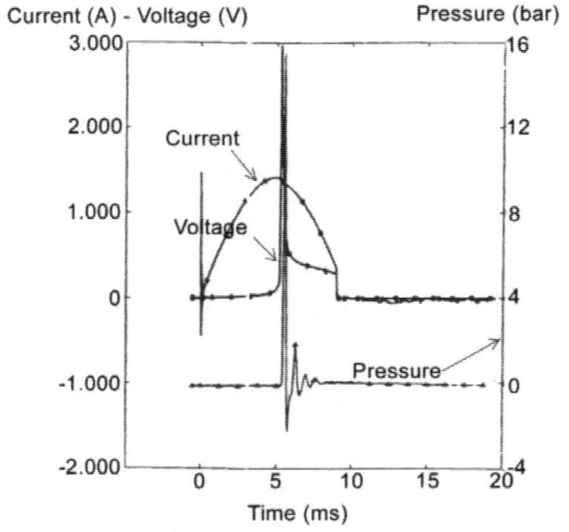

(a)

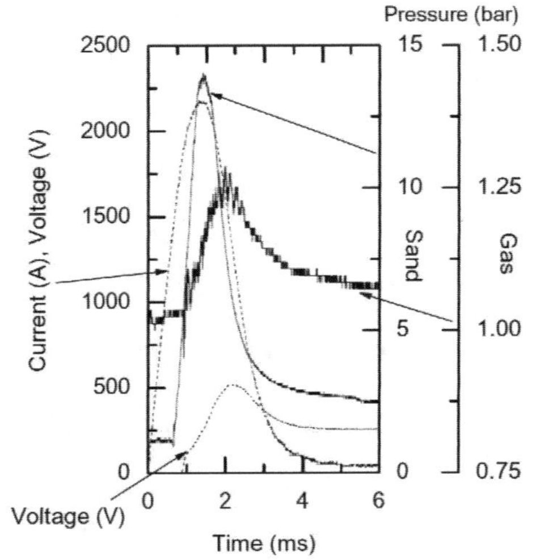

(b)

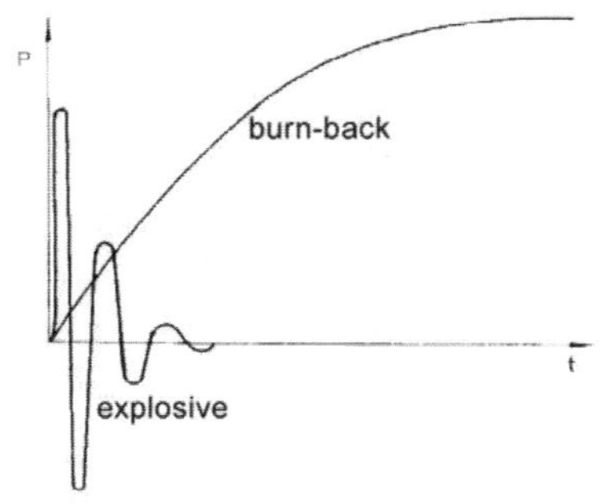

(c)

Figure 9. Experimental pressures of the arc plasma measured in (a) [19] (relative pressure) and (b) [20] (absolute pressure) and (c) idealized traces of pressure components, explosive and burn-back [21].

be reasonably assumed in thermal equilibrium once the erosion fronts have moved enough. To be depicted on the whole duration of the fuse operation the properties are calculated for various initial compositions of the plasma that is to say for a given mass of material (SiO_2 and Ag) which is initially present and fully vaporized to form the plasma. In figure 8 the same mass is put for SiO_2 and Ag from 0.05 kg·m^{-3} to 2.5 kg·m^{-3}. These values are chosen in such a way that the calculated properties fall in the same range as those observed experimentally.

For the pressure within the fuse arc plasma one obtains values varying from approximately the atmospheric pressure for 0.05 kg·m^{-3} around 5000 K up to ~ 10^6 Pa at 25 000 K. The pressure within the arc column is very sensitive to the amount of material actually vaporized because for 2.5 kg·m^{-3} one obtains ~ 4×10^6 Pa at 5000 K up to ~ 45×10^6 Pa at 25 000 K. From the experimental point of view it appears that the measured pressure (obtained by means of a piezoelectric pressure transducer located outside the fuse cartridge, the pressure developped inside the arc being transmitted through a high-temperature ceramic tube filled with glycerine) increases quickly and strongly right from the beginning of the *arcing* period if we refer to [19]: the *arcing* period begins as an explosive situation from the pressure recording as shown in figures 9(a) and 9(c). It is obtained by putting the piezoelectric pressure transducer as close as possible to the arc column. To avoid any possible damage the sensitive area was shielded by organic grease eventually responsible for the amplification of the explosive feature. On the contrary measurements from [20] clearly show a steadier increase especially for the pressure defined as the sand pressure. This pressure is also measured by means of a piezoelectric sensor but the sensitive area is put inside the granular filler at a given length from the initial position of the fuse element in such a way that it can not be in contact with the fulgurite at high temperature. Thus the measured pressure is the pressure due to the sand grains on the sensitive area (protected by silicon grease against point to point contact with the sand grains) and not the plasma pressure. The pressure curve follows the electric power curve and shows a maximum of $(12.3 \pm 2.4)\times10^5$ Pa (figure 9(b)). The measurement of the gas pressure is also given: the gas pressure is the pressure of the gas flowing in the interstices of the granular filler measured at the same position from the fuse element. Values are significantly lower (typically ~ 1.25 bar) pointing out the fact that the diffusion of vapours from the plasma column towards the surroundings – through the outer part of the plasma column mixing viscous silica and partly molten silica sand grains – is very weak or even insignificant. In fact the actual pressure is the result of two components as described in [21]. As it is shown in figure 9(c) there exists an explosive component whose features are linked to the number of reduced sections and the value of overcurrent. It is especially strong at the arc ignition when the reduced section is melted and vaporized. A second component arrives once the erosion of the fuse element process begins namely during the *burn-back*. One may add the ablation of silica sand grains in this last component [21]. The actual pressure is the superposition of these components. The second component is not visible in figure 9(a), and the first component can be seen in figure 9(b) but strongly reduced due to the granular filler between the fuse element and the sensor (and the small dimensions of the reduced section). To conclude about pressure one understands the important contribution of the calculations shown in figure 8(a) in order to help the analysis of experiments and the formulation of a physical modelling.

For the same SiO_2-Ag mixtures the electronic concentrations are shown in figure 8(b) for the temperature range 5000 K – 25 000 K. For the various mixtures the electronic concentrations increase from ~ 3×10^{15} cm^{-3} to ~ 2×10^{16} cm^{-3} at 5000 K to from ~ 2×10^{18} cm^{-3} to ~ 4×10^{19} cm^{-3} at 25 000 K. For the high temperature domain electronic concentration can be roughly identified to the electrical conductivity because electrons are the main chemical species. Thus electrical conductivity increases both with the temperature and the amount of vaporized material: this latter parameter can be linked with the morphometric properties of the silica sand which are carefully checked and controlled in industry. A specific study of the influence of these properties on electrical properties and the temperature and the electronic density of the arc plasma can be found in [22].

The internal energy of SiO_2-Ag plasma is given in figure 8(c). The plasma internal energy defines the energy for a given set of the thermodynamic amounts: temperature, volume (or pressure for example in [23]) and density (amount of SiO_2 and Ag). In other words for a given amount of vaporized material one obtains the maximum energy which can be accumulated within the plasma. The various curves show an increase with the temperature linked to the dissociation and ionization processes and sharp increases due to chemical reactions. These latters can be observed more precisely in figure 8(d). The heat capacity at constant volume results from the derivative of the internal energy. Each fluctuation is associated to a specific chemical reaction which can be isolated from the composition calculation given in figure 8(e). The first peak between 5000 K and 10 000 K corresponds to the dissociation of SiO in Si and O. The second peak between 8000 K and 15 000 K corresponds to the ionization of Si. The ionization of silver appears between 5000 K and 15 000 K and the ionization of Ag^+ for temperature higher than 20 000 K (see [17] for more details). Calculations in figure 8(e) are done for the same weight percentage (50% SiO_2 and 50% Ag) but as the molar weight of silver is higher the composition in figure 8(e) approaches the one of a pure SiO_2 plasma if we consider the main chemical species up to 25 000 K. Electrical neutrality is observed between e^- and Ag^+ up to 9800 K, between e^- and Si^+ from 9800 K to 21 800 K and e^- and O^+ up to 25 000 K.

The knowledge of pressure is also of very high interest both at the beginning and at the end of the *arcing* period. And one has to keep in mind that theoretical studies concerned by fuses have to provide useful informations for industrial purposes. Figure 10(a) shows the dependence between the vaporization temperatures of liquid silver (—) and liquid silica (– – –) and the air percentage in silver and silica for pressures up to 50 bars, and figure 10(b) shows the dependence between the vaporization temperature of liquid silver (—) and liquid silica (– – –) and the pressure (for various mixtures of SiO_2

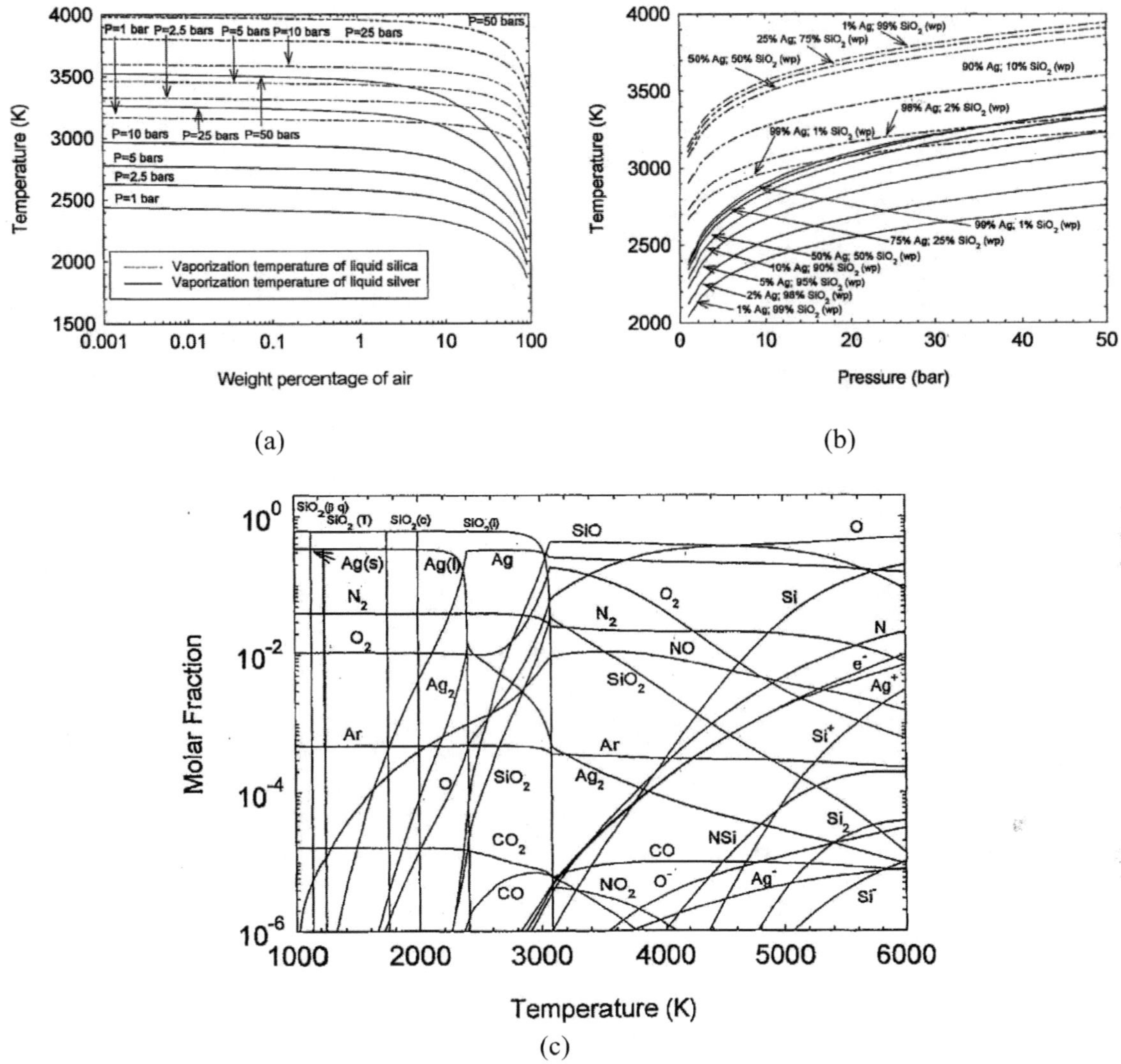

Figure 10. (a) Vaporization temperature of liquid silver versus weight air percentage in silver and vaporization temperature of liquid silica versus weight air percentage in silica for various pressures. (b) Vaporization temperature of liquid silver and liquid silica calculated for various pressures (weight percentage of air fixed at 2%, silver and silica proportion given in weight percentage). (c) Molar fractions of chemical species versus temperature calculated at 1 bar (weight percentage of air fixed at 2%, silver and silica proportion fixed at 50%-50% given in weight percentage): s for solid, l for liquid, βq for beta quartz, T for tridymite and c for cristobalite. From [23].

and Ag defined in weight percentage, the weight percentage of air being set to 2%) [23]. First for a given pressure the influence of the weight air percentage on the vaporization temperature is insignificant up to 2% air whatever the material. On the contrary pressure plays a great role: between 1 and 50 bars the vaporization temperature of silver increases by 1.5 and those of silica by 1.3. These results can be linked to industrial considerations, especially the mean granulometry of the granular silica sand. If the porosity or volume of the void space is kept constant one understands that the characteristic length of the pore will vary with the mean equivalent diameter of the silica sand grains. For example in the case of sphere bead instead of silica sand grains bead with a porosity ~ 0.4 [24], the

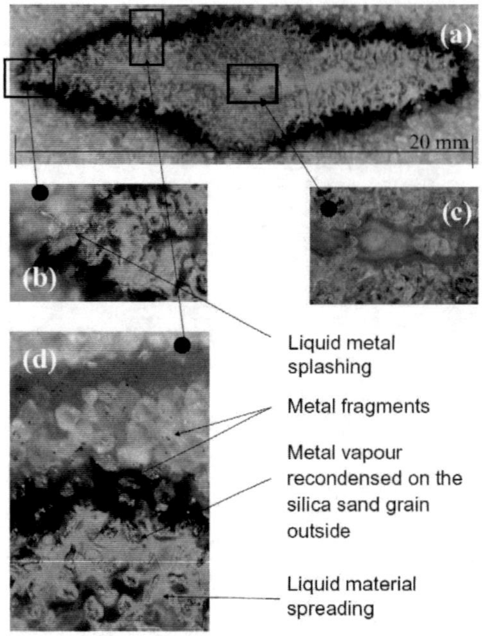

Figure 11. Views of the section of a fulgurite [71]. (a) Section view of the inter-electrode gap (20 mm), the fuse element strip width being initially perpendicular to the plan of the picture. (b) Magnifying of the erosion front. (c) Magnifying of the central part of the plasma column. (d) Magnifying of the edge of the fulgurite.

equivalent pore diameter is found ≈ 116 μm for 275 μm-mean diameter spheres, ≈ 172 μm for 375 μm-mean diameter spheres and ≈ 2086 μm for 5 mm-mean diameter spheres (the granulometric distribution around the central quoted value being different for each mean granulometry). The void space volume depends directly on the grain size which is an industrial parameter and can modify the vaporization temperature of silver and silica thus changing the plasma composition. Moreover the mean granulometry also governs the pressure. The plasma composition is the result of many dissociation and ionization processes whose temperature of appearance depends on the plasma pressure: higher the pressure, higher the temperature for which these chemical processes appear.

Figure 10(c) gives the plasma composition for the temperature range from 1000 K to 6000 K: this temperature range is of special interest both for the beginning of the *arcing* period – ignition of the plasma, increase of the temperature and evolution of the plasma composition initially dominated by silver vapours and modified because of the ablation of silica sand grains – and for the end of the *arcing* period (but without air). A single example is shown for the silver-silica proportion (50% Ag, 50% SiO_2) and 2% air (all in weight percentage). First whatever the material – silver or silica – it is observed in the solid and liquid states before to be dissociated and/or ionized. For the low temperature domain (< 1500 K) the gaseous species are issued from air only, namely Ar, CO_2, N_2 and O_2. For silica (< 3000 K) we observe the following phase transitions in the solid state (at 1 bar): from β-quartz to tridymite at 1140 K, from tridymite to cristobalite at 1743 K and from cristobalite to liquid at 1996 K. Secondly for high temperature the charged neutrality is mainly made between Ag^+ and electrons with a smaller contribution of Si^+. But it depends strongly on the proportion of silver in the silver-silica mixture. For the mixture (99% Ag, 1% SiO_2) the neutrality is made between Ag^+ and e^-, for the mixture (1% Ag, 99% SiO_2) the neutrality is made between Si^+ and e^-. Thirdly the behaviour of liquid silver vaporization and liquid silica vaporization depends on the relative proportion of silver with silica [23].

Various section views of the fulgurite obtained after the fuse operation are shown in figure 11. The width of the fuse element strip is initially perpendicular to the picture. The whole of the fulgurite is shown in figure 11(a), the interelectrode length being ~ 20 mm. The thickness of the fulgurite is more important in the centre which corresponds to the reduced section area where the arc is ignited. Due to the vaporization of silver first and hence silica on one hand and the very high temperature on the other hand, the pressure is maximum in the center of the fulgurite volume: for a fuse element strip of ~ 0.1

IOP Conf. Series: Materials Science and Engineering **29** (2012) 012001 doi:10.1088/1757-899X/29/1/012001

mm in thickness, the plasma column thickness is about 50 times this initial thickness. During the fuse operation the fuse strip is melted and only a part of the liquid silver is vaporized: figure 11(b) shows a liquid metal splashing which is observed around one of the two erosion fronts. The central part of the plasma column where the reduced section is initially present is shown in figure 11(c): the plasma column is surrounded by silica sand grains partially fused. The surrounding area of the fulgurite is shown in figure 11(d) with: metal fragments and especially some silver droplets which are solidified in the interstices of the silica sand grains which are not connected to the fulgurite volume ; some silver vapours which are condensed in the edges of the plasma column ; and liquid material spreading mainly composed of fused silica which has been solidified between the partially eroded silica sand grains. Thus the depiction of the fulgurite formation is a complex task because silver and silica can be observed in the three phases simultaneously but in different areas in the whole of the fulgurite volume, from the core of the plasma column to the extreme edges in the silica sand grains. Among the transport coefficients the viscosity appears to be very important. The volume of the fulgurite – especially in the thickness direction – depends on the viscosity of the fused silica which spreads through the silica sand grains. In the case of an industrial fuse the merging of the fulgurites formed on the whole length of a fuse element could imply the explosion of the fuse which leads to a dangerous situation from the safety point of view.

2.5. Pre-arcing period depiction

The *pre-arcing* period is studied in section 3 independently of the *arcing* period which is the main topic of the Part 2 of the current review. Up to know many measurements and modellings have been published for the *pre-arcing* period, with a good correlation between measurements and calculations. In Part 2 modellings based on thermal-electrical analogy [41] or equivalent electric circuits [40,61] or black box modellings [25] or Cassie-Mayr arc model [62] are not discussed. On the contrary the *arcing* period is discussed in Part 2 by refering to various types of studies – empirical and physical modellings, experiments – because the situation is more complex.

As a consequence it is not possible to build a fully physical modelling taking into account the whole of the industrial parameters as input data (features of the fuse element strip and of the quenching material, nominal electrical parameters, features of the electric *fault*, electromagnetic emissions from the surroundings or 3-phase disturbance [28], properties of the ambiant air, coupling with neighbouring electrical sources or fuses or various switchgears [29], …) and providing the whole set of electrical and physical output data (fuse current waveform, actual voltage drop, Joule intergal, fuse electrical resistance, dielectric rigidity of the fulgurite *i.e.* of the fuse [30], …). The works dealing with the *arcing* period are discussed in Part 2 with a comparison with experimental results when available.

3. Pre-arcing phenomena in electric fuses

3.1. Pre-arcing time and time-current characteristics

Pre-arcing periods can be classified into three main groups according to their duration as shown in table 2: short time, medium time and long time [33]. The simplest case is the short time where adiabatic assumptions are often sufficient to obtain valuable calculations. The two more complex situations are medium and long *pre-arcing* times. For these two latter cases the influence of the morphometric properties becomes significant and makes more difficult the treatment of the thermal diffusion. Up to now and at our knowledge it is not possible to consider the thermal diffusion in the intersection areas of adjacent grains precisely and without long time consuming calculations – inability to define the intersection area: shape and value – but only the transfer in the arc quenching material with the help of mean features such as the mean granulometry, the mean packing density, the mean characteristic length of the void space.

Time-current characteristics are essential curves which allow the prediction of the *pre-arcing* duration namely in this case the time necessary to melt the fuse element. It has to be pointed out that most of

Table 2. Definition of the *pre-arcing* period according to its duration [33].

	Duration	Features
Short time	Up to 10 ms	Up to 1 ms : adiabatic assumption is valid, *i.e.* the heat transfer takes place in the reduced section mainly, the heat transfer towards the remainder fuse element parts and other material is negligible. From 1 ms to 10 ms : heat transfer by thermal diffusion between the reduced section and the remainder fuse element parts is significant and governs the temperature increase within the reduced section.
Medium time	10 ms to 1 s	In addition to the heat transfer in the remainder parts of the fuse element, the heat transfer towards the surroundings silica sand grains becomes significant: this latter contribution depends strongly on the morphometric properties of the silica sand grains.
Long time	More than 1 s	The whole of the outer components – end-caps, electrodes, connecting cables, cartridge, arc quenching material – play a more or less role in the effective heat transfer.

the published calculations are done up to the melting temperature of the constitutive metal of the fuse element. Thus the phase transition from liquid to vapour is disregarded. A typical example of time-current curve is given in figure 12(a) in the case of fuselinks used for semiconductor protection and typical curves of industrial technical instructions are shown in figure 12(b). In the past such data were obtained by experiments only. Some calculated results have been obtained mostly in the case of plain cylindrical wire elements up to 35 ms for various element forms and at extremely high currents which allowed to neglect the heat transfer [3]. But for most of the actual or modern industrial fuselinks a *pre-arcing* time of 3 ms or even 0.3 ms in the case of fuselinks used for semiconductors protection appears to be the high limit for a close agreement between experiments and calculations. And nowadays industrial laboratory tests remain the only way to proceed in regard to the validation of the operating conditions set by standards such as those from the IEC. The treatment of the *pre-arcing* period in adiabatic conditions is discussed in section 3.2.

Because of mathematical and numerical improvements [3, 32] some valuable calculated results have been obtained in the recent past for both severe and not so severe short-circuit currents by taking into account the heat transfer in the whole of the fuse element (and not only in and around the reduced section) but not towards the various components such as the quenching material, the end-caps and the cartridge. Close agreement was found between measurements and calculations for *pre-arcing* duration up to around 5 ms to 10 ms.

3.2. Pre-arcing time modelling

3.2.1. Adiabatic assumption with the Meyer's coefficient. This approach corresponds to the simplest case where strong overcurrents imply short pre-arcing period durations. The adiabatic assumption considered within the framework of the Meyer's coefficient can be defined in a simple way. The Joule integral is assumed to be proportional to the conduction section (s) of the electric current: $\int_{t_0}^{t_{prearc}} i^2(t)\cdot dt = K\times s^2$, where t_0 is the time of application of the fault current, t_{prearc} is the duration of the pre-arcing period where it is assumed that the pre-arcing period ends once the melting temperature is reached. The constant of proportionality K is usually given in $A^2\cdot s\cdot mm^{-4}$. This

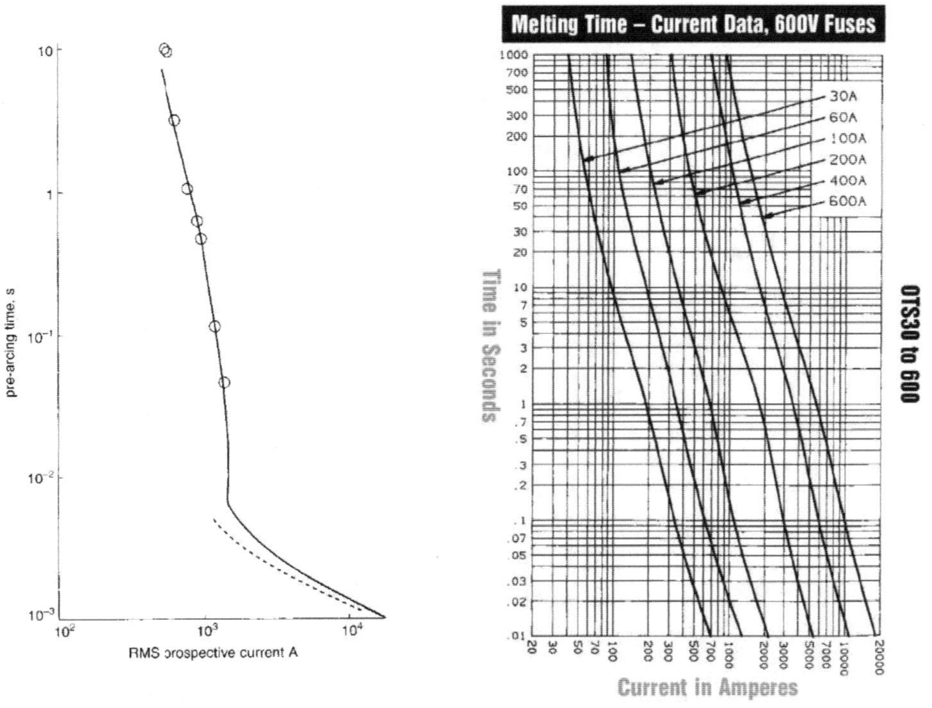

(a) (b)

Figure 12. Pre-arcing time-current characteristics: (a) for a 200 A semiconductor-protection fuselink carrying 50 Hz symmetrical sinusoidal current (—, predicted results ; ○, test results ; – – –, curve from adiabatic formula) [3] ; (b) for OTS general purpose fuse series [31].

approach has been first developped by Meyer [35]. More recently the coefficient K has been studied experimentally by Deshayes [34] for various metals and especially silver and copper.

For only one material different values for K can be defined. From the voltage curve the time corresponding to the ignition of the arc can be measured very accurately if it is identified to the strong and quick increase of the voltage across the fuse. Then the Joule integral is calculated up to this time from the measured current curve. Even if the end of the *pre-arcing* period is still not well understood from the physical point of view the quoted voltage reference refers to the ignition of the arc *i.e.* the splitting of the fuse element into two parts which leads at least to the total cathode-anode voltage fall. So the coefficient K includes the various phase changes without any differentiation.

Another approach consists in calculating the theoretical value of K within the adiabatic assumption for each following case [15]: i) up to the end of the heating of the solid state once the melting temperature is reached, ii) or once the energy necessary for melting has been provided, iii) or at the end of the heating of the liquid up to the vaporization temperature, and finally iv) once the energy necessary for vaporization has been provided. After calculation the value of K is obtained after each step: i) 6.13×10^4 $A^2.s.mm^{-4}$, ii) 7.11×10^4 $A^2.s.mm^{-4}$, iii) 8.89×10^4 $A^2.s.mm^{-4}$ and iv) 28.5×10^4 $A^2.s.mm^{-4}$. Because of the adiabatic approximation these values have to be compared with experimental results obtained in the case of strong overcurrents *i.e.* short *pre-arcing* period duration. Experimental values of K lie between 8.89×10^4 $A^2.s.mm^{-4}$ and 28.5×10^4 $A^2.s.mm^{-4}$ and are mainly close to the smallest value. It shows that the fuse element is splitted before the reduced section is fully vaporized.

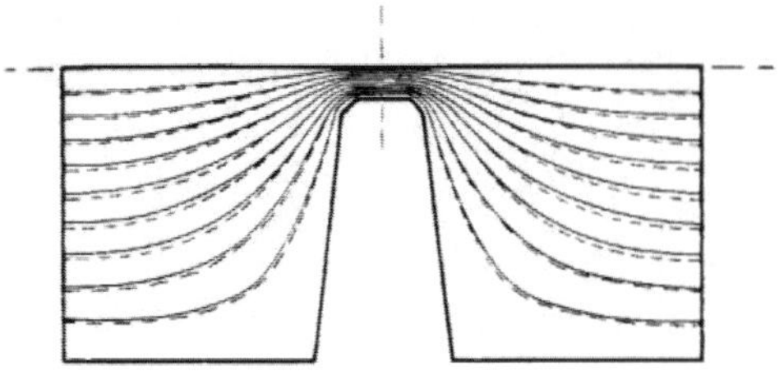

(a)

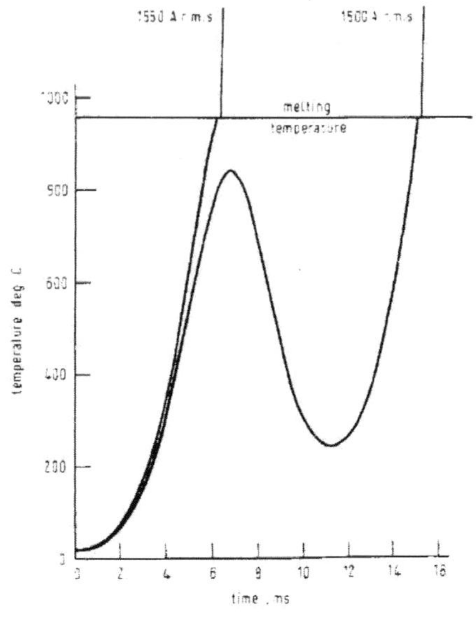

(b)

Figure 13. Contribution of a 2D-finite-difference modelling dedicated to the calculation of the *pre-arcing* period duration [32].
(a) Current-flow lines for uniform temperature distribution (—, initial temperature) and once the reduced section is at melting temperature (– – –).
(b) Temperature versus time curve in the case of a semiconductor-protection reduced section illustrating the sensitivity of the modelling in the case of two sinusoidal currents: at 1550 A *rms* and 1500 A *rms*.

In industry this coefficient is of very high interest. For a given set of reduced section or fuse element features or material the coefficient can be experimentally assessed with a significant accuracy. Once it is known the *pre-arcing* period duration can be estimated for various electric fuse configurations. The coefficient K based method provides a highly efficacious tool to design new electric fuses for novel electrical safety features.

3.2.2. Physical modelling dedicated to the fuse element. Various modellings have been published by taking into account heat transfer in the fuse element only [42], the influence of the filling material being disregarded. Of course such assumption is valid for strong overcurrent but for smaller overcurrent also. In fact one has for the thermal condutivities at ambiant temperatrue k_{Ag} >> k_{SiO_2} >> k_{Air}. In the case where the silica sand grains are uniformely spread (especially the contact areas between the grains and the fuse strip) along the whole length of the fuse strip they play a minor role in the total heat transfer. It is particularly true when the *pre-arcing* period lasts one or two half-periods of the 50-Hz that is to say for sufficiently high overcurrents.

Numerical results have been obtained in [32] by means of finite-difference methode. First (figure 13(a)) the modelling shows that the current lines are not sensitive to the temperature distribution in the reduced section: uniform temperature (the initial temperature in the example) leads to a close result as those obtained in the case of the reduced section at the melting temperature. Secondly it possible to reproduce real fuse operation thanks to a high sensitivity on the *rms* current value. Figure 13(b) shows the temperature versus time curve for two symmetrical *fault* currents in the case of a semiconductor-protection reduced section: 1550 A *rms* in the first case, and a 3% lower *rms* value in the second case (~ 1500 A *rms*). The melting temperature is respectively obtained ~ 6 ms and ~ 14.5 ms. At 1550 A *rms* the Joule integral and the consecutive heating of the reduced section is enough to reach the melting temperature in less than one halfcycle. At 1500 A *rms* and during the first halfcycle the temperature reaches its maximum ~ 850 °C < 962 °C = T_m between 6.5 ms to 7 ms which corresponds to the peak current during the first halfcycle. During the second part of the halcycle the current decrease implies the Joule integral decrease. Therefore the reduced section temperature decreases

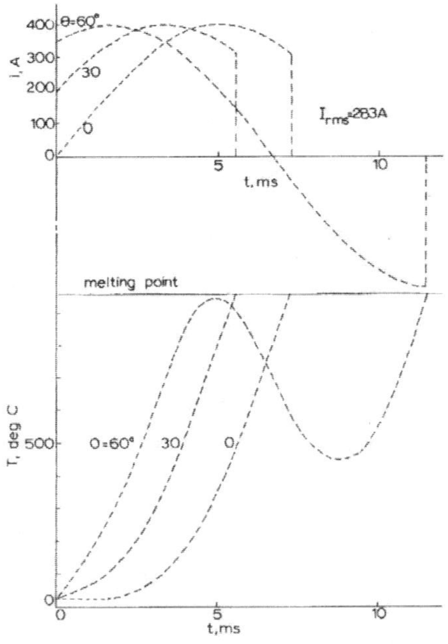

(a)

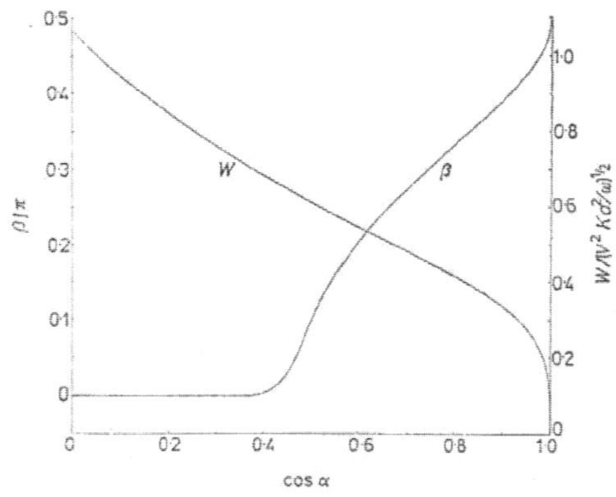

(b)

Figure 14. Influence of the closing angle during the *pre-arcing* period.
(a) Effect of the closing angle on melting of notched element (power factor $\cos\varphi = 1.0$, prospective current set at 283 A, temperature observed in the centre of the reduced section) [37].
(b) Maximum pre-arcing energy W and optimum closing angle ($\beta \equiv \theta$, rad) as a function of circuit power factor ($\cos\alpha \equiv \cos\varphi$) [38].

down to ~ 250 °C between 11.5 ms to 12 ms, the temperature decrease being intensified by the heat transfer by conduction in the two parts of the fuse element. The melting temperature is only reached around 15.5 ms *i.e.* for approximately the current peak during the second halfcycle. These results are close to those obtained in practical tests. Such results are consistent with more recent ones [36] obtained for various fuse elements and operating conditions: the calculation is done by taking into account the phase changes up to the vaporization.

Similar calculations are done in [37] where the effect of the closing angle θ – equivalent to the time delay between the start of the *fault* current and the voltage source – is studied. From figure 14(a) and for the closing angle values 0° and 30° the melting point is respectively obtained after 7.25 ms and ~ 5 ms. For the 60° theta value the melting point is not reached by the centre of the reduced section during the first halfcycle and the temperature falls by ~ 500°C as the current falls and passes through zero. The melting point is reached only in the second halfcycle ~ 11.5 ms. The heat transfer by conduction in the outer parts of the fuse element combined with both the fuse element geometry and the *fault* features – peak value of the sinusoidal current and closing angle for a given power factor – play a great role in the resulting temprature of the reduced section temperature and thus in the time necessary to reach the melting temperature. These calculated trends are reinforced by practical tests whose results are close to the calculations as quoted in [37]. These two main electrical features – the closing angle and the power factor – have been investigated in [39] by means of an experimentally approach and in [38] by considering an adiabatic heating of the fuse element, the *pre-arcing* period being depicted from a set of common electrical equations. From this latter we show in figure 14(b) the dependence between the maximum pre-arcing energy and the circuit power factor with the corresponding optimal value for the closing angle.

As said before the *pre-arcing* period in modelling is defined in most of the studies as the time necessary to reach the melting point of the constitutive material of the fuse element. In practice – experimental tests in laboratory or industrial tests for normalized qualification for example – the end of the *pre-arcing* period is identified to the sharp voltage drop linked to the arc ignition or splitting of

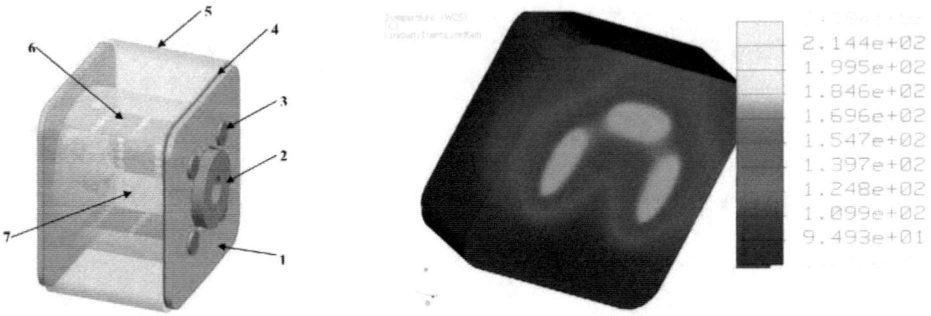

(a) (b)

Figure 15. 3D thermal modelling including the arc quenching material in the heat transfer resolution [46]. (a) Sketch of the fast fuse dedicated to semiconductor protection: 1, outer cap ; 2, end tag : 3, rivet ; 4, inner cap ; 5, ceramic bodic ; 6, fuselink ; 7, granular quartz. (b) Temperature distribution through the fuse at 50% cross section.

the fuse element into two parts. But in practice many works show that the time to vaporize the partly liquid reduced section is very short. This can be easily understood: once the core of the reduced section is liquid, electrical conductivity is strongly increased because of the phase change and the Joule effect is accordingly increased thus shortening the delay to reach the vaporization ($\int_{t'}^{t''} r_f(t)\cdot i^2(t)\cdot dt$, where $t' < t'' \leq t_b$). Some additional works attempted to depict the liquid to vapour phase change on one hand and the role played by the quenching material in the total heat transfer on the other hand. First these attempts are done to improve calculations for long *pre-arcing* periods and secondly they may be considered as a preliminary step before to model *arcing* period in its whole by means of a physical-electrical approach only *i.e.* without any empirical factors.

3.2.3. Physical modelling dedicated to the fuse element and the filler. There exists only a few numbers of such attempts for the *pre-arcing* period. In fact the treatment of the filler is a complex task and as mentionned before main of the valuable modellings work quite well even for some medium *pre-arcing* times (small delay to reach the arc ignition from the melting point) because thermal conduction in the fuse element is the main heat transfer.

Most of the modellings dealing with the arc quenching material are designed for industrial applications in order to predict the thermal volumic cartography (figure 15). This is of high interest in order to know the thermal behaviour of the fuse especially the temperature of the fuse caps or of the cartridge. In [43-45] the filler is considered in its whole and without any phase change and concerns mostly the heat transfer during *arcing* period where the temperature in the plasma column reaches its maximum. Such calculations are also done for the *pre-arcing* period but are of lesser interest because the temperature increase is small compared to those for the *arcing* period except in the case of long *pre-arcing* time where the temperature of the fuse components can be modified. In [2] the high limits in temperature for each of the electric fuse components are quoted in the case of the nominal current. More attention is dedicated to the pressure increase and spreading towards the ceramic body linked to the arc ignition at the end of the *pre-arcing* period.

The difficulties met in the assessment of the thermal properties of the arc quenching material are discussed in Part 2 for the *arcing* period. But these properties can play a non negligible role even during the *pre-arcing* period. The problem with the thermal problematics for the filler lies mainly in the lack of basic data such as the knowledge of thermal conduction for silica – in the solid, liquid and vapour phases. Another question is concerned with the formal depiction of the contact area between the silica sand grains. Especially during the filling of HBC fuses the silica sand grains are continuously vibrated to obtain the escounted value of the final packing density typically ~ 1.6 g·cm^{-3}.

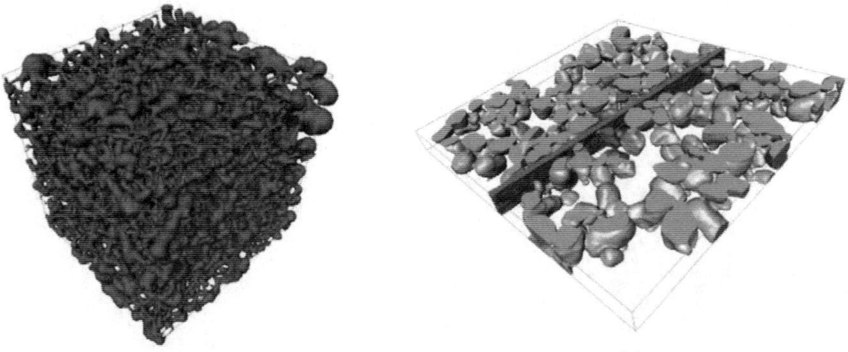

(a) (b)

Figure 16. Characterization of the void space defining the porosity of the electric fuse [47]. (a) 3D network showing the void space channels. (b) Focus around the fuse element strip (in blue) showing a single plan of compacted silica sand grains (in brown).

This operation results in a complex void space network which is otherwise disturbed by the fuse element strip. In figure 16(a) is shown the network corresponding to the void space channels between the silica sand grains [47]. Even with silica sand grains with a mean granulometry centred on a narrow granulometric interval the resulting network is very complex and is characterized by a wide range of subvolumes. In figure 16(b) is shown a section view of the silica sand grains around the fuse element strip. The homogeneous distribution of the grains is altered because the fuse strip has the same effect as a wall for the closest grains.

The most valuable works dealing with heat transfer in granular media are mainly concerned with spherical particles *i.e.* symetrical particles. The coordination number *i.e.* the number of contacts of one sphere made by the surrounding spheres is linked to the void fraction of the packed bed and plays a major role in the evaluation of the effective thermal conducitivity [48]. Efficient formulas are built to predict the actual effective thermal conductivity on one hand and the effective conductivity of a wet bed with void fractions ~ 0.36 to 0.54 on the other hand. A synthesis of the main contributions is given in [49] both for experimental and theoretical approaches: many results are concerned with spherical packed beds but particles with various shape factors are also discussed. Sand packed beds have been widely studied but comparative works show noticeable discrepancies between experiments and calculations [50]. This is due to the wide variety of sands from the particle shape point of view: from one to another quarry the mean shape of the particles can vary together with the mean granulometry and thus the resulting packing density or equivalent void fraction. Such scattering in the basic morphometric properties makes more difficult the establishment of a coherent physical depiction valid for each type of sand. Thus many discrepancies appear when predicting formulas or empirical modellings are compared with measurements obtained for packed beds built on particles issued from different quarries. The numerical attempt published in [51] based on an axi-symetric fuse model gives an actual value of the filler thermal conductivity which lies in the range from ~ 5.06×10^{-3} $W\cdot cm^{-1}\cdot°C^{-1}$ to ~ 3.77×10^{-3} $W\cdot cm^{-1}\cdot°C^{-1}$ for the electric fuse concerned in the study. But in this numerical assessment the filler is considered in its whole namely the result depends on the filler morphometric properties and the way to proceed for packing. At the end this numerical values are helpful when one needs to assess the temperature of the outer parts of the electric fuse but are inadequate to explain the thermal processes close to the arc ignition point or the close surroundings mostly because of the phase changes. In [60] the thermal conductivity (k ,$W\cdot m^{-1}\cdot°C^{-1}$) was assessed experimentally for a sand of one chemical composition but various particle size distributions. Results are given under the form of a

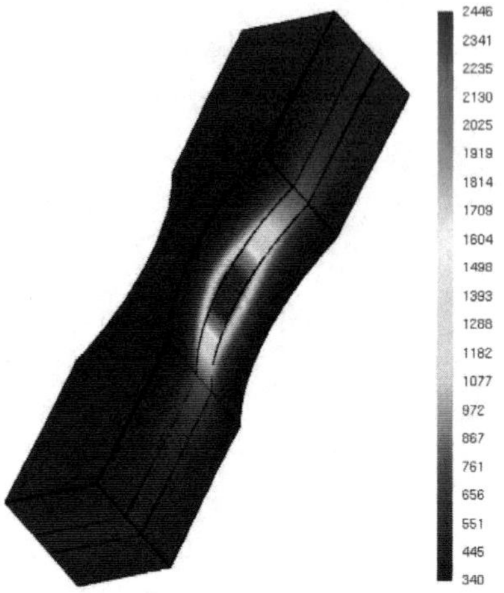

Figure 17. Temperature cartography in Kelvin of a single fuse element strip with circular reduced sections (see [15] for more details) at the vaporization point.

simple expression taking into account the form factor (f) and the mean grain size (D, mm) of the particles and the density of compactness (δ , g·cm^{-3}): $k = 0.535\times f^{1/1.55}\times D^{1/5.5}\times \delta^{1/1.6}$.

In [15] the treatement of the silica sand effective conductivity is considered by means of two typical schemas, namely a parallel shema (high limit of the effective thermal conductivity) and a series schema (low limit of the thermal conductivity). As a result the temperature cartography is calculated as shown in figure 17.

3.3. Fuse element disintegration dedicated to pre-arcing understandings

The *pre-arcing* period ends with the arc ignition that is to say with the fuse element disintegration. This step is of very high interest because it is the former step of the following operation time namely the *arcing* period. This intermediate step between the *pre-arcing* and the *arcing* steps is not still fully understood because many processes occur in the same time with a strong discontinuity in the main electrical and physical properties. Many experimental works have been done in order to explain the disintegration of the fuse element equipped with one or several reduced sections. The time-voltage curve shown in figure 18(a) is obtained for a single reduced section fuse element (see [4] for operating conditions) at the onset of the arc initiation. Two steps are observed: a sudden increase (U_{in}) followed by a more gradually increase up to a limit value (U_{top}). The explosive component observed for the pressure (figure 9) is linked to the first step and indicates a violent disintegration of the fuse element such as in the case of a simple wire. If two reduced sections equip the fuse element as in figure 18(b) two sudden voltage increases are observed. These voltage peaks correspond to the consecutive disintegrations of each reduced section, this step being immediately followed by a more gradually increase corresponding to the erosion of the fuse element during the *arcing* period. The complex structure of typical fuse elements such as those of figure 2 makes more difficult the study of the fuse element disintegration. As a consequence in many experimental works typical fuse elements are replaced by cylindrical fuse wires without any reduced sections [53-57]. Moreover this phenomenon has been widely investigated within the framework of plasma guns [58]. Even if these latters work are concerned with very high value of di/dt many shared problematics are in common. A comprehensive work dedicated to electric fuses problematics is published in [53]. In figure 18(c) we show the influence of various fault currents on especially the voltage curve up to the beginning of the *arcing* period thus including the transition between the *pre-arcing* and the *arcing* periods. For the investigated range of currents this influence is rather weak: it is due to the fuse wire diameters and the dissipated energy. In their paper the authors explain that the whole length of the fuse wire is disintegrated leading to metal chunks and droplets in suspension in the interaction volume. If the fuse

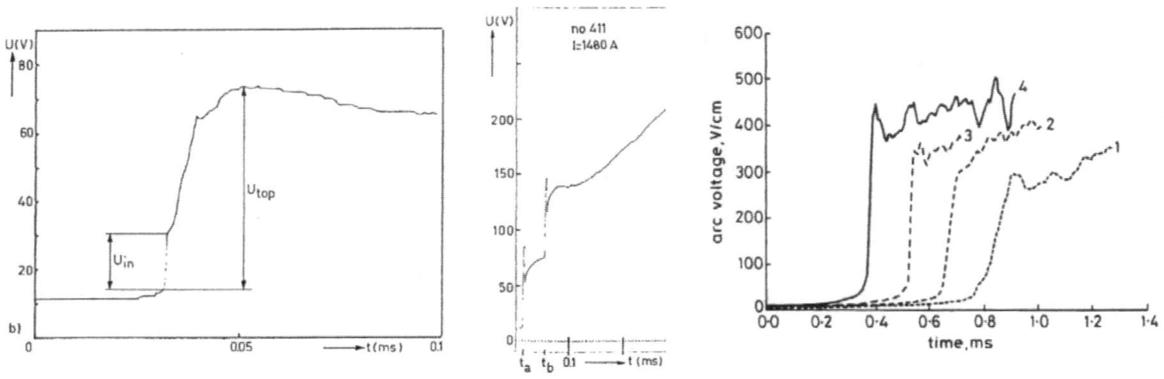

(a) (b) (c)

Figure 18. Influence of the reduced section disintegration on the arc voltage [4]: (a) for a single reduced section fuse element and (b) for two reduced sections fuse element. (c) Influence of the fault current on the arc ignition voltage in the case on narrow slots (1: 0.6 kA, 2: 1.6 kA, 3: 4.4 kA, 4: 12 kA) [53].

wire diameter is too small, the constitutive material is systematically destroyed in its whole and the internal energy of the resulting metallic plasma produced during the transition is at its maximal value depending on the pressure, temperature and vaporized material amount.

4. Conclusion

In the fifty most recent years experimental and theoretical works intended to increase the knowledge of the oldest electrical safety device namely electric fuse. From 1883 and the patent published by *Boys* and *Cunyngham* [3] many improvements have been carried out both by industry and academics. For both the final results of these should be to build a physical and comprehensive modelling of the two operation periods, namely the *preacing* and the *arcing* periods without disregarding the normalization tests to check the features imposed by international standards.

Most of the knowledge about the fuse operation comes from experiments dealing with the various processes occuring during the fuse operation.

The *pre-arcing* period is at the present time quite well understood especially as the *pre-arcing* end is identified to the melting of the fuse element. Within this limit powerful numerical tools have been produced especially those based on finite elements schema. Thermal transfer within the fuse element is correctly depicted but the thermal transfer through the granular filler is more roughly modeled by means of the effective thermal conductivity. In modelling some of the current works are focussed on the actual depiction of the contact areas between the silica sand grains and the fuse element because they can play a non negligible role in the whole heat transfer. Main difficulties lie in the accurate depiction of the silica sand grain shapes. The end of the *pre-arcing* period or arc ignition is not still fully understood. Many unknowns remain especially about the electrode phenomena – even in the case of electric fuse for not so high overcurrents some differences have been pointed out between the anode side and the cathode side – directly linked to the fuse element erosion or *burn-back* and about the disruption of the fuse element (partial vaporization of the melted reduced section).

Aknowledgments

The author is grateful to industrial partners and academic partners for their useful discussions and their help during continued collaborations (Alstom, Electricité de France, Schneider Electric, Mersen, ACOIFF, Direction Générale des Entreprises) and also to the institutions for valuable conversations and for allowing us to reproduce some materials (SIMAP, Grenoble INP, Centre SPIN Ecole Nationale Supérieure des Mines de Saint-Etienne). The author would like to thank particularly the colleagues from LAEPT.

References

[1] Gelet J L 2007 To the origins of fuse (Clermont-Ferrand: Blaise Pascal Academic Press, IEEE Catalog Number: 07EX1870) pp 1-8

[2] IEC 60282-1 2002 ed 5: High-voltage fuse – Part 1: Current-limiting fuses *International Electrotechnical Commission Technical Committees & Subcommittees* http://www.iec.ch/dyn/www/f?p=103:6:0

[3] Wright A and Newbery P G 2004 Electric fuses *ed IEE Power & Energy Series* **49** 3rd edition (London)

[4] Daalder J A and Schreurs E F 1983 Arcing phenomena in high voltage fuses *EUT Report* **83-E-137** (Eindhoven: EUT Dept. of Electrical Engineering)

[5] Bouilliez O and Perez Quesada J C 2002 Conception et utilisation de fusibles limiteurs MT *Cahier Technique Schneider Electric* **n°128**

[6] International Electrotechnical Commission (IEC) http://www.iec.ch

[7] Rakhovskii V I 1976 Experimental study of the dynamics of cathode spots development *IEEE Transactions on Plasma Science* **4** N°2 81-102

[8] Sen Sn, Gantait M and Jana DC 1988 Voltage current characteristics of low current arcs in air with metal electrodes *Indian J. Phys.* **62B** 78-87

[9] Saqib M A and Stokes A D 1999 Characteristics of fuse arcing in different fillers *Proc. of the 6th Int. Conf. on Electric Fuses and their Applications* (ICEFA Torino Italy) pp 275-278

[10] Bahrin I 1995 *Thermochemical data of pure substances* 3rd ed Vol 1 (VCH Weinheim NY)

[11] Touloukian Y S, Powell R W and Ho C Y 1970 *Thermal conductivity – metallic elements and alloys, Thermophysical properties of Matter* Vol 1 TPRC Data Series ed IFI/Plenum (NY Washington)

[12] Matula RA 1979 Electrical resistivity of Copper, Gold, Palladium and Silver *J. Phys. Chem. Ref. Data* **8** N°4 1147-1298

[13] Rochette D, Touzani R and Bussière W 2007 Numerical study of the short pre-arcing time in high breaking capacity fuses via an enthalpy formulation *J. Phys. D: Appl. Phys.* **40** 4544-4551

[14] Cagran C and Pottlacher G 2007 Normal spectral emissivities of liquid copper,liquid gold and liquid silver at 684.5 nm *J. of Non-Crystalline Solids* **353** 3582-3586

[15] Memiaghe S 2010 Modélisation du régime de préarc dans les fusibles *Ph.D Thesis* **DU n°2033 EDSF 646** (in French, Clermont-Ferrand)

[16] Stephan K and Laesecke A 1985 The thermal conductivity of fluid air *J. Phys. Chem. Ref. Data* **14** N°1 227-234

[17] Bussière W and André P 2001 Evolution of the composition, the pressure, the thermodynamic properties and the monoatomic spectral lines at fixed volume for a SiO_2-Ag plasma in the temperature range 5000-25 000 K *J. Phys. D: Appl. Phys.* **34** pp 1657-1664

[18] André P 1995 Etude de la composition et des propriétés thermodynamiques des plasmas thermiques à l'équilibre et hors d'équilibre thermodynamique *Ph. D Thesis* (in French) Université Blaise Pascal **DU 771** EDSF **85**

[19] Saqib MA, Stockes AD and Seebacher PJ 1999 Pressure inside the arc channel of a high-voltage fuse *Proc. of the 6th Int. Conf. on Electric Fuses and their Applications* (ICEFA, Torino, Italy) pp 83-88

[20] Rochette D and Bussière W 2004 Pressure evolution during HBC fuse operation *J. Phys. D: Appl. Phys.* **13** 293-302

[21] Lipski T 1985 Generation and propagation of the pressure due to the fuse-element disintegration in HBC fuses *Proc. of the 8th Int. Conf. on Gas Discharges & their Applications* (Oxford, UK) pp 87-90

[22] Bussière W 2001 Influence of sand granulometry on the electrical characteristics, temperature and electron density during the high-voltage fuse arc extinction *J. Phys. D: Appl. Phys.* **34** 1-11

[23] Rochette D, Bussière W and André P 2004 Composition, enthalpy, and vaporization temperature calculation of Ag-SiO_2 plasmas with air in the temperature range from 1000 to 6000 K and for pressure included between 1 and 50 bars *Plasma Chemistry and Plasma Processing* **24** **N°**3 475-492

[24] Bussière W, Rochette D, Latchimy T, Velleaud G and André P 2006 Measurement of Darcy and Forchheimer coefficients for silica sand beads *High Temp. Mat. Processes* **10** 55-78

[25] Cividjian GA, Klejch J and Georgescu D 2003 Black box arc model after current zero *Proc. of the XVth Symp. on Physics of Switching Arc* (Czech Republic Sept. 22-26) pp 48-51

[26] Garrido C and Cidras J 1995 Study of different materials as fuse element *Proc. of the 5th Int. Conf. on Electric Fuses and their Applications* (ICEFA, Ilmenau, Germany) pp 201-206

[27] Ehrhardt A, Rother W, Schumann K and Nutsch G 1995 Dielectric reignition of electric fuses at small overcurrents *Proc. of the 5th Int. Conf. on Electric Fuses and their Applications* (ICEFA Ilmenau Germany) pp 265-272

[28] Lipski T and Partyka R 1995 Protection of disturbance arc on bus-bars by means of LV fuses *Proc. of the 5th Int. Conf. on Electric Fuses and their Applications* (ICEFA Ilmenau Germany) pp 171-175

[29] Aftyka W, Jakubiuk K and Lipski T 1995 An analysis of coordination of low-voltage circuit-breakers and HBC fuses *Proc. of the 5th Int. Conf. on Electric Fuses and their Applications* (ICEFA Ilmenau Germany) pp 209-213

[30] Semenowicz B and Partyka R 1999 Dielectric strength of the ultra-short fuse *Proc. of the 6th Int. Conf. on Electric Fuses and their Applications* (ICEFA Torino Italy) pp 99-102

[31] Ferraz Shawmut, North American Power Fuses visited the 30th of October 2011 http://stevenengineering.com/tech_support/PDFs/76NAF.pdf

[32] Leach JG, Newbery PG and Wright A 1973 Analysis of high-rupturing capacity fuselink pre-arcing phenomena by a finite-difference method *Proc. of IEE* **120** N°9 pp 987-993

[33] Beaujean D A, Newbery P G and Jayne M G 1995 Modelling fuse elements using a C.A.D. software package *Proc. 5th Int. Conf. on Electrical Fuses and their Applications* (ICEFA Ilmenau Germany) pp 133–42

[34] Deshayes R 1970 Mise au point et contribution à l'étude du fonctionnement d'un coupe-circuit à fusible pour courant continu *Mémoire d'Ingénieur CNAM* (in French) ed CNAM Paris

[35] Meyer GJ 1907 Theoretisches und Praktisches über Abschmelzsicherungen *Elektrotechnische Zeitschrift Heft* **17 18 47**

[36] Rochette D, Bussière W, Touzani R, Memiaghe S, Velleaud G and André P 2007 Modelling of the pre-arcing period in HBC fuses including solid-liquid-vapour phase changes of the fuse element *Proc. of the 8th Int. Conf. on Electric Fuses and their Applications* (ICEFA, Clermont-Ferrand, France) ed Blaise Pascal University Press pp 87-93

[37] Wilkins R and McEwan PM 1975 A.C. short-circuit performance of notched fuse elements *Proc. IEE* **122** N°3 289-292

[38] Wheeler CB 1976 Optimum conditions for testing electrical fuses: I. Maximization of pre-arcing energy *J. Phys. D: Appl. Phys.* **9** 1809-1816

[39] Bussière W, Rochette D, Velleaud G, Latchimy T, Gelet JL, Gentils F, Perez-Quesada JC, Rambaud T and André P 2008 Experimental study of HBC fuses working at short and medium pre-arcing times *J. Phys. D: Appl. Phys* **41** 105210

[40] Bokor A and Gonda G 1991 Modelling of filaments warming before breaking in high voltage fuses *Elektrotechn. Cas.* **42** 3-4 199-206

[41] Cheim LAV and Howe AF 1991 Calculating fuse pre-arcing times by trasmission-line modelling (TLM) *Proc. of the 4th Int. Conf. on Electric Fuses and their Applications* (ICEFA Nottingham UK) pp 162-167

[42] Fogorosi A and Blaabjerg F 2004 The analysis of pre-arcing period and modeling of a High Breaking Capacity fuse-link in SABER *Proc. of the IAS* ed IEEE pp 2337-2344

[43] Torres E, Mazon AJ, Fernandez E, Zamora I and Perez JC 2007 New FEM model for thermal

analysis of medium voltage fuses *Proc. of the 19th Int. Conf. on Electricity Distribution* (Vienna Austria) Session **1** Paper N°0576

[44] Kawase Y and Miyatake T 2000 Heat analysis of a fuse for semiconductor devices protection using 3-D finite element method *IEEE Trans. on Magnetics* **36** N°4 1377-1380

[45] Ishikawa Y, Hirose K, Asayama M, Yamano Y and Kobayashi S 2007 Dependence of current interruption performance on the element patterns of etched fuses *Proc. of the 8th Int. Conf. on Electric Fuses and their Applications* (ICEFA Clermont-Ferrand France) ed Blaise Pascal University Press pp 51-56

[46] Plesca A 2007 A complete 3D thermal model for fast fuses *Proc. of the 8th Int. Conf. on Electric Fuses and their Applications* (ICEFA Clermont-Ferrand France) ed Blaise Pascal University Press pp 79-85

[47] Olmos L *et al.* 2010 *Private Communication* Laboratoire SIMAP Grenoble INP Centre SPIN Ecole des Mines de Saint-Etienne

[48] Okazaki M, Ito I and Toei R 1997 Effective thermal conductivities of wet granular materials *AichE Symposium Series* **163** N°73 pp 164-176

[49] Tsotsas E and Martin H 1987 Thermal conductivity of packed beds: A review *Chem. Eng. Process.* **22** 19-37

[50] Botteril JSM, Salway AG and Teoman Y 1989 The effective thermal conductivity of high temperature particulate beds – II. Model predictions and the implication of the experimental values *Int. J. Heat Mass Transfer* **32** N°3 595-609

[51] McEwan M P 1977 Numerical determination of the thermal conductivity of powered fuse filler *Proc. of the Switching Arc Phenomena (*Lodtz Poland) Part 1 pp 282-286

[52] Memiaghe S, Bussière W, Rochette D, Touzani R and André P 2010 Simulations and measurements of the pre-arcing times in HBC fuses under typical electric faults *JHTMP* **14** N°3 255-270

[53] Wolny A and Stokes A D 1994 Electrical arc ignition in narrow slots: high-voltage generation *IEE Proc.-Gener. Transm. Distrib.* **141 4** 263-269

[54] Psomopoulos CS and Karagiannopoulos 2002 Measurement of fusible elements during current interruption and interpretation of related phenomena *Measurement* **32** 15-22

[55] Psomopoulos CS and Karagiannopoulos 2002 Temperature distribution of fuse elements during the pre-arcing period *Electric Power Systems Research* **61** 161-167

[56] Psomopoulos CS, Diamadopoulos D, Karras C, Karagiannopolous P and Tsivlikas E 2007 Experimental investigation of the fusion in cylindrical exploding wires using high current densities with industrial frequency *Proc. of the 8th Int. Conf. on Electric Fuses and their Applications* (ICEFA Clermont-Ferrand, France) ed Blaise Pascal University Press p 17-22

[57] Arai S 1999 Geometrical model of discharge space in the state of the high arc voltage generation of capillary arcs *Proc. of the 6th Int. Conf. on Electric Fuses and their Applications* (ICEFA Torino Italy) pp 119-124

[58] Taylor MJ 2002 Formation of plasma around wire fragments created by electrically exploded copper wire *J. Phys. D: Appl. Phys.* **35** 700-709

[59] Wright A and Beaumont KJ 1976 Analysis of high-breaking-capacity fuselink arcing phenomena *Proc. IEE* **123** N°3 252-258

[60] Reineri CA, Gomez JC and Magnago F 1993 Experimental determination of fuse filler thermal conductivity *7th Int. Conf. on Switching Arc Phenomena* (Lodz Poland) **Part I** pp 220-223

[61] Beaujean DA, Newbery PG and Jayne MG 1995 Modelling fuse elements using a CAD software package *Proc. of the 5th Int. Conf. on Electric Fuses and their Applications* (ICEFA Ilmenau Germany) pp 133-142

[62] Baraboi A, Adam M and Leonte P 1995 Modelling of circuit breaking at the fuses working *Proc. of the 5th Int. Conf. on Electric Fuses and their Applications* (ICEFA Ilmenau Germany) pp 143-149

[63] Wilkins R and Gnanalingam S 1978 Burn-back rates of silver fuse elements *Proc. of the 5th Int.*

Conf. on Gas Discharges (University of Liverpool) pp 195-198
[64] Daalder JE and Hartings RM 1981 The burn-back rate of high voltage fuses *Proc. of the 4th Int. Symp. on Switching Arc Phenomena* pp 158-164
[65] Lakshiminarashima CS, Barrault MR and Oliver R 1978 Factors influencing the formation and structure of fuse fulgurites *Proc. IEE* **125** N°5 pp 391-399
[66] Saqib M and Stokes AD 1999 Time resolved spectrum of the fuse arc plasma *Thin Solid Films* **345** 151-155
[67] Cheim LAV 1992 Arc temperature measurements in high breaking capacity fuses *Ph.D Thesis* Univeersity of Nottingham Department of Elect. and Electronic Engineering
[68] Bussière W and Bezborodko P 1999 Measurements of time-resolved spectra of fuse arcs using a new experimental arrangement *J. Phys. D: Appl Phys.* **32** N°14 1693-1701
[69] Arai S 1999 Geometrical model of discharge space in the state of the high arc voltage generation of capillary arcs *Proc. of the 6th Int. Conf. on Electric Fuses and their Applications* (Torino Italy) pp 119-124
[70] Dou Z, Richardeau F, Sarraute E, Blaquière JM, Bley V and Flumian D 2011 Comparison of Short-Circuit Failure Capability of Epoxy Molded Package versus Silicone Gel Module for New Fault-Tolerant Inverter and Interleaved Chopper *Power Conversion and Intelligent Motion Conf.* (PCIM May 17th – 19th 2011 2011 Nuremberg)
[71] Bussière W 2000 Mesure des grandeurs (T, Ne, P) au sein du plasma d'arc des fusibles en moyenne tension *Ph. D Thesis* (in French) Université Blaise Pascal **DU 1258** EDSF **301**

IOP Conf. Series: Materials Science and Engineering **29** (2012) 012002 doi:10.1088/1757-899X/29/1/012002

Electric fuses operation, a review: 2. Arcing period

W Bussière

Clermont Université, Université d'Auvergne, LAEPT, BP 10448, F-63000 Clermont-Ferrand, France

E-mail: william.bussiere@univ-bpclermont.fr

Abstract. In the electric fuse operation the arcing period follows immediately the pre-arcing period depicted in Part 1 (*Part 1. Pre-arcing period*). The transition between these two operation steps is not fully understood at this time. To simplify the beginning of the arcing period can be identified with the electric arc ignition i.e. with the electrodes voltage drop. The consecutive plasma is of metallic type at the beginning of the arcing period and of metallic plus silica type with varying mixture up to the end of the arcing period. The energy brought by the fault current is withdrawn by means of the interaction between the electric arc and the arc quenching material (usually silica sand) whose morphometric properties influence the properties of the plasma column: composition, thermodynamic properties and transport coefficients of the plasma column depend on the porosity (and other morphometric properties) of the filler. The fuse element erosion also known as burn-back is responsible for the lengthening of the plasma column and the variations of the electric field. The whole of these processes is depicted by means of experimental results or modellings when possible.

1. Introduction

In spite of their apparent simplicity both for technical, geometrical and working considerations the physical processes involved in electric fuse operation are far to be fully understood. Moreover they are of very high interest for electrical safety purposes both for persons and systems and widely used in the Low and Middle Voltage systems.

Some general considerations linked to the *arcing* period from the electrical and physical point of views are given in the Part 1 (*Part 1. Pre-arcing period*) of the current review. Especially the *arcing* period in its whole is described from the chemico-chemical point of view.

During the *arcing* period the energy brought by the *fault* current has to be dispersed on one hand. And the electric field induced by the plasma column has to shorten the *arcing* period duration on the other hand. A long *arcing* time is problematic because the electrical conductivity of the plasma column is high enough to allow the flow of the *fault* current and can induce irreversible damage in the devices behind the point of the *fault* current appearance. The electric field (E) level of the plasma column necessary to ensure a short *arcing* time is the result of the compromise between the increase of the voltage across the fuse (v_f) and the relative limitation of the plasma column lengthening or inter-electrodes gap (d_{ie}). In other words the most favourable situation is to quickly increase the electric fuse voltage so that it can not be maintained any more by the supply thus forcing the current to fall down. If one considers that $v_f \approx E \times d_{ie}$ it follows two main remarks. First the electric field of the plasma column is the result of the chemical species issued from the melting and the vaporization of the metallic (mainly silver) fuse element and from the melting and the vaporization (and the dissociation of various chemical species produced at low temperature in the plasma) of the silica sand. The density of each of these chemical species varies inside the plasma volume and also during the *arcing* period. Consequently the electric field varies on the whole duration of the electric fuse operation and can be strongly inhomogeneous from the centre to the edges of the plasma column because of material (silver and silica) injection. Secondly the length of the plasma column which is roughly the gap between the two erosion fronts of the fuse element increases during the *arcing* period. The erosion of the fuse element is the result of various parameters. The fuse element features (material, thickness, shape of the reduced section, number and distribution of the reduced sections on the whole length of the fuse element), the *fault* current level, the level of di/dt at the *fault* current appearance, the energy transfer from the plasma column later during the *arcing* period are some of the main parameters. The selected set of these parameters has to be chosen in order to increase the length of the plasma column at the maximum. But there is a high limit: one has to prevent the merging of each of the plasma columns appearing around each reduced section. If not the electric fuse can be considered as a single plasma column on the whole length of the fuse element which could lead to the electric fuse destruction.

It is not possible to give a full depiction of the physical processes occuring during the *arcing* period be means of a single tool such as a modelling especially as these processes are not still well known or identified. Section 2 is organized as follows. The problematics of the *arcing* period depiction is discussed in section 2.1. Phenomena occuring around the erosion fronts are addressed in section 2.2. Section 2.3 is dedicated to the study of the plasma column, especially the *burn-back* rate, the assessment of some plasma column properties such as the plasma composition, the plasma radiation, the temperature and the electron density, and the *fulgurite*. The comparative analysis of these properties is done in section 2.4. Section 2.5 is dedicated to the trends in electric fuse modelling for the *arcing* period.

2. Arcing phenomena in electric fuses

2.1. Which depiction for which phenomenon?

Arcing period understanding is a very complex task. Many approaches are carried on with plus or less success depending on the expected level of physical depiction.

Difficulties lie in various problematics: the simultaneity of various physical mechanisms occuring in the whole of the interaction volume or are confined either in the plasma column either around the close surroundings of electrodes (with a differentiation between cathode and anode phenomena) ; the very strong gradients – in temperature, electron density and pressure – from the core of the plasma column to the sourroundings causing fluctuations in the dependent transport coefficients such as electrical conductivity, thermal conductivity, viscosity and radiation ; the two main materials - silver and silica – in the solid, liquid and vapour phases ; a continuously varying plasma composition resulting from material ablation – erosion of the fuse element also called *burn-back* and ablation of the silica sand grains – and material removal – spreading of liquid silica through the filler interstices, condensation of metallic vapours in the plasma surroundings – from the plasma core.

To be understood each of these problematics was first studied separately if possible and mainly by experiments. With these experimental databases empirical formulations or modellings were produced but often with significant limitations implying new experiments to obtain the value of adjustable parameters. The next step consisted in semi-empirical modellings or physical modellings dedicated to *arcing* period in its whole. In this latter case the contribution of experimental results is required for comparison purposes.

2.2. Cathode and anode phenomena

As for another types of arc discharges one has to take into account phenomenas occuring at the arc roots. General considerations about arc roots can be found in [1, 2], and cathode and anode phenomena are reviewed respectively in [3, 4] and [5-7] with a detailed review of experimental and

IOP Conf. Series: Materials Science and Engineering **29** (2012) 012002 doi:10.1088/1757-899X/29/1/012002

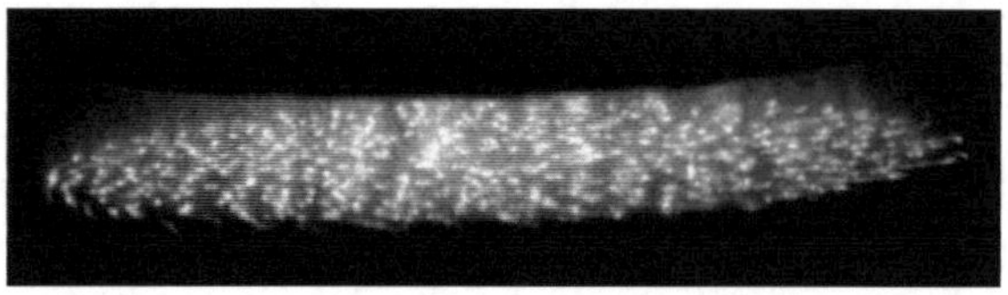

(a)

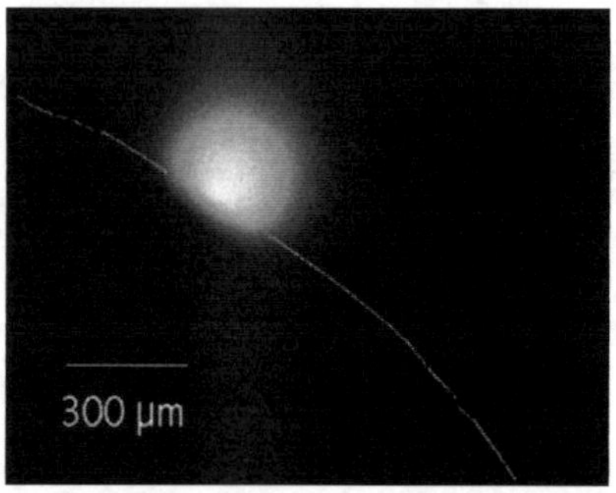

(b)

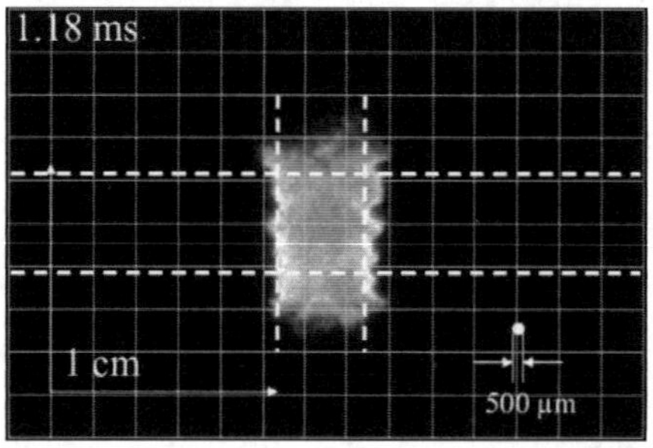

(c)

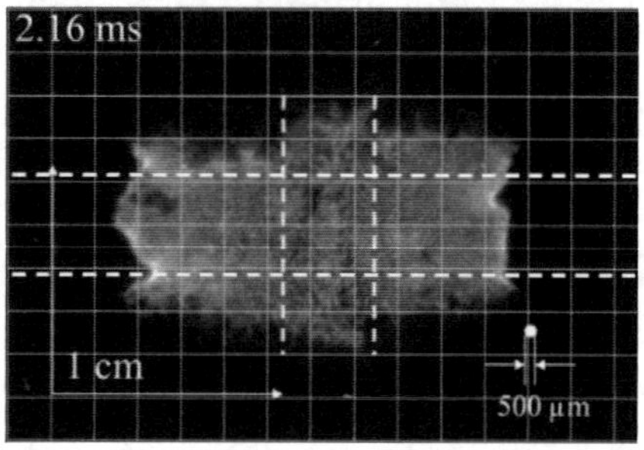

(d)

Figure 1. Pictures of arc roots obtained by high-speed imagery (see quoted references for full details). (a) Cathode spots view in the case of high-current (11 kA) developed vacuum arc stabilized by axial magnetic field (0.25T), exposure 25 µs (copper-chromium electrodes) [6]. (b) Side-on view of a cathode spot on a hot tungsten cathode (thin curved line) at 1.3 bar argon and 1 A current, exposure 500 µs (band-pass filter at 4008.nm suppressing argon lines) [3]. (c), (d) Silver fuse element strip (~ 5 mm-width, 2 circular reduced sections) view focussed on the reduced sections area, di/dt ~ 2×10^{6} $A\cdot s^{-1}$, exposure 3.8 µs (band-pass filter centered on Ag I 520 and Ag I 546 nm) respectively obtained 1.18 ms and 2.16 ms after the start of the *fault* current, *pre-arcing* time ~ 0.89 ms (dashed lines symbolize the intial position of the fuse element strip) [12].

theoretical studies. There exist only few works dealing with cathode and anode phenomena dedicated to electric fuses [8-10]. These latters include the cathode and anode regions in a theoretical approach with simplying assumptions.
No specific experimental studies are dedicated to arc roots or more exactly to erosion fronts in the case of electric fuses: one of the main difficulties lies in the shifting of the two erosion fronts during the *arcing* period. To illustrate the very high difficulty to assess experimentaly these phenomena even in the case of motionless electrodes we refer to figures 1(a) and 1(b) for cathode views. Figure 1(a) refers to a high-current vacuum arc whose ignition takes place on cold electrodes. As mentionned in [6] during the whole duration of an arc (~ ms) the cathode spots (bright points in figure 1(a)) appear, growth and hence disappear many times. During its life each cathode spot is the area where electrode material is removed (inside the cathode spot the temperature is higher than the boiling temperature) to form the plasma by means of plasma jet and microdroplets. The significance of these processes depends on the electrode material, the current and the duration of the arc. In figure 1(b) is shown a side-on view of a spot on a hot cathode in the case of an argon arc with a hot tungsten cathode. As quoted in [3] the bright spot can be seen on the surface of the cathode and its position changes by

IOP Conf. Series: Materials Science and Engineering **29** (2012) 012002 doi:10.1088/1757-899X/29/1/012002

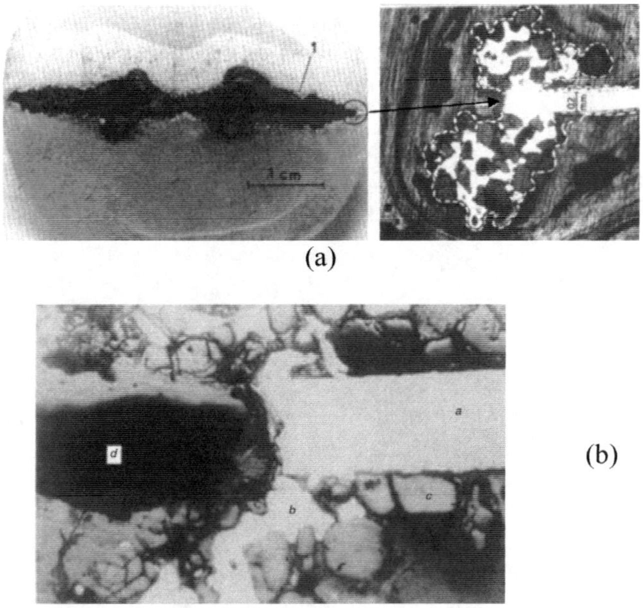

Figure 2. (a) Cross-section of a fulgurite cut along the arcing channel axis, copper fuse element with two reduced sections : 1, copper droplets ; circle fuse element end directly in contact with the arc column (metal droplets in white, sand grains in grey) [13]. (b) Microscopic picture of fulgurite section from *Wright et al.*: a, electrode ; b, electrode material which has flowed aways in liquid form ; c, grains of filler ; d, arc cavity which has been filled with araldite to enable the section to be made through the fulgurite [9].

jumps of the order of 100 ms. As aconsequence of these two latter quotations one understands that the erosion fronts in a fuse can be seen as a combination of all these features.

An example of the erosion fronts in a silver fuse strip is shown in figures 1(c) and 1 (d). These two frames are extracted from a high-speed film for the times 1.18 ms and 2.16 ms, the *pre-arcing* duration ~ 0.89 ms, the time of measurable intensity ~ 3.61 ms, $di/dt \sim 2\times10^6$ A·s^{-1}, Joule integral ~ 6.2×10^3 A^2·s, and the observed spectrum is restricted to those including the two neutral silver lines at 520 nm and 546 nm (to be disturbed the less possible by the radiation issued from silicon lines). In the used camera pictures are focussed on a black and white photographic film. Figure 1(c) shows that ~ 300 µs after the arc ignition (sharp increase of the voltage across the fuse) which is close to the peak fuse current (~ 2.1 kA) the inter-electrodes gap is ~ 4.6 mm and that the two erosion fronts are erratic with roughly half-circular shapes for the irregularities distributed along the fuse element strip width. It has to be noticed that metallic vapours spread in the width direction further away from the intial position of the fuse strip. Figure 1(d) is observed ~ 1.3 ms after the arc ignition *i.e.* during the decrease of the fuse current. Erosion fronts are more regular without showing a clear trend. Some gradients in intensity are observed in the whole of the interaction volume with strong inhomogeneities right in the centre. Especially at the cathode side (left part in figure 1(d)) it can be seen a highly irregular erosion front.

Keeping in mind the previous remarks – especially the considerations given to explain the observed fluctuations – and the pictures shown in figures 1(c) and 1(d) we understand that a thorough investigation of the arc roots in a fuse or at least of the erosion fronts is made more complex by the instability of the erosion fronts and the time-changing shape. So there remains a non negligible lake of knowledge which could be very helpful to understand better the transition between the *pre-arcing* and the *arcing* periods.

2.3. Burn-back and growth of the plasma column

2.3.1. Burn-back rate. The erosion of the fuse element also called *burn-back* has been widely studied within the electric fuse community. Many experimental studies and modellings can be found in the literature [14].

The initial point of the *burn-back* approach can be seen in figure 2(b) and also in figure 11(b) (*Part 1. Pre-arcing period*): at the solid fuse element end some liquid silver splashings can be seen starting from the liquid-vapour erosion front. It is interesting to note that in the two pictures the liquid splashing is seen flowing through the interstices which would lead to assume a high level of pressure linked to the vaporization: it is clearly shown in figure 2(a) close to the fuse element end. This indicates that only a part of the fuse element is vaporized to form the plasma column: this would seem to confirm the assertion quoted in section 2.2 about *microdroplets* but at another scale and raises the problem of the vaporized part of the total fuse element material. *Wright et al.* [9] gives the following relation for the vaporized mass (m_V) as a function of the total mass (m_t): $m_V/m_t = 40\%$ which results from many actual fuse tests. To assess the inter-electrodes gap or *burn-back* length equality is assumed between the energy provided for melting and vaporisation: $E_{MV} = 2\times(V_{af} + V_{wf} + V_T)\times\int_0^{t_{arc}} i(t)\cdot dt$ (V_{af} : voltage of anode fall, V_{wf} : work function equivalent voltage of the fuse element material, V_T : voltage associated with the thermal energy of the electrons which enter the anode), and the energy necessary to remove electrode material: $E_{MV} = m_t \times L_m + m_v \times L_b + m_t \times (T_m - 200)\times c_s$ where 200°C is the bulk temperature raise of the electrode material during the *pre-arcing* period. So doing the energy necessary for the temperature increase from melting to boiling point is neglected.

From these bases two other main calculations are done: the cross-sectional area of the positive column and its electrical conductivity. Many simplifying assumptions are done and justified keeping in mind that the complexity of physical processes in a fuse prevents a full modelling of the fuse arc [9]. We highlight two problematics.

First both for electrons and atoms the Maxwell distribution for velocity is assumed valid. Respectively for the kinetic energy (E_{ai}) of atoms/ions (n_{ai}) which are scattered out of the column and for the kinetic energy (E_e) of the electrons (n_e) scattered out of the column *Wright et al.* write: $E_{ai} = 3/2\times n_{ai}k_BT$ and $E_e = 3/2\times n_ek_BT$, where T is the mean temperature defined for a step time in the calculation. These two latter formulas imply the same temperature for electrons (T_e) and heavy particules (T_h) that is to say that thermal equilibrium is checked [15]. This assumption can be reasonably accepted for the core of the plasma column *i.e.* far from the edges where material is injected by means of liquid silica vaporization and where the pressure is the highest. On the contrary it is rather questionable for volumes close to the silica liquid/vapour interface (plasma column surroundings) and close to the silver liquid/vapour interface namely the fuse element ends.

Secondly electrical conductivity (σ) is assessed from the Spitzer's formula written for ionisation levels of more than 0.01%, the electron density being deduced from Saha's equation assuming that $n_e = n_i$. As previously this can be properly assumed for the plasma core. But it is more questionable for the periphery of the plasma core because of the strong gradients for temperature, pressure and density moreover as material injection changes the plasma composition during the whole of the fuse operation. As a matter of fact we recall that σ follows roughly the electron density as n_e^2 [16]: small variations of the metallic material amount vaporized to form the fuse plasma can imply strong variations of the resulting electron density [17].

Empirical formulas deduced from many experiments have been published in [18]. The *burn-back* rate is deduced by assuming that the fuse element erosion is mainly due to local electrode effects. The temperature of metal droplets is obtained from the comparison between experimental *burn-back* rate and empirical values calculated from the latter consideration: 1700 K for silver droplets and 1356 K for copper droplets. From [19] a useful formulation is given for the *burn-back* rate (figure 3(a)): $dx/dt = 10^{-4}\times(4.6 + 0.236\times i^{0.6})\times i/S$ where $i = i(t)$ (A) and S (cm^2) is the cross-section area. This formulation improves the older reliability obtained with linear approximation. For the volumetric erosion rate *Wilkins et al.* give: $dv/dt = 10^{-4}\times(4.6\times i + 0.236\times i^{1.6})$ $cm^3\cdot s^{-1}$. It is compared in figure 3(b) with the results of *Turner et al.* [20] where the volumetric erosion rate is studied in the case of

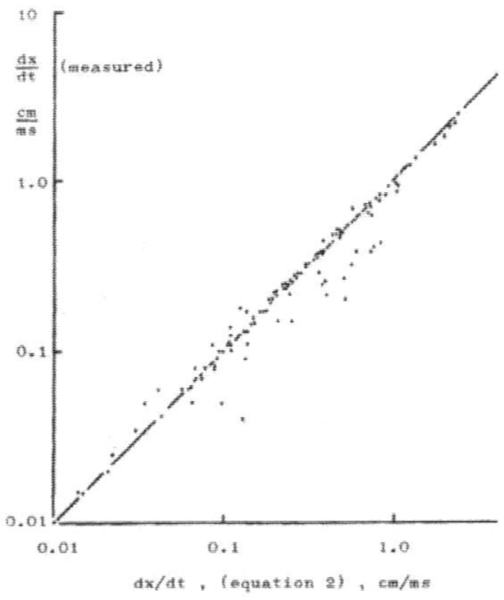

(a)

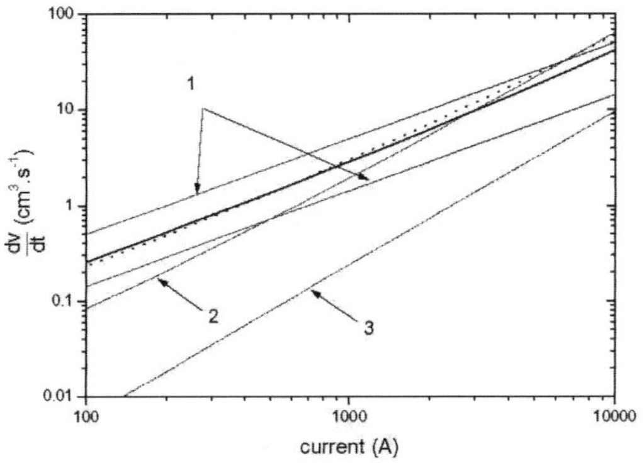

(c)

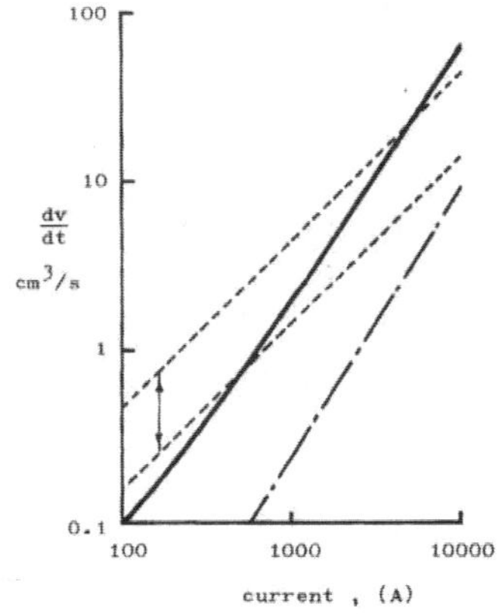

(b)

Figure 3. (a) Erosion rate: correlation between experimental results and $dx/dt = 10^{-4} \times \left(4.6 + 0.236 \times i^{0.6}\right) \times i/S$ [19]. (b) Volumetric erosion rate versus current: – – –, range of linear approximation ; ▬, $dv/dt = 10^{-4} \times \left(4.6 \times i + 0.236 \times i^{1.6}\right)$ [19] ; — - — - , approximation from [20] deduced from contact erosion. (c) Comparison of volumetric erosion rate [12]: at intemediate current levels for silver (- - -) and copper (▬), limits deduced from the linear approximation (1), *Wilkins et al* [19] and *Turner et al.* [20].

contact erosion. There is a significant shift between the results because the heat transfer is different for the two situations, heating of contacts and heating of the fuse element erosion fronts. In the first situation a large part of the incoming energy is used in the heating of the contact, the remaining part being used for erosion. In the second situation only a small part of the incoming energy is used in the heating of the fuse element and the significant remaining energy is used for the fuse element erosion. [19]. Another limitation in the trend curves has to be pointed out: the level of current linked to the level of the overcurrent that has to be cut by the fuse. Correlatively to the type of *pre-arcing* period (short, medium and long) one defines small overload current (typically some I_n, the nominal current), medium overload current and high overload current. These definitions depend on the nominal current of the electric fuse, the features of the *fault* current and supply regarding the power factor, the phase angle of the electrical *fault*, current peak and are resumed in [21] (test series at I_1, I_2 and I_3). In figure 3(c) experimental volumic erosion rates from [12] fit well with the approximation from [19] for currents higher than ~ 1000 A. For smaller currents down to ~ 100 A there is a significant gap. For smaller overload currents the role played by the filler and the solid parts of the fuse element can not be considered as negligible in the actual thermal transfer: thus discrepancies appear because the Joule integral contribution is divided in the global heat transfer and in the fuse element erosion.

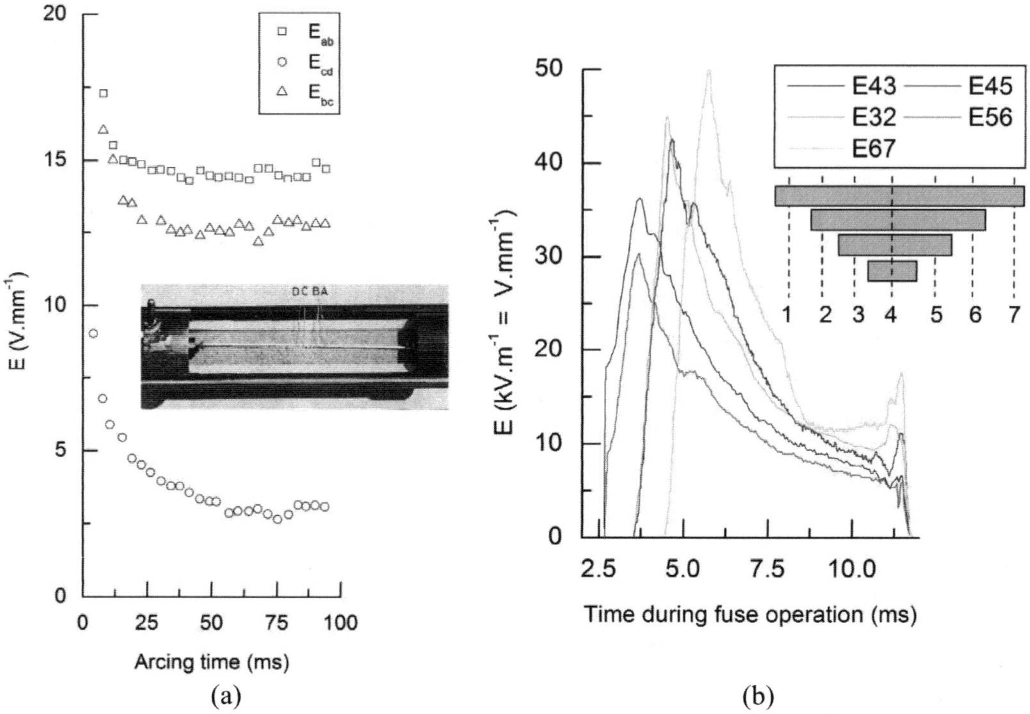

Figure 4. (a) Axial electric field values measured in the case of d.c. arcs in current-limiting fuses [62]. (b) Electric field values measured in the case of HBC fuse tested with high fault current level (1: cathode side, 2: anode side, each probe wire being at 3 mm from adjacent ones) [63].

*2.3.2. Plasma column.*The formation and the evolution of the plasma column geometry (inner part of the fulgurite obtained after the fuse operation) and of the whole interaction volume (which is assimilated to the whole fulgurite) has been studied experimentaly mainly [13, 18, 22] with some attempts to formalize these experimental results in the form of more or less empirical modellings dedicated to the whole of the fulgurite or focussed on the plasma column section also called lumen [9, 10, 18]. This task is of very high difficulty even in the case of experimental results: in an actual fuse the plasma column is enclosed by liquid silica itself enclosed by silica sand grains more or less melted themselves enclosed by solid silica sand grains. As a consequence each experimental attempt can strongly modify the true geometry of the plasma column. In [9] the column is assumed cylindrical for simplifying purposes and the output current and voltage curves obtained by means of the modelling are satisfactory.

The electric field assessment is shown in figure 4 with two different measurements (voltage probes are referenced to earth in both measurements). In figure 4(a) axial electric field values range from ~ 2.5 V·mm^{-1} to ~ 17.5 V·mm^{-1} at the most. These measurements correspond to the average axial electric fields obtained between the probes (nickel alloy probe wires) at positions A, B and C in the case of d.c. arcs in current-limiting fuses. This technique has been fitted to assess the time evolution of average electric field both at the cathode side and at the anode side as shown in figure 4(b) for a total *arcing* period duration ~ 9 ms. Around the centre of the fuse element and on the whole duration of the measurement the evolution of E_{43} is similar to those of E_{45}, the main difference being observed for the peak values at the beginning of the *arcing* period with $E_{43} > E_{45}$. On the contrary as the distance from the centre of the fuse element increases the discrepancies become smaller. The peak values of the

Table 1. Spectroscopic lines from various chemical species observed in the plasma column during the fuse operation, from [25].

3a. Spectrocopic lines

Line	Wavelength (nm)
Ag I	487.410, 520.908, 546.550
Si II	504.101, 505.598, 587.376, 595.756, 597.893, 634.710, 637.136
Si III	455.262, 456.782, 573.973
O II	418.546, 418.979, 431.714, 433.686 434.556, 434.943, 436.690, 439.595 441.491, 441.698, 444.821, 445.238 446.628, 446.783, 459.097, 463.885 464.181, 464.914, 465.084, 466.164 469.921, 470.536, 492.460, 494.306

3b. Duration of lines in the arc

Wavelength (nm)	Line	Durations (ms)
441.491	O II	1.924
441.698	O II	1.765
455.262	Si III	2.310
456.782	Si III	2.150
466.164	O II	1.820
469.921	O II	1.765
470.536	O II	1.870
487.410	Ag I	0.760
504.103	Si II	3.300
505.598	Si II	3.420
520.908	Ag I	3.004
546.550	Ag I	3.180
573.973	Si III	2.150
587.376	Si II	3.004
595.756	Si II	3.100
597.893	Si II	3.300
634.710	Si II	3.498
637.136	Si II	3.498

experimental average electric field range from ~ 30 V·mm^{-1} to ~ 50 V·mm^{-1}. Whatever the couple of probe wires or the location of the measurement a regular decrease is observed from the peak value showing that the steady plasma column assumption sometimes made appears to be really questionable. The comparison between the measurements from [62] and [63] can not be done because the fuse elements have different features and are tested in different conditions about the electric *fault*.

The plasma column is now discussed preferentially from the physico-chemical point of view [23]. The first paper dealing with the study of the electric fuse arc plasma has been published by *Chikata et al.* [24]. The main aim of the work is the spectroscopic study – especially Optical Emission Spectroscopy (OES) – of the radiation escaped from the fuse arc plasma: identification of the spectral lines escaped from the plasma column, evolution of the intensity of these lines on the whole duration of the fuse operation and assessment of electron temperature, electron density and ratios of particle densities. This work was done by means of a specific fuse body and a spectrograph equipped with three photomultipliers to observe simultaneously different spectral intervals (use of bandwidth filters) on the whole duration of the fuse operation. A wavelength calibration and a relative intensity calibration were performed. From the time integrated spectra over the 330-560 nm spectral range many lines from different radiated species are identified: Ag I, Si I-II-III (I: neutral, II: first ion, …) but no lines corresponding to O I-II-III which is due to the studied spectral interval and the value of the energy levels of the excitated states for these latter species. It wasfound a typical temperature ~ 20 000 K and a typical electron density ~ 10^{18} cm^{-3}. But it has to be noticed that no variation during the fuse operation has been found. Because of the continuously changing radiating volume no Abel inversion was done to obtain the local values of the temperature and the electron density: thus the measurements lead to a mean value or integrated value. On the basis of LTE a set of Saha's equations was built from the experimental relative intensities of the quoted radiated species to calculate the composition and the pressure of the pressure. Stark broadening of ionized silicon lines was also used

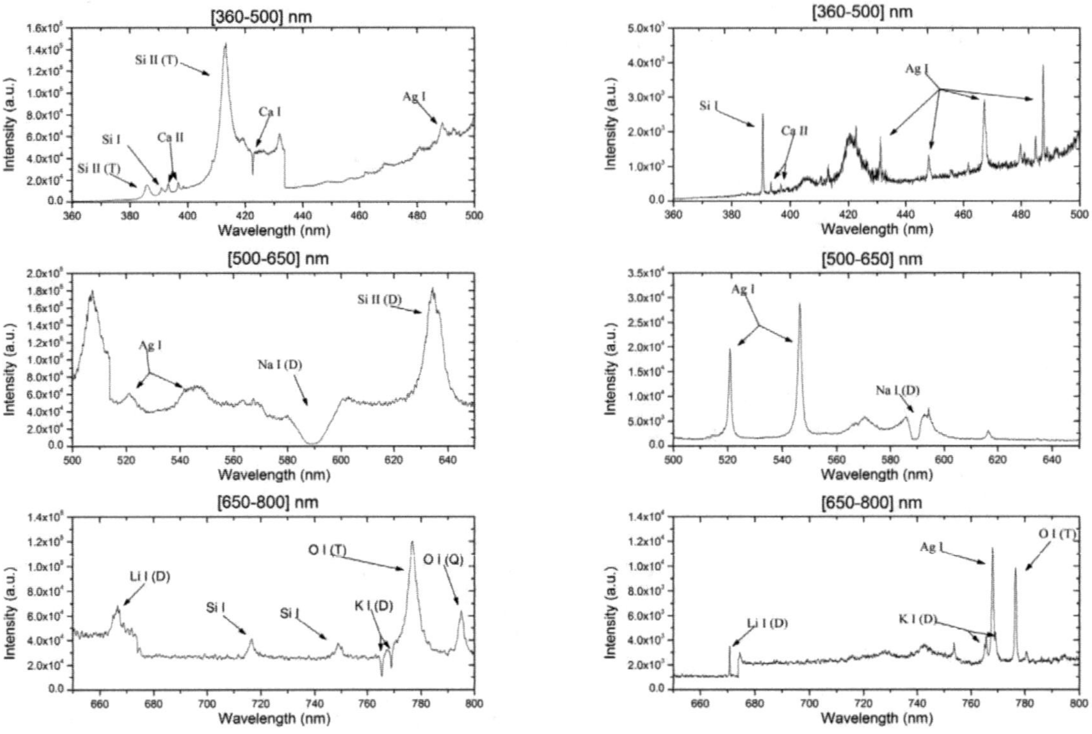

(a) (b)

Figure 5. Optical emission emission spectroscopy dedicated to the fuse arc plasma spectroscopic lines identification [34]: (a) at the beginning of the *arcing* period (~ 0.9 ms) (b) at the end of the arcing period (~ 3.6 ms) in the case of a silver fuse element tested by means of a capacitor bank discharge.

to assess electron density which was compared with values otherwise obtained from composition calculation and experimental line intensity ratios.

This work represents the reference of the later spectroscopic studies: *Stokes AD* and co-workers [25-30] where the *wall-stabilized* interaction dedicated to the fuse arc plasma was tested experimentaly and *Cheim* [31, 32]. From these works and [33-36] it appears that the modified fuse body or way to collect the light escaped from the plasma column can influence the results concerning the observation of the ionization degree radiating lines especially as many features – the overload current, the fuse element (geometry and features of the reduced sections and position within the fuse body) and the arc quenching material (chemical features, mean granulometry and packing density) – play a significant role in the resulting temperature, electron density and pressure of the plasma column during the fuse operation [35]. Main trends and results are resumed in the following paragraphs.

We show in table 1(a) the spectroscopic lines observed in [25], the light radiated by the plasma column being directly collected via an optical fibre whose end is inserted very close to the plasma. On the contrary to [24] many O II lines are observed and even some Si III lines which may be attribute to the proximity between the plasma and the optical fibre and electric features. The dependence between the duration of observation, the ionization degree of the chemical species and the implied excitation level is shown in table 1(b). In figure 5 we show the spectra observed by means of a direct integration of the escaped light without any optical fibre inserted in the plasma column, the light being collected by means of a lens from the glass window close to the fuse element to the entrance slit of a spectrometer via an optical fibre [34]. The same chemical species are observed but with a lower ionization degree. By comparison between figures 5(a) and 5(b) the discrepancies between the

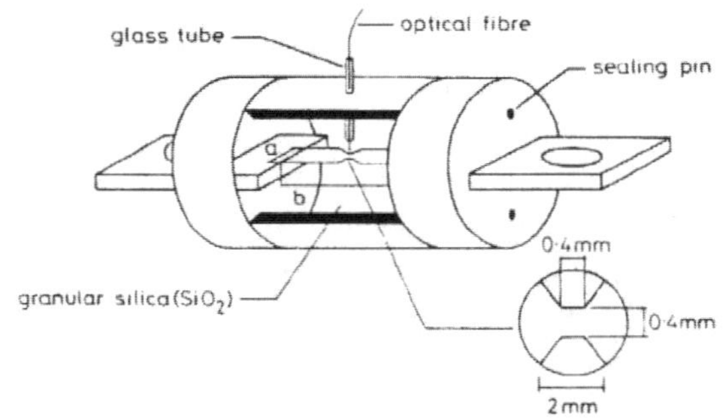

(a)

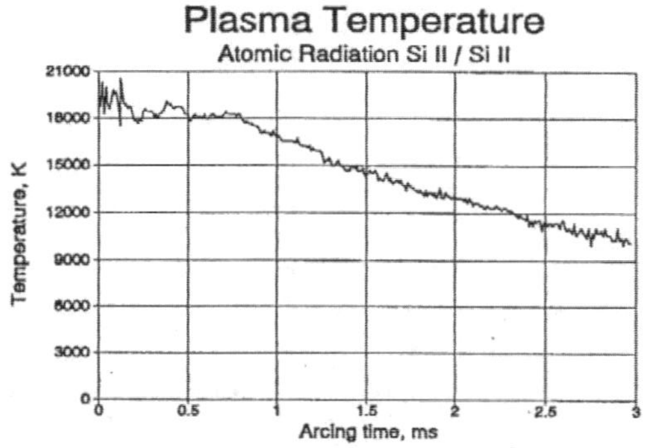

(b)

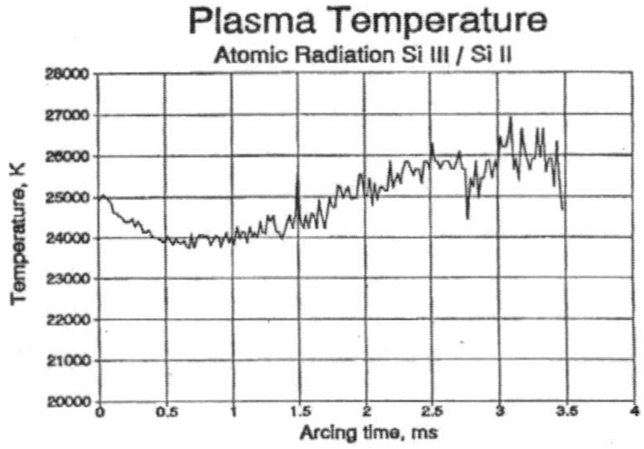

(c)

Figure 6. Arc temperature measurement by *Cheim* [31,32]. (a) Sample fuse with the insertion of the optical fibre (silica sand with mean granulometry ~ 100 μm). Plasma temperature versus *arcing* time deduced (b) from Si II 505 nm and Si II 634 nm lines and (c) from Si II 413 nm and Si III 457 nm.

radiation observed at the beginning of the *arcing* period (~ 0.9 ms) and at the end of the *arcing* period (~ 3.6 ms) are clear: Si II lines – especially the multiplets Si II (1), Si II (2) and Si II (3) – and O I – especially the triplet ~ 777 nm – show a high level of intensity for ~ 0.9 ms ; on the contrary for ~ 3.6 ms ionized silicon lines are not visible any more, and the spectra are dominated by neutral silicon and silver lines whose excitation energy levels are lower than those of Si II and O I.

The Na I doublet seen ~ 589 nm is due to the nature of the glass window (the same for Li I, Ca I and K I). It was used in [37,38] to study the colder periphery of the plasma column, especially the temperature: in the case of the capillary [37] a temperature ~ 4300 K is found, in the case of the electric fuse [38] a temperature ~ 6600 K is found. To limit the number of chemical species in the fuse arc plasma the glass window can be replaced by a quartz window [35].

The specific fuse body with optical fibre inserted used by *Cheim* is shown in figure 6(a), the optical fibre being close to the fuse element reduced section (≈ 3 mm) and in front of its centre. From the measurement of the ratio between radiation intensities from various lines the temperature is deduced versus the *arcing* time. Temperature is obtained from the ratio of the intensities (ξ'/ξ) of emissions of radiation [39]: $\xi/\xi' = \left(\lambda'^3\, gf / \lambda^3 g' f'\right) \times \exp[(E'-E)/k_B T]$ where λ is the wavelength, g the statistical weight of the line, f the oscillator strength, E the energy level. This formulation can also be written for lines corresponding to two consecutive ionization stages. From these formulations the temperature T identified to the plasma temperature is obtained for the whole duration of the *arcing* period: from Si II 505 nm and Si II 634 nm lines in figure 6(b) and from Si II 413 nm and Si III 457 nm in figure 6(c). From Si II lines the temperature decreases from ~ 20 000 K down to ~ 9000 K, and from Si II and Si III lines the temperature decreases from ~ 25 000 K down to ~ 24 000 K during the first millisecond of the *arcing* time, and increases weakly and steadily up to ~ 26 000 K for ~ 3 ms. As it is mentioned by the authors themselves: the trend in figure 6(b) appears to be coherent with the fact that during the *arcing* period energy in the fuse decreases which justifies the cutting role played the fuse ; but the trend in figure 6(c) shows a mean value ~ 25 000 K during the whole of the *arcing* period and even during the last millisecond of operation which seems to reinforced the measurements of *Chikata*. During this last millisecond current in the fuse falls quickly down to zero which appears to be in contradiction with the temperature measurement. Authors explain that the radiations emitted in the hot plasma core may be absorbed by colder layers located at the edges of the plasma column. These layers develop between the emission area of the studied transitions and the end of the optical fibre close to the edges of the interaction volume. They finally argue that for lines corresponding to high quantum levels the influence of absorption is negligible and that these lines can be considered as thin lines. First, narrow bandwidth filters are used to record only the part of radiation corresponding to the lines studied. The silicon lines used (Si II, and Si II and Si III) are sensitive to Stark broadening so the resulting full width at half maximum of these lines can be many nanometers of even tens of nanometers. This can be increased by the high level of continuous radiation, the observed spectrum being the result of the convolution of the Stark broadened profile with the continuum. Secondly, we can imagine that once the electric power is around its maximal value which falls approximately in the same time as maximum pressure, there exists a very strong gradient in material density from the core to the surroundings of the plasma column. As once more the Abel inversion is impossible to go back to local emissivities, the effect of these strong gradients can not be understood. Finally it is evident that the lack of the radial profiles of the silicon lines is a strong limitation which makes more difficult the temperature analysis from the recording of lines intensities.

Reliable comparisons are sometimes difficult to perform mainly because of the spectroscopic considerations linked to the various devices or whole experimental set-ups. For example in the past interferential filters were often used to obtain the spectral selectivity. At the present time CCD matrix with high sensitivity and high spectral resolution is commonly used. Considerations linked to intensity calibration are also of high interest but they are poorly discussed in the bibliography [40] and even less in the case of the electric fuse plasma analysis. On the contrary main of the electrical properties can be easier compared in so far as electrical features and fuse element features are accurately depicted. As a conclusion about the reliability and the reproductivity of the measurements dedicated to electric fuses one can conclude that reproducible measurements can be performed (especially owing to the improvement of technological solutions), the more questionable step lying in the analysis of the experimental results and the corresponding assumptions made to apply classical diagnostic methods such as in spectroscopic analysis.

The fuse arc plasma properties such as the temperature and the electron density i.e. microscopic properties can be directly linked to macroscopic properties such as the morphometric properties of the silica sand.

- Temperature of the fuse arc plasma is obtained by assuming a Boltzmann distribution for the energy levels [35, 38, 40] thus leading to the excitation temperature. If thermal equilibrium conditions are assumed one can identify the excitation temperature with the electron temperature (electron velocities following a Maxwell distribution). This method can be applied with integrated intensities (I_l) for the studied line (with spectral profile $P_l(\lambda)$) instead of local emissivities ($\varepsilon_{l\lambda}$) at the condition that the plasma can be considered as homogeneous along the line of sight (L) namely: $I_l = \int_\lambda I_{l\lambda} \cdot d\lambda = L\int_\lambda \varepsilon_{l\lambda}(\lambda) P_l(\lambda) d\lambda = L \times (1/4\pi) \times (hc/\lambda_{j,ul}) \times A_{j,ul} \times n_{j,k}$ (h : Planck constant, c : speed of light in vacuum, $\lambda_{j,ul}$: wavelength of line emitted by the j-chemical species where u and l symbolize the upper E_u and lower E_l energy levels of the transition, $A_{j,ul}$: transition probability of the line, and $n_{j,k}$ is the part of excited particles of the j-chemical species with $u,l \equiv k$). In the case of fuse arc plasma one can only work with integrated intensities because the set of radial measurements necessary to obtain $\varepsilon_{l\lambda}$ from I_l (by Abel inversion) is impossible. The hypothesis for which homogeneity is considered valid along the line of sight is highly questionable. As a consequence it is better to consider the experimental temperature as an average temperature. In the case of the fuse arc plasma the assumption of the Boltzmann distribution for the energy levels and more generally about the Maxwell-Boltzmann distribution linked to equilibrium considerations is a complex task. Even in the case of apparent easier situations – such as plasma produced by ICP torch – some key points are still under consideration (see [40] for discussions about the accuracy of classical spectroscopic methods used to assess excitation temperature, electron temperature and the thermal disequilibrium). The high pressure level in the fuse arc plasma column is a favourable condition: in fact in the more general case of plasmas at least close to atmospheric pressure the number of collisions is sufficient for efficient energy transfer. As a consequence the Maxwellian distribution function is obtained for particles of the same species. As high level of electron density is obtained in the fuse arc plasma especially when the electric power is maximum [35], typically ~ 10^{18} cm^{-3}, we can assume that the transfer of energy between electrons (e) and heavy particles (h) is efficient. Thus the unicity of temperature is assumed in order to define the plasma temperature by $T = T_e = T_h$. This unicity appears to be valid for the inner part of the plasma core where gradients (in temperature, density and pressure) are small (figure 8(c)) compared to the edges of the plasma. In the plasma surroundings the quoted gradients are strong and the composition of the plasma is continuously changed because of material injection (around the two erosion fronts and around the inner wall of the *fulgurite* or interaction volume). As a consequence the equilibrium assumption is difficult to assume for these areas. Local measurements could be helpful to analyse these areas, but typical methods such as Abel inversion can not be used. More details about the whole of these considerations linked to equilibrium (thermal, chemical) can be found in [64].

• Electron density of the fuse arc plasma is mainly obtained by means of two different approaches. First, it can be deduced by applying the Saha's equation [24] and using the experimental variation of the spectral intensities of various chemical species with different ionization degrees. This way is questionable in so far as the plasma composition has to be defined and LTE has to be assumed. Secondly, it is deduced from the spectral line shapes (broadening and sometimes shift) of ionized silicon lines by assuming that the observed spectral width at half maximum is mainly due to Stark broadening (splitting and shifting of the energy levels of the emitters) that is to say by pressure broadening [24, 30, 34, 35, 38]. So doing the assumption of LTE is not necessary. Main limitations lie in the substraction of the continuous contribution and of the apparatus function. But in the analysis of the results one has to consider that even with ionized silicon lines the observed profile can be more or less broadened by absorption because of the radial gradients (in density, pressure and temperature).

2.3.3. Fulgurite. The formation of the *fulgurite* has been discussed in section 2.3 with quoted references in which more details can be found. This section is more concerned by the electric rigidity of the *fulgurite* once the *arcing* period is ended. It is of very high interest because the *fulgurite* has to ensure the highest dielectric rigidity possible to avoid any restriking linled to the appearance of a new electric *fault*.

In figure 7 we briefly illustrate some results from [41-43]. First in figure 7(a) is shown the dependence between the resulting resistance and the initial value of the copper fuse element thickness: higher values are obtained for thiner fuse elements. Secondly, as it has been previously observed in figures 5 and 11 (*Part 1. Pre-arcing period*), there is a bulge located right at the initial position of the reduced section (linked to the high pressure which is the result of the metal vaporization). Each part of the fulgurite contributes in the resulting resistance. From figure 7(b) we see that the resistance obtained for the whole length of the fulgurite is especially the result of the contribution of the bulges: if the fuse element is equipped with many tens of reduced sections in series (typical case of an industrial fuse element) the resistance of the fulgurite will depend directly of these. Figure 7(c)

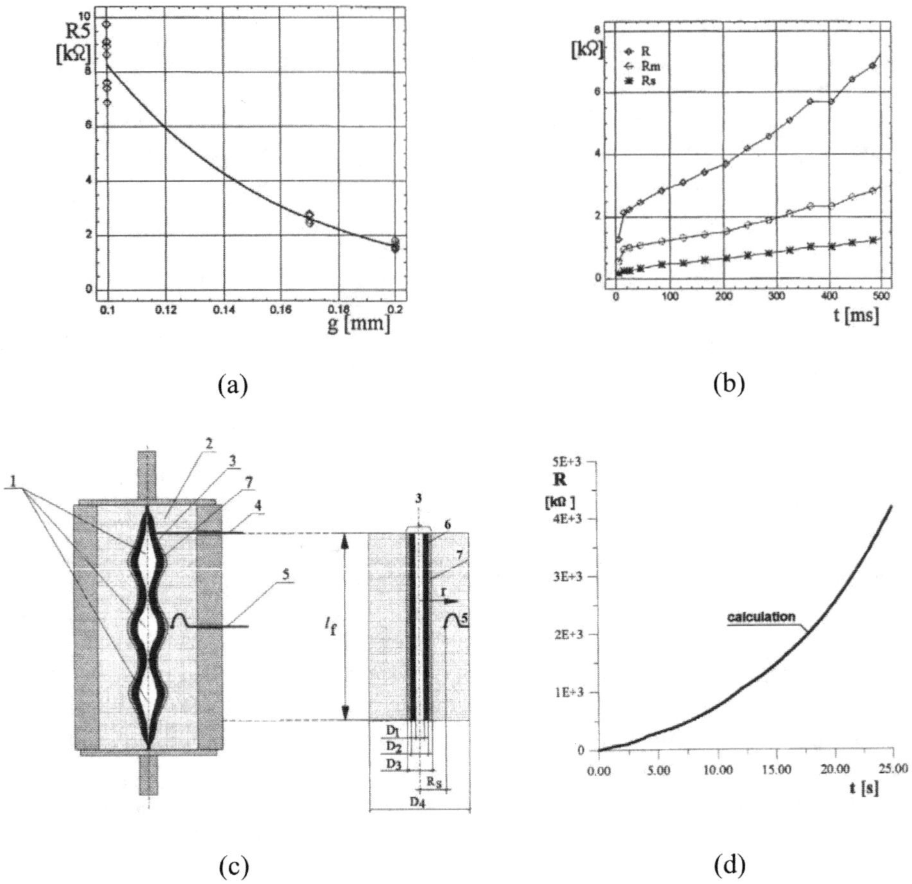

Figure 7. Considerations about the post-arc fulgurite resistance [41-43]. (a) Experimental post-arc resistance versus Cu element thickness (source voltage 500 V). (b) Distribution of post-arc resistance alonf different parts of fulgurite of Cu elements (resistances: R, for the whole length of the fulgurite ; R_m, for only one metal part ; R_s, for a swell of the fulgurite). (c) Sketches of the experimental fulgurite (on the left) and of the model put in the calculation: 1, swell observed at the initial place of the reduced section ; 2, quartz sand ; 3, fulgurite shell ; 4, voltage probe ; 5, thermocouple ; 6, fully melted layer ; 7, partially melted layer. (d) Profile of the fulgurite resistance versus time up to 25 s.

illustrates the simple schema retained to build the modelling: the actual fulgurite volume with bulges is simplified in an equivalent cylindrical volume. Especially fitted layers are deduced from a simple temperature cartography established by means of thermocouples inserted the filler volume, the temperature measurements being used to set the temperature profiles to put in the calculation. As a result we show in figure 7(d) the evolution of the post-arc resistance versus time. After few seconds a post-arc fulgurite resistance of some kΩ is obtained. Thus this simple tool can be helpful to design roughly the number of reduced sections and the gap between each of them.

2.4. Properties of the fuse arc plama

Among the studies published up to know there are only a few numbers of papers dealing with the physico-chemical properties of the fuse arc plasma in spite of the fact that microscopic properties *i.e.*

those of the fuse arc plasma are of high interest in order to understand the macroscopic properties *i.e.* those induced by the fuse.

• At the microscopic scale. The transport coefficients of the fuse arc plasma – electrical conductivity, thermal conductivity, viscosity and radiation – are difficult to assess in an experimental way. From theoretical calculations [16, 17, 44-46] one knows that small variations of the metallic material to silica sand ratio can imply strong variations. Among them the fuse arc plasma radiation is the most studied even if few studies have been done. In that case the high difficulty lies in the analysis of the radiative properties especially for temperature and electron density measurements. To be obtained these evaluations require experimenters to apply classical spectroscopic methods whose assumptions and applicability are sometimes questionable. Some examples are detailed in the following sections.

• At the macroscopic scale. The analysis of the fuse resistance can be improved by use of the calculated electrical conductivity of the fuse arc plasma but many calculations are necessary to depict the whole duration of the *arcing* period and the whole of the plasma volume (for example of metallic type at the beginning and hence with an increasing percentage of silica). The instantaneous mesurement of the pressure can be helpful at the condition that the measured pressure is the plasma pressure and that the plasma column shape is known at the time of evaluation [47]. The dimensional features of the fulgurite can be an indirect way to check the trends calculated for the viscosity but in this case the correlation has to be established from post-operation features (and the task is made more difficult because of the role played by the plasma pressure).

For physical modelling purposes of the *arcing* period one has to increase the knowledge of the physical processes occuring inside the plasma column. In the case of fuse arc plasma OES methods have led to the most significant contributions. In addition to the identification of the chemical species of the fuse arc plasma the temperature and electron density are obtained. As the fuse arc plasma radiation is assessed by means of an integration over the whole interaction volume thickness according to a single line of sight – whatever the way: either by direct insertion of an optical fibre either through a quartz window – the measured spectrum has to be considered as an average over this line of sight for the given integration time. This way to proceed raises many unknowns linked to fluctuations during the integration duration: about the changes in the plasma composition, about the optical thickness of the layers met by the light from the core of the plasma column to the surrounding layers where the light is collected, about the subvolume from which the studied spectroscopic line is emitted, about the validity of the necessary assumptions for the use of classical spectroscopic methods [40].

As example we show in figure 8(a) the spectra observed during the *arcing* period (*pre-arcing* time ~ 0.9 ms) for the spectral interval [360-440] nm. Neutral and ionized silicon lines are observed together with the continuum whose level changes significantly: sharp increase up to ~ 1.7 ms (for which electric power is maximum) followed by a regular decrease down to the end of the *arcing* period. The pressure evolution follows the electric power during *arcing* period [48]: thus for the time interval where the pressure level is high the lines used for temperature and electron density assessments are strongly broadened and the line wings are strongly disturbed. In the same time the level of the continuum increases and even becomes higher than the discret radiation level: as a consequence discret radiation becomes weaker compared to continuous radiation and can not be used any more for diagnostic purposes. In figure 8(b) we show a comparison between an experimental spectrum focussed on the ionized silicon multiplet Si II (2) and the calculated profiles for the two lines of the multiplet, namely ~ 634.7 nm and 637.1 nm and assuming a 100% SiO_2 plasma. This spectrum if observed ~ 3.24 ms after the application of the fault current, that is to say at the end of the *arcing* period for which the continuum level is low. Calculated profiles are obtained by means of the radiative transfer equation [36] written for only one dimension comparable to the line of sight. The observed intensity I_λ is defined by: $I_\lambda(R)=\int_0^R \varepsilon_\lambda(x,T).\exp\left(-\int_x^R k_\lambda(x',T)dx'\right).dx$, where R indiquates the

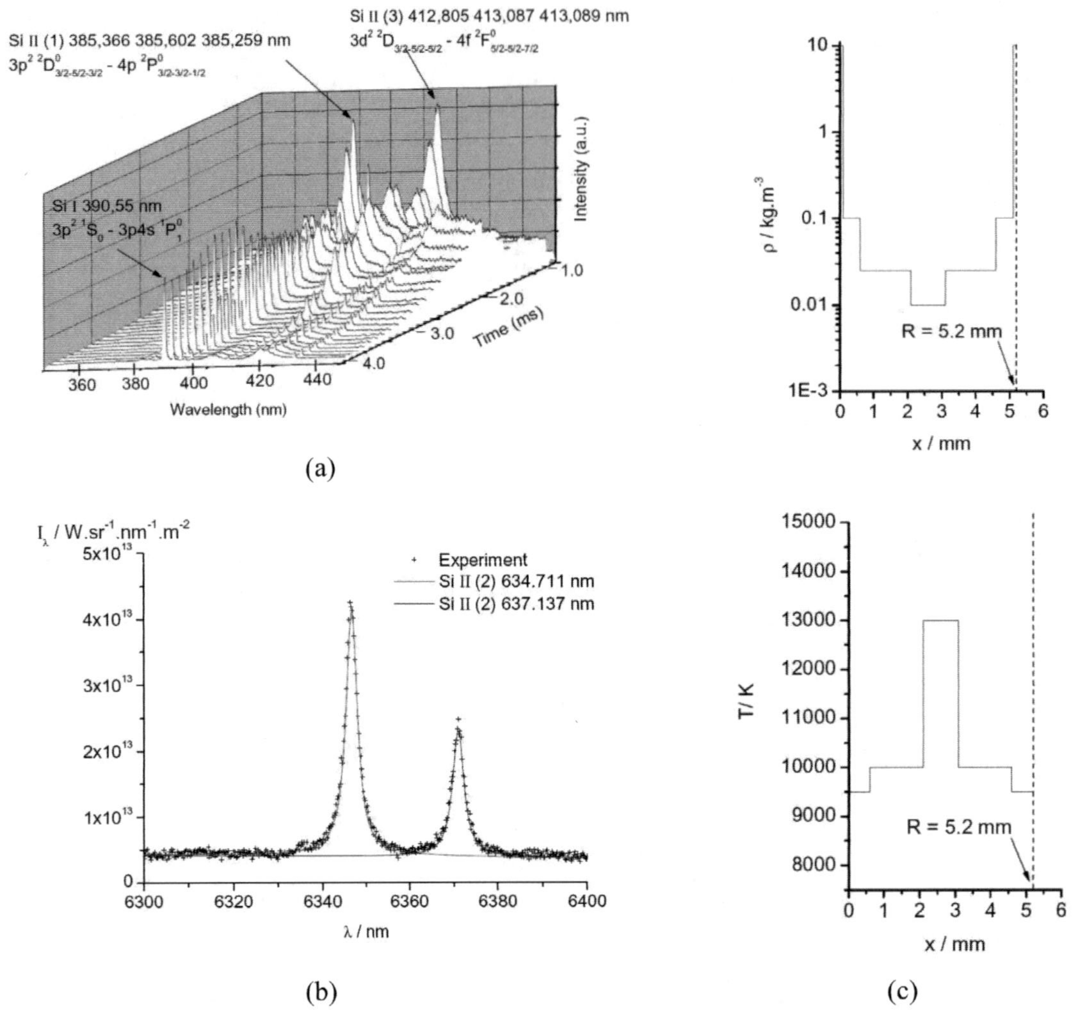

Figure 8. Fluctuations of the fuse arc plasma radiaition [36]. (a) Evolution versus *arcing* time of the neutral silicon line Si I 390 nm and the two ionized silicon multiplets Si II (1) and Si II (3). (b) Experimental spectrum at 3.24 ms focussed on the Si II (2) ionized silicon multiplet. (c) Radial density and temperature deduced from the Radiative Transfer Equation.

edge of the integration domain, k_λ is the spectral absorption coefficient (necessary to calculate the optical thickness of the plasma at the wavelength λ) and the total spectral emissivity is: $\varepsilon_\lambda(\lambda,T)=\varepsilon_{j,ul,\lambda}(\lambda,T)+\varepsilon_{fb,\lambda,j}(\lambda,T)+\varepsilon_{ff,\lambda,j}(\lambda,T)$, where ε_{fb} and ε_{ff} are respectively the *free-bound* and the *free-free* contributions of the continuum. For this calculation the radial density and temperature of the plasma are shown in figure 8(c): the plasma density is weak in the center ~ 0.01 kg·m^{-3} and increases up to ~ 1 to 10 kg·m^{-3} at the edges ; the maximum temperature is ~ 13 kK in the plasma core and decreases down to ~ 9.5 kK at the edges.

Spectroscopic considerations on one hand, and composition and transport coefficients calculations [46, 49-51] on the other hand are currently improved. For the latters the non equilibrium plasmas are studied in the low temperature domain namely from 1000 K to 6000 K. In addition to these novel calculation tools some experimental works have been done to study chemistry considerations linked to the *arcing* period in a fuse [23, 52, 53]. Especially in [23] experiments performed after the device

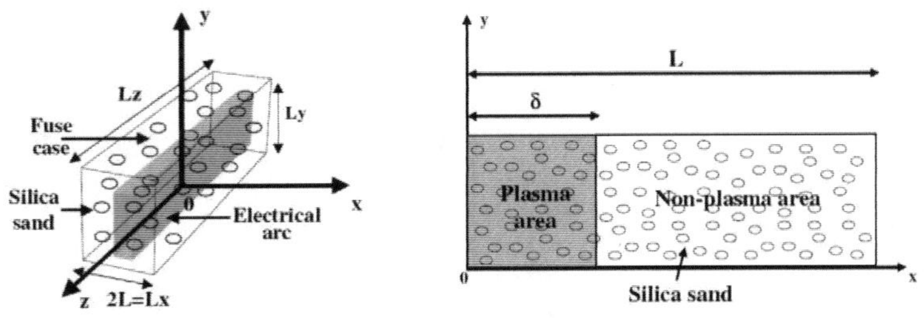

Figure 9. (a) Sketch of the geometry of the studied fuse. (b) Sketch of the one-dimensional geometry of the fuse process. From [57].

operation do not show any significant evidence of the lasting chemical change. The devices concerned are fast-acting HBC fuses with silver fuse elements designed to protect semi-conductor devices. Experiments include calorimetry, 'wet' analysis of the fuse fulgurite, electron spectroscopy for chemical analysis and determination of gas produced. On the whole of the results and within the range of experimental error no specific trend is pointed out from the chemical point of view and the results doesn't show any permanent chemical change due to the fuse operation. Moreover the black regions that can be seen in figure 11 (*Part 1. Pre-arcing period*) have been analysed also and three main conclusions are given [23]: (a) the presence of silver oxide is not proved in significant quantities, (b) the central black area of the fulgurite corresponds to silicon dioxide (SiO_2) and silver, and hence (c) there is no evidence of lasting chemical change.

The distribution of silver within the fulgurite has been studied in [53] by means of local neutron activation. Results have been averaged over tens of fulgurites obtained after fuse operation and show that an axial flow of silver exists from the centre of the fulgurite towards the extremities, the anode side flow being stronger than the cathode side flow.

2.5. Trends in arcing period modelling

Most of the published works up to now are digital simulations [9, 10, 54] or more or less empirical calculations especially the one of *Daalder* [18] which is very comprehensive. Among digital simulations reliable calculations have been produced to predict current and voltage induced by the fuse but they are dependant on fitting paramaters *i.e.* specific to one type of fuse or based on simplifying geometrical considerations espcecially for the plasma column. One of the main difficulty to build a physical modelling is first to identify the physical processes that have to be included. This task is complex because the whole of the processes are not known or not known with acceptable accuracy to be depicted properly. Secondly one has to integrate the three phases for the materials with different silver to silica proportions, the ratio between each of these phases varying during the *arcing* period (due to the localization within the interaction volume). Thirdly the chemical species forming the fuse arc plasma can vary with the proportion of the material and with the level of pressure (the increase of pressure shifts the chemical reactions to the high temperatures). Fourthly the whole interaction volume is not delimited by a homogeneous and steady boundary with accurate limits. On the contrary the thickness of the melted silica layer surrounding the plasma column is inhomogeneous in temperature and most probably in density and viscosity. The diffusion of the melted silica and of the vapours produced in the plasma column has to be depicted because it corresponds to energy withdrawal and it modifies the whole electrical conductivity of the plasma column and of the electric fuse. And fifthly one has to find experimental and calculated or modelised quantities to establish reliable comparisons.

A physical modelling of the *arcing* period in a HBC fuse has been published in [55-58]. We invite to refer to these works to obtain the full mathematical and physical considerations included in themodelling. In the next parts only the quantities which can be compared with experiments are discussed.

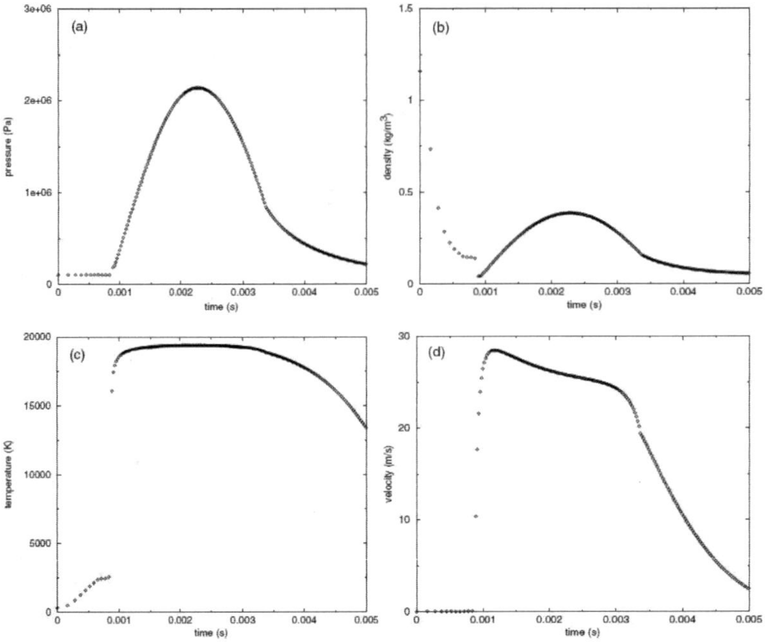

Figure 10. Pressure (a), density (b), temperature (c) and velocity (d) distributions in the plasma core during the fuse operation of 5 ms duration [57].

One of the contributions of these latter works is the approach developped to take into account the fluid flow through the porous granular filler. The sketch of the studied electric fuse is shown in figure 9(a) together with the one-dimensional geometry retained for the modelling. To simplify the fuse domain is divided into two main areas, the plasma area and the silica sand area. To model the HBC fuse the compressible Euler equations for perfect gas are used. The basic flow model is coupled with a model of the porous medium which integrates a mechanical interaction between the gas and the silica sand and the heat transfer between the hot gas and the cold silica sand grains. From the numerical point of view a finite volume scheme is used and each cell is defined according to the material involved depending on the area: especially in the silica sand the void space filled with air is set by the porosity ϕ whereas the silica sand correspond to $(1-\phi)$. The fluid flow through the porous filler is described by means of the Darcy and Forchheimer coefficients experimentally assessed by means of a modified Ergun's equation [59,60] where the pressure drop (ΔP) through the porous layer of given thickness (L) is given as a function of the fluid velocity (v) by: $\Delta P/L = (\mu v)/k_1 + (\rho v^2)/k_2$, k_1 is the viscous term linked to the Darcy coefficient and k_2 is the inertial term linked to the Forchheimer coefficient. Mass, momentum and energy conservations are written and the heat transfer is described by taking into account the effective thermal conductivity of the filler. In figure 10 we show the evolutions versus time of the pressure, the density, the temperature and the velocity in the plasma core for total fuse operation of 5 ms, the *pre-arcing* period duration is ~ 0.9 ms. The pressure increases quickly up to ~ 2.2 MPa at ~ 2.3 ms before to decrease steadily down to nearly atmospheric pressure. The plasma temperature shows a sharp increase at the beginning of the *arcing* period up to ~ 20 000 K and remains nearly unchanged up to ~ 3 ms, the temperature at 5 ms being ~ 13 000 K. The density normally falls from ~ 1 kg·m^{-3} at the beginning of the *pre-arcing* time to less than 0.1 kg·m^{-3} at the beginning of the *arcing* time. Next it increases due to vaporization of material and high temperature and reaches ~ 0.4 kg·m^{-3} when the pressure is maximum and finally decreases down to less than 0.1

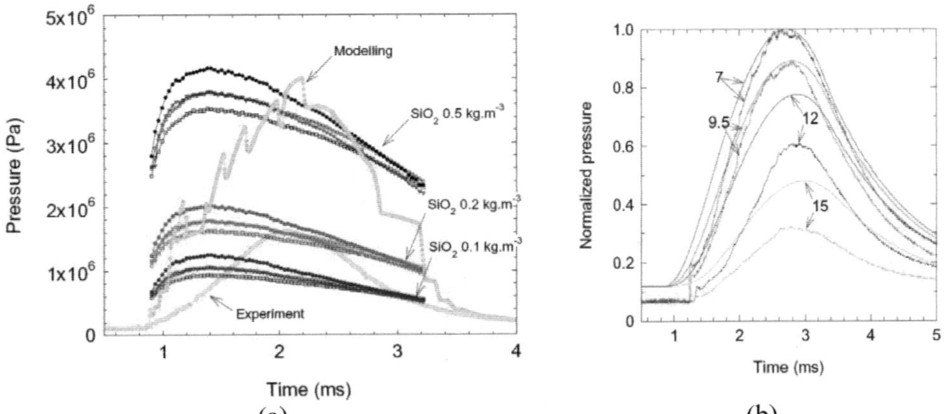

Figure 11. (a) Evolution of the pressure during the fuse operation [36]: comparison between experiments [48], simulations from modelling [57] and calculations (for three initial SiO_2 amounts fully vaporized to form the fuse arc plasma) [17]. (b) Comparison between experiments and simulations given in normalized pressure: at 7 mm, 9.5 mm, 12 mm and 15 mm from the intial position of the fuse strip [61].

kg·m^{-3} at the end of the *arcing* period. More simulations can be found in [57] at different positions (at the plasma periphery ~ 1 mm and in the filler) which exhibit coherent trends.

The fuse arc plasma pressure is one of the physical quantities which can be compared (figure 11). Experimental curves shown in figure 11(a) are similar to those showed in figure 9(b) (*Part 1. Pre-arcing period*): they correspond to the pressure measured the closest as possible of the plasma column within the granular filler for a given experimental test [48]. Simulated pressure is the arc plasma pressure obtained in the same way as in figure 10 but for different conditions corresponding to the experimental tests. Calculated pressures are given for three silica plasma compositions namely 0.1 kg·m^{-3}, 0.2 kg·m^{-3} and 0.5 kg·m^{-3}: for each case three curves are given depending on the temperature scale which is deduced from the spectroscopic measurements [36]. The experimental and simulated pressures follow a similar trend, the simulation (arc plasma pressure) showing logically higher values compared to experiments (pressure within the solid silica sand but the closest as possible to the plasma edges). On the contrary and whatever the composition, calculated pressures exhibit a significantly different trend. These latters are obtained from the temperature/time correlation for each of the plasma compositions. The absorption of the radiation emitted in the plasma core by the colder surrounding plasma layers justifies partially these different trends.

From the calculation of the Darcy and Forchheimer contributions the total flow resistance can be calculated [48, 57, 58] and the variation of the pressure during the fuse operation can be deduced at various positions from the fuse element and compared with measurements [61]. The comparison is shown in figure 11(b) in terms of normalized pressure for the following positions from the initial position of the fuse element: 7 mm, 9.5 mm, 12 mm and 15 mm (absolute measurements can be found in [61]). For the smallest value *i.e.* 7 mm the sensitive area of the quartz pressure transducer is very close to the edges of the fulgurite (a closer position could imply irreversible damages for the piezolectric transducer). A good agreement is found about the time variation especially for the closest positions 7 mm and 9.5 mm. For 12 mm and 15 mm there are some reasonable discrepancies between measurement and simulation. This is due to the higher silica sand thickness. The pressure wave ignited in the plasma column is transferred via the silica sand grains intersection points: higher the silica sand thickness, higher the number of intersection points and stronger the reduction of the pressure wave magnitude.

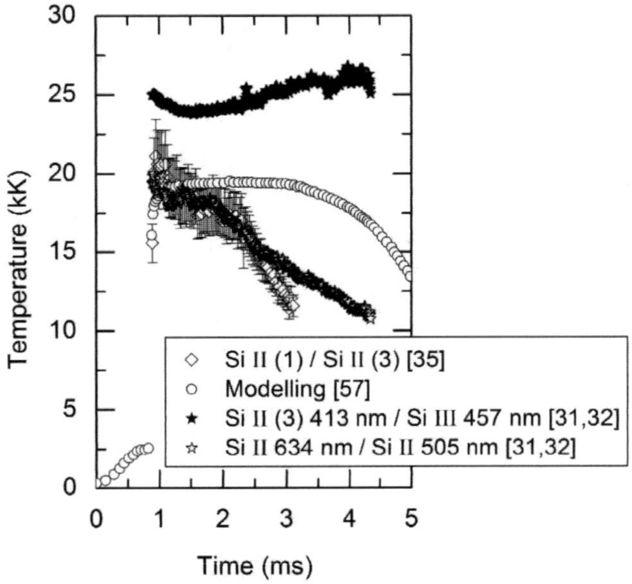

Figure 12. Comparison of various fuse arc plasma temperature values from experiments and modelling: M, average excitation temperature with uncertainties bar from Si II (1) and (3) multiplets [35] ; ○, simulated temperature [55,57] ; ψ, plasma temperature from Si II (2) 634 nm / Si II 505 nm ratio [31,32] ; ξ, plasma temperature from Si II (3) 413 nm / Si III 457 nm ratio [31,32]. For ψ and ξ, the initial time scale refering to the *arcing* time only is shift ~ 0.9 ms to be compared.

In figure 12 we show a comparison between the fuse arc plasma temperatures obtained in various ways. For a given mean granulometric interval [35] the mean excitation temperature is obtained from the relative intensity ratio method with the multiplets (1) and (3) of ionized silicon. The temperature increases from ~ 15.5 kK (the first spectrum during the *arcing* period is obtained ~ 30 μs after the arc ignition) to ~ 21 kK at the most. Hence the temperature decreases less or more regularly down to ~ 12 kK for ~ 3.1 ms (spectra integrated up to the cancellation of current can not be used for spectroscopic diagnostic). Up to ~ 2 ms these measurements are close to the simulated temperature [55, 57] especially as uncertainty bars are considered. Significant discrepancies appear after 2 ms of operation. Simulated temperatures are roughly unchanged up to ~ 3.2 ms and hence decrease steadily down to ~ 13.5 kK at 5 ms. First noticeable inhomogeneity appears from the plasma core towards the plasma edges thus changing the optical thickness of the ionized silicon lines along the line of sight and increasing the broadening of the ionized silicon lines by absorption. Secondly the spreading of melted silica through the pores of the filler is omitted in the modelling thus an energy loss term is omitted and this remaining energy corresponding to the whole hot plasma energy is overestimated. The temperatures obtained by means of spectroscopic measurements with Si II lines by *Cheim* [31, 32] are consistent with those of [35]. On the contrary the temperatures deduced from the relative intensity ratio method with Si II and Si III lines are quite stable with a rough mean value ~ 25 kK. A significant increase is observed after 2 ms which is inconsistent with the decrease of the energy provided to the fuse plasma. Considerations linked to optical thickness and absorption due to radial gradients could explain this trend.

From temperature [24, 29] and electron density [24, 30, 61] measurements the electron density versus temperature curves can be obtained for the whole duration of the fuse operation. An example is shown in figure 13 and compared with electron densities calculated for various SiO_2 plasmas (volume is kept constant): 0.1 $kg{\cdot}m^{-3}$, 0.2 $kg{\cdot}m^{-3}$, 0.5 $kg{\cdot}m^{-3}$ and 1 $kg{\cdot}m^{-3}$. It can be noticed that in the temperature range from ~ 16.5 kK to ~ 18.5 kK the assessment of electron density is not possible. This temperature domain is obtained when the electric power is at its maximum level. For these conditions the level of continuous radiation is the highest observed during the fuse operation and it is not possible to observe the ionized silicon lines any more: the radiation emitted in the plasma core crosses dense layers in the radial direction towards the plasma edges as shown in figure 8(c). From the calculated curves shown in figure 13 one notes that a single calculation is not sufficient to describe the fuse

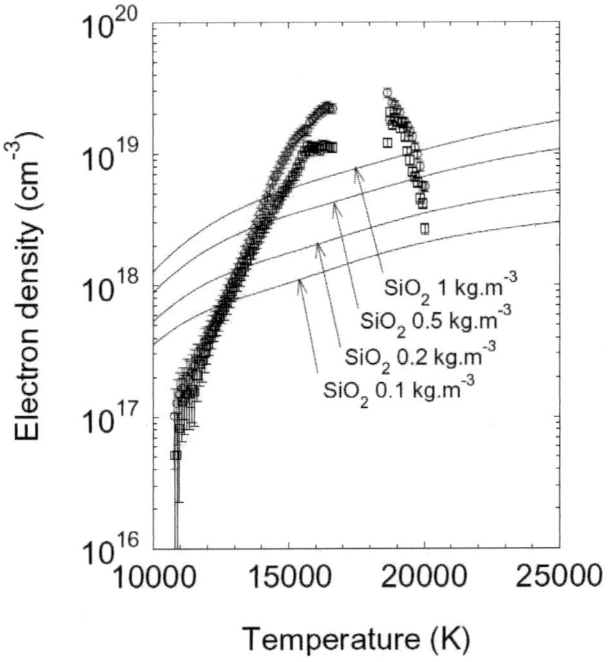

Figure 13. Electron density curves $n_e(T)$ [61]: comparison between experiments (Stark broadening of Si II (2) mutiplet: ∀, 634 nm and –, 637 nm) and composition calculation for SiO_2 plasma.

operation. Actually the evolution of the fuse arc plasma electron density is the result of the variation of the plasma composition: of metallic type at the beginning of the *arcing* period and of silica type for most of the *arcing* period duration. The electron density shown in figure 13 lies varies from ~ 10^{17} cm^{-3} to ~10^{19} cm^{-3}. These values and the trends during the *arcing* period depend strongly on the morphometric properties of the silica sand.

3. Conclusion

The *arcing* period is clearly more complex to describe and to model compared to the *pre-arcing* period (*Part 1. Pre-arcing period*). The assessment of the fuse arc plasma properties is mainly done by means of classical spectroscopic diagnostics whose applicability is sometimes difficult to justify. Especially conditions to obtained LTE within the plasma core can be reasonably assumed. On the contrary for the area where the arc is ignited and for the edges of the fuse arc plasma the LTE assumption appears questionable and the changes in the plasma composition (both close to the erosion front and close the plasma edges) are still unknown. The plasma core – high temperature, electron density and pressure and low density – is surrounded by colder layers whose composition changes during the *arcing* period and with distance from the plasma core. These various temporal and spatial gradients imply strong inhomogeneities and absorption of the radiation issued for the hotest areas of the fuse arc plasma. The knowledge of the plasma properties has to be increased to identify the physical processes in order to build a comprehensive modelling.

Aknowledgments

The author is grateful to industrial partners and academic partners for their useful discussions and their help during continued collaborations (Alstom, Electricité de France, Schneider Electric, Mersen, ACOIFF, Direction Générale des Entreprises) and also to J Aubreton and MF Elchinger from SPCTS Limoges. The author would like to thank particularly the colleagues from LAEPT.

References

[1] Lefort A 1998 Questions about arc roots ? *Invited lecture. XIIIth Symp. on Phys. Switching Arc* (Brno 7-11 Sept. 1998)

[2] Abbaoui M and Lefort A 2009 Arc root interaction with the electrode: a comparative study of 1D-2D axi symmetric simulations *Eur. Phys. J. Appl. Phys.* **48** N°1 11001

[3] Jüttner B 2001 Cathode spots of electric arcs *J. Phys. D: Appl. Phys.* **34** R103-123

[4] Benilov MS 2002 Theory and modelling of arc cathodes *Plasma Sources Sci. Technol.* **11** A49-54

[5] Heberlein J, Mentel J and Pfender E 2010 The anode region of electric arcs: a survey *J. Phys. D: Appl. Phys.* **43** 023001

[6] Shkol'nik SM 2011 Anode phenomena in arc discharges: a review *Plasma Sources Sci. Technol.* **20** 013001

[7] Lefort A and Andanson P 1985 Characteristic parameters of the anode spot *IEEE Trans. on Plasma Science* **13** N°5 296-299

[8] Wright A and Newbery P G 2004 Electric fuses ed IEE Power & Energy Series **49** 3rd edition (London)

[9] Wright A and Beaumont KJ 1976 Analysis of high-breaking-capacity fuselink arcing phenomena *Proc. IEE* **123** N°3 pp 252-258

[10] Gnanalingam S and Wilkins R 1980 Digital simulation of fuse breaking tests *Proc. IEE* **127 C 6** pp 434-440

[11] Reineri CA, Gomez JC and Magnago F 1993 Experimental determination of fuse filler thermal conductivity 7th *Int. Conf. on Switching Arc Phenomena* (Lodz,Poland) **Part I** pp 220-223

[12] Bussière W 2001 Estimation of the burn-back rate in high breaking capacity fuses using fast imagery *J. Phys. D: Appl. Phys.* **34** 1007-1016

[13] Jabubiuk K and Lipski T 1993 Dynamics of fulgurite formation during arcing in HRC fuses *J. Phys. D: Appl. Phys.* **26** 424-430

[14] Sloot JGJ, Kalasek VKI and Sikkenga J 1987 *A one dimensional mathematical model for the dynamical burnback velocity of silver strips Proc. of the 3rd Int. Conf. on Electric Fuses and their Applications* (ICEFA Eindhoven Netherlands) pp 72-77

[15] Fauchais P 2005 *Plasmas thermiques : aspects fondamentaux* ed Techniques de l'Ingénieur **D 2810**

[16] André P, Bussière W and Rochette D 2007 *Transport coefficients of Ag-SiO2 plasmas* Plasma Chem. Plasma Process **27** pp 381-403

[17] Bussière W and André P 2001 *Evolution of the composition, the pressure, the thermodynamic properties and the monoatomic spectral lines at fixed volume for a SiO_2-Ag plasma in the temperature range 5000-25 000 K* J. Phys. D: Appl. Phys. **34** pp 1657-1664

[18] Daalder J A, Schreurs E F 1983 *Arcing phenomena in high voltage fuses* EUT Report **83-E-137** (Eindhoven: EUT Dept. of Electrical Engineering)

[19] Wilkins R and Gnanalingam S 1978 *Burn-back rates of silver fuse elements* Proc. of the 5th Int. Conf. on Gas Discharges (University of Liverpool) pp 195-198

[20] Turner HW and Turner C 1977 *Calculation of the burn-back rate of a fuse element and its relation to contact erosion* Proc. of the 3rd Int. Symp. on Switching Arc Phenomena (Lodtz, Poland) pp 334-337

[21] IEC 60282-1, ed 5: High-voltage fuse – Part 1: Current-limiting fuses, International Electrotechnical Commission, 2002

[22] Lakshiminarashima CS, Barrault MR and Oliver R 1978 *Factors influencing the formation and structure of fuse fulgurites* Proc. IEE **125 5** 391-399

[23] Barrow DR, Howe AF and Cook N 1991 *The chemistry of electric fuse arcing* IEE Proc.-A **138 1** pp 83-88

[24] Chikata T, Ueda Y, Murai Y and Miyamoto T 1976 *Spectroscopic observations of arcs in current limiting fuse through sand* Proc. Int. Conf. On Electric Fuses and their Applications (Liverpool, UK) 114-121

[25] Saqib M and Stokes AD 1999 *Time resolved spectrum of the fuse arc plasma* Thin Solid Films **345** 151-155

[26] Stokes AD, Sibilski H and Kovitya P 1989 *Ablation arcs. II. Arcs in plastic materials and in boric acid* J. Phys. D: Appl. Phys. **22 11** 1702-1707
[27] Cao LJ and Stokes AD 1991 *Ablation arc. III. Time constants of ablation-stabilized arcs in PTFE and ice* J. Phys. D: Appl. Phys. **24 9** 1557 1562
[28] Cao LJ and Stokes AD 1992 *Ablation arc. IV. Spectrum analysis of the ablation-stabilized arcs in ice* J. Phys. D: Appl. Phys. **25 4** 669-676
[29] Saqib MA, Stokes AD, James BW and IS Falconer 1999 *Arc temperature measurement in a high-voltage fuse* Proc. of the 6th Int. Conf. on Electric Fuses and their Applications (Torino, Italy) 107-110
[30] Saqib MA, Stokes AD, James BW and Falconer IS 1999 *Measurement of electron density in a high-voltage fuse arc* Proc. of the 6th Int. Conf. on Electric Fuses and their Applications (Torino, Italy) 129-132
[31] Cheim LAV 1992 *Arc temperature measurements in high breaking capacity fuses* Ph.D Thesis, Univeersity of Nottingham, Department of Elect. and Electronic Engineering
[32] Cheim LAV and Howe AF 1994 *Spectroscopic observation of high breaking capcity fuse arcs* Proc. IEE Sci. Meas. Technol. **141** 123-128
[33] Barrow DR and Howe AF 1991 *The use of optical spectroscopy in the analysis of electric fuse arcing* Proc. of the 4th Int. Conf. on Electric Fuses and their Applications (ICEFA, Nottingham, UK) pp 221-225
[34] Bussière W and Bezborodko P 1999 *Measurements of time-resolved spectra of fuse arcs using a new experimental arrangement* J. Phys. D: Appl Phys. **32 14** 1693-1701
[35] Bussière W 2001 *Influence of sand granulometry on electrical characteristics, temperature and electron density during high-voltage fuse arc extinction* J. Phys. D: Appl. Phys. **34 6** 925-935
[36] Bussière W, Rochette D, André P, Velleaud G and Memiaghe S 2007 *Study of the SiO_2 plasma radiation. Application to the fuse arc plasma* Proc. of the 8th Int. Conf. on Electric Fuses and their Applications (ICEFA, Clermont-Ferrand, France) ed Blaise Pascal University Press p 109-119
[37] Arai S 1999 *Geometrical model of discharge space in the state of the high arc voltage generation of capillary arcs* Proc. of the 6th Int. Conf. on Electric Fuses and their Applications (Torino, Italy) 119-124
[38] Bussière W 2000 *Mesure des grandeurs (T, Ne, P) au sein du plasma d'arc des fusibles en moyenne tension* Ph. D Thesis (in French) Université Blaise Pascal **DU 1258** EDSF **301**
[39] Griem H 1964 *Plasma spectroscopy* ed McGraw-Hill
[40] Bussière W, Vacher D, Menecier S and André P 2011 *Comparative study of an argon plasma and an argon copper plasma produced by an ICP torch at atmospheric pressure based on spectroscopic methods* **20 4** 23pp
[41] Cwidak K and Lipski T 1993 *Post-arc resistance in HBC fuses* Proc. of the 7th Int. Conf. on Switching Arc Phenomena (TU Lodz, Poland) pp 202-205
[42] Cwidak K and Lipski T 1995 *New results on the post-arc fulgurite resistance of HBC fuses* Proc. of the 5th Int. Conf. on Electric Fuses and their Applications (ICEFA, Ilmenau, Germany) pp 219-222
[43] Jakubiuk K and Cwidak K 1995 *Analytical model of post-arc fulgurite resistance of HBC fuses* Proc. of the 5th Int. Conf. on Electric Fuses and their Applications (ICEFA, Ilmenau, Germany) pp 223-228
[44] André P 1995 *Etude de la composition et des propriétés thermodynamiques des plasmas thermiques à l'équilibre et hors d'équilibre thermodynamique* Ph. D Thesis (in French) Université Blaise Pascal **DU 771** EDSF **85**
[45] Rochette D, Bussière W and André P 2004 *Composition, enthalpy, and vaporization temperature calculation of Ag-SiO_2 plasmaswith air in the temperature range from 1000 to 6000 K and for pressure included between 1 and 50 bars* Plasma Chemistry and Plasma Processing **24 3** pp 475-492

[46] André P, Aubreton J, Bussière W, Elchinger MF, Memiaghe S and Rochette S 2007 *Basic data: composition, thermodynamic properties and transport coefficients applied to fuse* Proc. of the 8th Int. Conf. on Electric Fuses and their Applications (ICEFA, Clermont-Ferrand, France) ed Blaise Pascal University Press pp 95-100

[47] Bussière W, Duffour E, André P, Pellet R and Brunet L 2004 *Experimental assessment of temperature in plasma wall interaction* Eur. Phys. J. D **28** pp 79-90

[48] Rochette D and Bussière W 2004 *Pressure evolution during HBC fuse operation* J. Phys. D: Appl. Phys. **13** pp 293-302

[49] Koalaga Z and Zougmore F 2003 *Composition of CxHyOzNt plasmas out of thermal equilibrium using quite different modified forms of Saha and Guldberg-Waage equations* IEEE Trans. on Plasma Science **31 1** pp 82-93

[50] André P and Koalaga Z 2010 *Composition of a thermal plasma formed from PTFE with copper in non-oxidant atmosphere. Part I: definition of a test case with the SF6* JHTMP **14 3** pp 279-288

[51] André P and Koalaga Z 2010 *Composition of a thermal plasma formed from PTFE with copper in non-oxidant atmosphere. Part II: comparison of a test case with nitrogen* JHTMP **14 3** pp 289-296

[52] Wasserman AA, Melosh HJ, Lauretta DS 2002 *Fulgurites: a look at transcient high temperature processes in silicates* Proc. of the 33rd Lunar and Planetary Sc. Conf. (LPSC, Texas, US) **1308** 2pp

[53] Murin-Borel V, Lieutier M, Parizet MJ, Barrault M, Melquiond S, Rambaud T and Vérité JC 2000 *Silver mass flow in high-voltage fuses* J. Phys. D: Appl. Phys. **33 6** pp 753-756

[54] Wilkins R and McEwan PM 1975 *A.C. short-circuit performance of notched fuse elements* Proc. IEE *122 3* 289-292

[55] Rochette D 2002 *Modélisation et simulation de la décharge d'arc électrique dans un fusible moyenne tension* Ph. D Thesis (in French) Université Blaise Pascal **DU 1400** EDSF **372**

[56] Rochette D and Clain S 2003 *Numerical simulaiton of Darcy and Forchheimer force distribution in a HBC fuse* Transport in Porous Media **53** pp 25-37

[57] Rochette D and Clain S 2004 *Mathematical model and simulation of gas flow through a porous medium in high breaking capacity fuses* Int. J. of Heat and Fluid Flow **25** pp 115-126

[58] Rochette D and Clain S 2004 *Local heat transfer of compressible fluid in porous media: application to the HBC fuse* Int. J. of Heat and Fluid Flow **26** pp 322-333

[59] Bussière W, Rochette D, Latchimy T, Velleaud G and André P 2006 *Measurement of Darcy and Forchheimer coefficients for silica sand beads* High Temp. Mat. Processes **10** pp 55-78

[60] Ergun S 1952 *Fluid flow through packed columns* Chem. Eng. Progress **48 2** pp 89-94

[61] Bussière W, Rochette D and Pellet R 2003 *Inluence of the filling material properties on the pressure in HBC fuse. Comprison between experimental results and simulation* Proc. of the 7th Int. Conf. on Electric Fuses and their Applications (ICEFA, Gdansk, Poland) pp 111-118

[62] Ranjan R and Barrault MR 1980 *Axial electric-field measurements of d.c. arcs in current-limiting fuses* IEE Proc. **127 C 3** pp 199-202

[63] Murin-Borel V 1999 *Fusibles en moyenne tension : détermination des mécanismes physiques de fonctionnement* Ph. D Thesis (in French) Université Blaise Pascal EDSPIC

[64] Rat V, Murphy AB, Aubreton J, Elchinger MF and Fauchais P 2008 *Treatment of non-equilibrium phenomena in thermal plasmas* J. Phys. D: Appl. Phys. **41** 28pp

IOP Conf. Series: Materials Science and Engineering **29** (2012) 012003 doi:10.1088/1757-899X/29/1/012003

Transport Coefficients of Ar/C/H/O/N Systems Thermal Plasma at Atmospheric Pressure

A Sanon[1,2] and J M Baronnet[2]

[1]Institut de Recherche en Sciences Appliquées et Technologies – CNRST, BP 2393 Bobo Dioulasso, Burkina Faso
[2]Laboratoire de Chimie des Plasmas, Université de Limoges, 123 avenue Albert Thomas, 87060 Limoges, France

Email : amadou_sanon@hotmail.com

Abstract. The transport coefficients of thermal plasma composed of Ar/C/H/O/N elements are presented in the temperature range from 1000 to 15000 K. Several chemical composition calculations with different proportions of the elements lead us to retain 35 major species. We have characterized 595 interactions the between these species and derived the collision integrals necessary for the calculation of transport coefficients by Chapman-Enskog method. The following approximation orders have been used: 3rd for electrical conductivity and electrons translation thermal conductivity, 2nd for heavy particles translation thermal conductivity and 1st for viscosity, reaction and internal thermal conductivity. The development of thermal plasma processes lead to the use of plasma gases or gases mixtures more and more complexes. If data concerning the transport coefficients of simple gases like argon, hydrogen, oxygen or nitrogen or some binary mixtures like argon-hydrogen, argon-oxygen or oxygen-nitrogen can be early found in literature, those concerning ternary mixtures or more are not available. Ar/H/O and N brought together and C is added permitting to take into account hydrocarbons. First ours results for simple gases or gases mixtures are compared with recently published data of the literature and finally we present the transport coefficients of oxygen-nitrogen-methane, and argon-steam mixtures.

1. Introduction

During the last twenty years, news processes using thermal plasmas have widely penetrated in many industries: iron and steel Metallurgy with heating of air blowed into the nozzle, metallurgy with fusion or re-fusion of metals and alloys, chemistry with acetylene synthesis; materials treatments with ceramic powder synthesis, plasma spraying, wastes treatments….

The knowledge of transport coefficients and thermodynamic properties at high temperature for gases and gases mixtures is necessary for modelling of plasma processes and the design of processes components.

Thermodynamic properties such as enthalpy, specific heat, density are generally well defined and relatively easy to calculate. The fundamental data upon which calculations are based are accurately known. But data concerning transport coefficients, for which data are rarely available, nevertheless thermal conductivity, viscosity and electrical conductivity of simple gases (Boulos [1]) such as argon, hydrogen, oxygen, nitrogen or some mixtures argon-hydrogen (Aubreton [2]), argon-oxygen (Mexmain [3]), nitrogen-oxygen (Bacri [4])… can be found.

However, it must be noticed that at high temperature and high velocity the surrounding atmosphere is sucked in by the plasma jet. For example, in the case of an argon plasma jet in free air, at only a few centimetres from the nozzle exit, the proportion of air in the gaseous phase is close to the argon one. Likewise, when heating poor gases of hydrocarbons, we find in the plasma jet all the species able to be formed with the elements C/H/O/N. Moreover, plasma spraying using pure gases or more and more complex mixtures is in development.

In this work, we have characterized the interactions between the major species in an Ar/C/H/O/N plasma jet at atmospheric pressure and temperatures from 1000 to 15000 K, and calculated the corresponding collision integrals. With the hypothesis of local thermodynamic equilibrium, we have calculated the chemical composition and deduced the transport coefficients.

The system developed allows the determination of transport coefficients of every mixture composed of Ar/C/H/O/N. Our results are compared with some recently published studies in the case of simple gases. Finally transport coefficients of methane, oxygen-methane, air-methane and argon-steam mixtures are presented.

2. Method of calculation

As part of kinetic theory, the state of a system formed of discrete particles is characterized by a function called distribution function depending on position and velocity of all the particles and on the time. With the hypothesis of binary collisions and interactions between charged particles described by Coulomb potential screened and Debye length, the velocity distribution function is obtained from Boltzmann equation.

We have used the well-known Chapman-Enskog method [5] of resolution of Boltzmann equation for the determination of transport coefficients. This method is adapted for systems not far from equilibrium.

The calculations have been developed at the 1st order of Enskog approximation with development in Sonine polynomials. The number of terms of the development in Sonine polynomials defined a new approximation order inside the 1st order solution of Enskog, and fixes the number of collision integrals to be calculated. Subsequently, it is this approximation order we will refer to and call "Chapman-Enskog approximation order".

Due to the mass difference between an electron and the other particles and following the results of Devoto [6], we have separated the contribution of electrons and heavy particles to the transport coefficients.

With these hypotheses, we have calculated the diffusion coefficients and the viscosity depending essentially of the heavy particles at the 1st order approximation of Chapman-Enskog. The electrical conductivity depending in contrary of electrons has been determined at the 3rd order. The thermal conductivity appears as the contribution of 3 terms: Translation, internal and reaction thermal conductivity.

$$\lambda = \lambda_{trans} + \lambda_{\text{int}} + \lambda_{reac} \quad (1)$$

The reaction and internal components have been calculated at the 1st order using respectively Butler-Brokaw theory and Eucken theory extended to partially ionized gases.

Following Devoto [6], the translation component has been divided into 2 parts, respectively for electrons and heavy particles.

$$\lambda_{trans} = \lambda_{tr}^{\ e} + \lambda_{tr}^{\ h} \quad (2)$$

The electrons translation thermal conductivity has been calculated at the 3rd order and those of heavy particles at the 2nd order approximation.

2.1. *Collision Integrals and Interaction Potentials*

In the Chapman-Enskog formalism (Hirschfelder [7]) appear the so-called collision integrals which are averaged over Maxwellian distribution of the collision cross-section for pairs of particles. The collision integral $\Omega_{i,j}^{(l,s)}$ for interaction between particles *i* and *j* is defined by:

$$\Omega_{i,j}^{(l,s)} = \left(\frac{kT}{2\pi\mu_{i,j}} \right)^{1/2} \int_0^\infty e^{-\gamma^2} \gamma^{2s+3} Q^{(l)}(g) d\gamma \qquad (4)$$

Where $\gamma^2 = (1/2)\mu_{i,j} g^2 / kT$

$\mu_{i,j}$ and g are the reduced mass and relative velocity of the two colliding particles i and j respectively, k is the Boltzmann constant and $Q^{(l)}(g)$ is the collision cross section defined as :

$$Q^{(l)}(g) = 2\pi \int_0^\infty (1-\cos^l \chi) b db = 2\pi \int_0^\pi \sigma(g,\chi)(1-\cos^l \chi) \sin\chi d\chi \qquad (5)$$

Here $\sigma(g,\chi)$ is the differential collision cross-section, b is the impact parameter and χ is the deflection angle:

$$\chi(g,b) = \pi - 2b \int_{r_m}^\infty (1 - \frac{b^2}{r^2} - \frac{V_{i,j}(r)}{\mu_{i,j} g^2/2})^{-1/2} r^{-2} dr \qquad (6)$$

Where $V_{i,j}(r)$ is the interaction potential, r the distance between particles i and j and r_m the distance of closest approach during the collision.

Except the case of electron-neutral interactions where collision integrals can be deduced from momentum transfer cross-section $Q^{(l)}(g)$ using only one integration, collision integrals are generally determined from interaction potentials. An exhaustive list of interaction potentials can be found in Aubreton [2]. In the table 1 below are reported the analytic form of potential used and the references where the tables of collision integrals can be found.

Table 1. References of collision integrals tables

Potential name	Analytic Expression	Reference
Attractive Power	$V = -V_0 r^{-n}$	Kihara [8]
Repulsive Power	$V = V_0 r^{-n}$	Kihara [8]
Attractive Exponential	$V = -V_0 \exp(-Br)$	Brokaw [9]
Repulsive Exponential	$V = V_0 \exp(-Br)$	Monchick [10]
Morse	$V = D_e[1-\exp(-\beta(r-r_m))]^2$	Smith [11]
Morse – corrected	$V = D_e[\exp(-2\beta(r-r_m) - 2\exp(-\beta(r-r_m))]$	Samuilov [12]
Buckingham	$V = D_e(\alpha-6)^{-1}[6\exp(\alpha(1-r/r_m)) - \alpha(r_m/r)^6]$	Mason [13]
Lennard – Jones	$V = 4\varepsilon[(r_0/r)^{12} - (r_0/r)^6]$	Neufeld [14]
Stockmayer	$V = 4\varepsilon[(r_0/r)^{12} - (r_0/r)^6 + \delta(r_0/r)^3]$	Monchick [15]

Two cases frequently appear for interaction potentials between particles:

The analytic form of potential is not adapted to the direct determination of collision integral.

Absolutely no data is available from the literature to characterize the interactions.

In the 1st case, analytic form can be adapted according to Hirschfelder-Eliason rule (Hirschfelder [16]) which gives the interaction energy domain appropriated to the determination of transport coefficients:

- For repulsive interactions, the important part of potentials curves is for energy between 0.9 and 1.7 *kT*.
- For attractive interactions, potential energy around -0.6 *kT*

Thus, for a given interaction, the data have to be fitted as accurately as possible in the intervals defined by this rule, with an analytic form of potential adapted to the direct determination of collision integrals.

In the 2nd case, especially when the interactions considered concern minor species, a simple potential can be built to describe the interaction. Here are presented some of the methods we have used.

2.1.1. *Dispersion potential.* For collisions between particles at low temperature such as polyatomic species, and some radicals, one can only take into account the zone of potential with weak interaction by using a long distance potential (Margeneau [18]). Long distance potential can be written as the contribution of 3 terms:

$$V_{LD} = V_{elec} + V_{disp} + V_{ind} \tag{7}$$

Where V_{elec} is the multi-polar electrostatic potential and depend on the dipole momentum (μ_d) of the molecule, V_{disp} is the dispersion term and V_{ind} the induction potential which contribution is generally neglected.

When the 2 colliding particles are non-polar molecules ($\mu_d=0$) and neglecting V_{ind}, the long distance potential reduces to only the dispersion term. Following the dispersion forces theory of London and the model of Drude, V_{disp} can be approached by (Hirschfelder [7]):

$$Vdisp = -\frac{3}{2}\left(\frac{E_{ia}.E_{ib}}{E_{ia}+E_{ib}}\right)\frac{\alpha_a\alpha_b}{r_{ab}^{\ 6}} \tag{8}$$

Where α_a, α_b are the static polarizability of species *a* and *b*, and E_{ia}, E_{ib} are empirical constants assimilated to ionization potentials.

2.1.2. *Polarization potential.* When molar fractions of molecules are high those of ions are low and vice versa. Ion-molecules interactions are not very important and can be taken into account by the simple model of polarizability:

$$V(r) = -\frac{\alpha(Ze)^2}{2r^4} \tag{9}$$

Where *Ze* is the charge of the ion, α is the static polarizability of the neutral molecule and *r* the distance between the 2 particles.

2.1.3. *Hulburt-Hirschfelder potential.* In the case of interactions between 2 mono-atomic species A and B, it is possible to build a potential using the spectroscopic data of the AB molecule. Hulburt-Hirschfelder [19] has proposed the following Morse like potential:

$$V = D[(1-e^{-x})^2 + cx^3e^{-2x}(1-bx)] \tag{10}$$

With $x = \dfrac{w_e}{2(B_e - D)^{1/2}}\left[\dfrac{r-r_e}{r_e}\right]$ $\quad c = 1 + a_1\left(\dfrac{D}{a_0}\right)^{1/2}$ $\quad b = 2 + \left[\dfrac{7}{12} - \dfrac{Da_2}{a_0}\right]/c$

a_0, a_1, a_2 are the Dunham coefficients given by:

IOP Conf. Series: Materials Science and Engineering **29** (2012) 012003 doi:10.1088/1757-899X/29/1/012003

$$a_0 = \frac{w_e^{\ 2}}{4B_e} \quad , \quad a_1 = \frac{-1-\alpha_e w_e}{6B_e^{\ 2}} \quad , \quad a_2 = \frac{5}{4}a_1 - \frac{2}{3}\frac{w_e x_e}{B_e}$$

w_e, x_e, $w_e x_e$, B_e, α_e are the spectroscopic constants, r_e the inter-atomic distance at equilibrium and D the dissociation potential.

2.1.4. *Combination of potentials.* It is possible to determine the potential parameters of interaction A-B between 2 different particles from the corresponding parameters of interactions A-A and B-B between identical particles. Several authors have proposed combination rules. Among them:

- Kramer [20] for Dispersion potentials
- Kong [21] for Morse potentials, and particularly:
- Smith [22] for exponential repulsive:

$$\rho_{AB} = \frac{1}{2}(\rho_A + \rho_B) \text{ With } \rho=1/B \qquad \text{and} \qquad \left(\frac{V_{0AB}}{\rho_{AB}}\right)^{2\rho_{AB}} = \left(\frac{V_{0A}}{\rho_A}\right)^{\rho_A}\left(\frac{V_{0B}}{\rho B}\right)^{\rho_B} \tag{11}$$

- Hirschfelder [7] for Lennard-Jones potentials:

$$\varepsilon_{AB} = (\varepsilon_A.\varepsilon_B)^{1/2} \text{ and } \qquad r_{AB} = \frac{1}{2}(r_A + r_B) \tag{12}$$

These methods use drastic approximations for the parameters derivation and also suppose that the same analytic form of potential can describe interactions between identical and different particles. Despite these limitations they are a great help when no data are available.

In the case of interactions between polar P and non-polar N molecule, Hirschfelder [7], has proposed the following Lennard-Jones potential parameters derived from the Stockmayer potential parameters of interaction between polar molecules and the Lennard-Jones potential parameters of interaction between non-polar molecules:

$$r_{NP} = \frac{1}{2}(r_N + r_P)\xi^{-1/6} \text{and} \quad \varepsilon_{NP} = (\varepsilon_N.\varepsilon_P)^{1/2}\xi^2 \tag{13}$$

$$\text{With} \qquad \xi = 1 + \frac{1}{4}\frac{\alpha_N}{r_N^{\ 3}}\frac{\mu_{dP}^{\ 2}}{\varepsilon_P r_P^{\ 3}}\left(\frac{\varepsilon_P}{\varepsilon_N}\right)^{1/2}$$

2.1.5. *Average of collision integrals.* At high temperature, the interactions between some mono-atomic species (atoms or ions) can follow different energy curves. Each of these curves corresponds to a different alignment of electronic spin of the 2 colliding particles [23]. The collision integrals must be calculated for each potential curve and the average is given by the expression:

$$\left\langle \Omega^{(l,s)} \right\rangle = \frac{\sum_n \rho_n \Omega_n^{\ (l,s)}}{\sum_n \rho_n} \tag{14}$$

Where ρ_n et $\Omega_n^{(l,s)}$ are respectively the statistical weight and the collision integrals corresponding to the nth curves.

3. Collision integral of Ar/C/H/O/N systems

After several calculation of chemical composition with different proportions of Argon, Carbon, Hydrogen, Oxygen, and Nitrogen elements, the following 35 major species have been chosen:

e-, Ar, C, H, O, N, H_2, O_2, N_2, CH_4, CO, CO_2, OH, H_2O, NO, C_2, C_3, CH, CH_2, CH_3, C_2H, C_2H_2, CN, C_2N, C_2N_2, NH, NH_3, HCO, HCN, Ar+, C+, H+, O+, N+.

These species are enough for a good description of Ar/C/H/O/N mixtures at atmospheric pressure and temperature range from 1000 to 15000 K.

It should be noticed that in the expression of transport coefficients appear terms such as "$x_i.x_j$" or "$n_i.n_j$" corresponding to products of molar fractions or density of species. Thus neglecting minor species (molar fraction < 10^{-4}) will not affect perceptibly the results.

3.1. *Neutral – Neutral and Neutral - Charged Interactions*

In Table 17 are presented the method used for the derivation of collision integrals concerning neutral-neutral interactions. Interaction potentials have been chosen according to the following criteria:

When data on a particular interaction are available in the literature, they have been used even if it has been necessary to fit with analytic form of potential adapted to the derivation of collision integrals. We have taken a special care for the atom-atom and atom-ion interactions which can follow several potentials curves.

Due to the fact that Buckingham potential described better the repulsive region of interactions, it has chosen rather than Lennard-Jones one when the 2 sets of parameters were available. Moreover, exponential repulsive potentials have been preferred to power repulsive potentials.

Stockmayer potentials have been used for interactions between polar molecules while those between polar and non-polar molecules have been described by Lennard-Jones potentials with parameters deduced from relation (13)

Most part of interactions between non-polar molecules has been described by Lennard-Jones potentials using the combination rule (11) from potential parameters between identical species found in the literature.

Interactions between molecules and ions have been described using polarization potential while those between atom and ion have been treated with care in the same way as for atom-atom interactions.

When there was absolutely no data available, we have used the dispersion potential or the rigid sphere model. So, interactions between some free radicals and molecules have been taken into account by dispersion potentials and some of those between atom and free radical by the rigid sphere model.

3.1.1. *Interactions described with Buckingham potentials.* Table 2 gives the parameters values and references for interactions described by Buckingham potentials. The corresponding collision integrals have been deduced from Mason [13] tables.

Table 2. Parameters for interactions described with Buckingham potentials

Interactions	ε/k (K)	r_m (A°)	α	References
H2 – O2	143.000	3.234	14.0	Chapman [5]
O2 – CO	110.000	3.733	17.0	Chapman [5]
O2 – CO2	255.000	3.616	17.0	Chapman [5]
O2 – CH4	220.000	3.612	17.0	Chapman [5]
N2 – CO2	184.000	3.782	17.0	Chapman [5]
CH4 – CH4	152.800	4.206	14.0	Hirschfelder [7]
CO2 – CO2	262.000	4.180	14.0	Chapman [5]

3.1.2. *Interactions described with Stockmayer potentials.* We have reported in Table 3 the parameters of interactions involving polar molecules which have been described with Stockmayer potentials.

Table 3. Parameters of interactions described with Stockmayer potentials

Interactions	ε/k	r_0 (A°)	δ	References
CO – CO	109.900	3.585	0.00983	Hirschfelder [7]
CO – H2O	227.273	3.090	0.11688	*Combination*
H2O – H2O	470.000	2.595	1.50000	Debbagh-Nour[23]
NO – NO	112.400	3.508	0.00337	Hirschfelder [7]
NH3 – NH3	146.800	3.441	1.24990	Hirschfelder [7]

3.1.3. *Interactions described with exponential potentials.* Table 4 gives the parameters of interactions described with exponential potentials. The asterisk (*) denote those for which the parameter have been obtained by adjusting the data of authors.

Table 4. Parameters of interactions described with Exponential potentials

Interactions	V_0 (eV)	B (A^{o-1})	References
Ar – H	138.300	2.783	Aubreton [2]
Ar – N	778.000	3.280	Mason [24]
Ar – N2	2030.000	3.240	Mason [24]
C – N	-25.330	1.710	Cubley [25]
C – H2	-7510.240	7.570	Herzberg [26]
C – CH	-118.988	1.802	Alexander [27]
C – C2	-559.650	2.750	Moiseyev [28]
H – H2	55.900	3.065	Aubreton [2]
H – CH	-9155.560	4.620	Herzberg [26]
H – C2	-1832.375	5.369	Alexander [27]*
O – N	348.000	3.410	Mason [24]
O – N2	840.000	3.320	Mason [24]
O – NO	1920.000	3.680	Mason [24]
N – H2	1740.000	4.520	Carter [29]
N – O2	1640.000	3.620	Mason [24]
N – N2	285.000	2.830	Mason [24]
N – NO	714.000	3.190	Mason [24]
H2 – H2	252.500	3.331	Aubreton [2]
N2 – O2	912.000	2.860	Mason [24]
N2 – N2	450.000	2.620	Mason [24]
N2 – NO	1950.000	3.130	Mason [24]
NO – O2	5120.000	3.560	Mason [24]
NO – NO	2150.000	3.320	Mason [24]

3.1.4. *Interactions described with Morse potentials.* We have reported in Table 5 the parameters of interactions described by Morse potentials. The corresponding collision integrals have been obtained from Samuilov tables [12].

Table 5. Parameters of interactions described with Morse potentials

Interactions	*De* (eV)	r_m (A°)	β (A^{o-1})	References
Ar – Ar	0.0121	3.760	6.279	Aubreton [2]
Ar – H2	7.019	2.615	1.895	Aubreton [2]
Ar – H+	4.100	1.330	1.792	Aubreton [2]
Ar – O+	0.670	2.219	1.910	Mexmain [3]
H – CH2	4.900	1.058	1.853	Capitelli [30]
H – CH3	4.795	1.086	1.880	Capitelli [30]
O – CH4	0.0094	3.780	1.667	Saxon [31]
O – CO	5.4528	1.160	3.461	Luiti [32]

3.1.5. *Interactions described with Lennard-Jones potentials.* Table 6 gives the Lennard-Jones potential parameters of interactions between identical species. The parameters for interactions between different species have been obtained from Hirschfelder combination rule (11).

Table 6. Lennard-Jones potentials parameters between identical species

Interactions	ε/k (K)	R_0 (A°)	References
Ar – Ar	93.300	3.542	Holmer [33]
H – H	145.000	2.050	Reid [34]
O – O	80.000	2.750	Reid [34]
N – N	71.400	3.298	Reid [34]
H2 – H2	33.300	2.968	Hirschfelder [7]
O2 – O2	113.000	3.433	Hirschfelder [7]
N2 – N2	71.400	3.798	Holmer [33]
CH4 – CH4	148.600	3.758	Holmer [33]
CO – CO	91.700	3.690	Holmer [33]
CO2 – CO2	195.200	3.941	Holmer [33]
OH – OH	82.360	3.392	Kee [35]
H2O – H2O	800.000	2.650	Khayar [36]
NO – NO	116.700	3.492	Holmer [33]
C2H2 – C2H2	231.800	4.033	Holmer [33]
C2H4 – C2H4	224.700	4.163	Holmer [33]
CN – CN	75.000	3.856	Reid [34]
C2N2 – C2N2	349.000	4.361	Reid [34]
NH – NH	80.000	2.650	Reid [34]
NH3 – NH3	558.300	2.900	Holmer [33]
HCO – HCO	498.000	3.590	Reid [34]
HCN – HCN	569.100	3.630	Holmer [33]

3.1.6. *Interactions described with power potentials.* Data from Capitelli [30] and Van-Dishoek [39] have been taken for respectively N2-H2 and Ar-C interactions.

- N2-H2 : $V_0 = 88.100$ eV.A^{on} n = 6.63
- Ar-C : $V_0 = 142.000$ eV.A^{on} n = 7.10

Dispersion potentials

We have reported in the table 7 the values of ionization potentials and static polarizability used for the determination of potential parameters of interaction taken into account by the simple model of dispersion.

Table 7. Values of Ionization Potential and Static Polarizabity used

Specie	Ionization Potential (eV)	Polarizability (A^{o3})	Specie	Ionization Potential (eV)	Polarizability (A^{o3})
Ar	15.755	1.640	CH	11.130	0.650
He	24.587	0.205	CH2	10.396	1.300
H2	15.427	0.767	CH3	9.843	1.950
O2	12.060	1.569	C2H	11.650	2.670
N2	15.576	1.760	C2H2	11.410	3.330
CH4	12.990	2.600	C2H4	10.500	4.260
CO	14.009	1.950	CN	14.300	1.960
CO2	13.769	2.650	C2N	13.600	5.010
OH	13.170	1.109	C2N2	13.600	3.000
H2O	12.600	1.480	NH	13.100	0.750
NO	9.250	1.740	NH3	10.200	2.260
C2	12.320	1.151	HCO	9.800	1.970
C3	12.900	1.318	HCN	13.800	2.590

Polarization potentials

As noted, ion-molecules interactions have been described by polarization model. Table VII give the values of static polarizability necessary to the calculation of power potential parameters according to expression (9)

3.1.7. *Interactions described with rigid sphere model.* By lack of data, some interactions (notably involving C, C2, C3 species) have been taken into account by the rigid sphere model. The collision radii used are reported in Table 8.

Table 8. Radii used for Rigid Sphere model

Specie	r (A°)	Specie	r (A°)	Specie	r (A°)
Ar	3.500	CO	3.590	C2H	3.500
C	0.9140	CO2	3.990	C2H2	3.800
H	0.6450	OH	3.390	C2H4	4.000
O	0.6775	H2O	2.650	C2N	3.800
N	0.6300	NO	1.700	C2N2	4.100
H2	2.500	C2	3.200	NH	2.500
O2	3.433	C3	3.600	NH3	3.000
N2	1.750	CH	2.600	HCO	3.500
CH4	3.700	CH2	2.800	HCN	3.600
		CH3	3.400		

3.1.8. *Multiples Interactions.* These concern atom-atom interactions which can follow several potential curves. For Ar/H and Ar/O systems we have used respectively the potentials given by Aubreton [2] and Mexmain [3]. The collision integrals given by Carter [29] have been directly used for N-N and N-N+ interactions.

H - C Interaction

We have admitted that this interaction can follow 4 potential curves corresponding to the states $^2\Pi$, $^4\Sigma^-$, $^2\Sigma^-$, $^4\Pi$. We have fitted the data from Amdur [37] to exponential potentials which parameters are given in Table IX.

Table 9. Exponential Potentials Parameters used for H-C Interaction

State	V_0 (eV)	B (A^{o-1})
$^2\Pi$	-257.206	2.806
$^4\Sigma^-$	-1614.024	4.402
$^2\Sigma^-$	-35.581	3.826
$^4\Pi$	56.0765	1.967

H - O Interaction

As Debbaagh-Nour [23], we have adjusted the data from Eaker [38] by exponential potentials.

- $X^2\Pi$ state V_0 = 111.535 eV B = 2.635 A^{o-1}
- $^2\Sigma^-$ state V_0 =68.578 eV B = 2.679 A^{o-1}

H - N Interaction

This interaction can follow 2 potentials curves corresponding to states $X^3\Sigma$ and $5\Sigma^-$. We have adjusted the data from Van-Dishoek [39]:

- $X^3\Sigma$ state by a Morse potential:
  D_e = 2.620 eV B = 2.660 A^{o-1} r_m = 1.100 A°
- $^5\Sigma^-$ state by a repulsive exponential potential:
  V_0 = 233.400 eV B = 3.290 A^{o-1}

H – C+ and C – H+ Interactions

Following the study of Graff [40], we have characterized the H-C+ interaction by 4 curves corresponding to states $X^1\Sigma^+$, $^3\Pi$, $^1\Pi$, $^3\Sigma^+$ of CH+ molecule. Data corresponding to the fundamental state $X^1\Sigma^+$ have been fitted by a power potential:

V0 = -651.670 eV.A^{on} n = 7.8,

and the other by exponential potentials which parameters are reported in Table 10:

Table 10. Exponential Potentials Parameters used for H-C+ Interaction

State	V_0 (eV)	B (A^{o-1})
$^3\Pi$	-266.850	2.780
$^1\Pi$	-5.464	1.280
$^3\Sigma^+$	475.140	3.195

For C-H+ interaction, the single data from Kusunoki [41] have been adjusted by an exponential potential:

V_0 = -118.830 eV B = 2.529 A^{o-1}

H – O+ and O – H+ Interactions

We have used the data from Chambaud [42] for these interactions. The potential curve of $X^3\Sigma^+$ and $^3\Pi$ states have been attributed to O-H+ interaction and those of $^5\Sigma^-$ and $^3\Sigma^-$ to H-O+ interaction. The data have been fitted to exponential potentials and Table 11 gives the parameter values found.

Table 11. Exponential Potentials Parameters used for H-O+ and O-H+ Interactions

State	V_0 (eV)	B (A°$^{-1}$)
$X^3\Sigma^+$	-3.699	1.310
$^3\Pi$	-15.332	1.447
$^5\Sigma^-$	602.000	1.830
$^3\Sigma^-$	610.920	1.830

C – C Interaction

This interaction can follow 6 potentials curves corresponding to the states $X^1\Sigma^+_g$, $a^3\Pi_u$ $b^3\Sigma^+_g$, $A^2\Pi_u$, $C^3\Sigma_u$, $d^3\Pi_g$.of C2 molecule. From Rosen [43] spectroscopic data, we have build Hulburt-Hirschfelder potentials. The results for $X^1\Sigma^+_g$, $b^3\Sigma^+_g$, $A^2\Pi_u$, $d^3\Pi_g$ states have been fitted by exponential potentials and those for $a^3\Pi_u$ et $C^3\Sigma_u$ by Morse potentials. The parameters values are reported in Table 12.

Table 12. Potentials Parameters used for C-C Interaction

	Exponential Potentials		
State	V_0 (eV)		B (A°$^{-1}$)
$X^1\Sigma^+$	-636.820		2.840
$B^3\Sigma^+_g$	-356.280		2.340
$A^2\Pi_u$	-162.000		2.240
$D^3\Pi_g$	-289.440		3.348
	Morse Potentials		
State	D_e (eV)	r_m (A°)	β(A°$^{-1}$)
$a^3\Pi_u$	6.000	1.317	2.200
$C^3\Sigma_u$	4.470	1.230	3.082

C – O Interaction

For this interaction, Data from Krupenie [44] have been fitted to exponential potentials. The parameters are given in Table 13.

Table 13. Potentials Parameters used for C-O Interaction

State	V_0 (eV)	B (A°$^{-1}$)
$X^1\Sigma^+$	-285.652	2.302
$^3\Pi$	-200.990	2.714
$^3\Sigma^+$	-134.959	2.113
$^3\Delta$	-171.684	2.403
$^3\Sigma^-$	-100.034	2.142
$^1\Sigma^-$	-109.400	2.189
$^1\Pi$	-232.432	3.139

C – N Interaction

Data from Cubley [25], which identify 2 potential curves corresponding $X^2\Sigma^+$ and $A^2\Pi$ states have been used for this interaction. Since the 2 potentials have the same values in the energy domain of interest, we have fitted the data with a single exponential potential:

V_0 = -25.330 eV B = 1.710 A^{o-1}

Ar – C+ and C – Ar+ Interactions

For these interaction, we have used data from Hillier [45] who assumed that Ar-C+ interaction follow potential curves corresponding to $^2\Pi$ and $^2\Sigma^+$ states and C-Ar+ interaction, potential curves corresponding to $^4\Sigma^-$, $^4\Pi$ and $^2\Sigma^-$ states. We have fitted data for $^2\Pi$, $^4\Sigma^-$ and $^4\Pi$ to Morse potentials and those for $^2\Sigma^+$ and $^2\Sigma^-$ state by exponential potentials. The parameter values are given in Table 14.

Table 14. Potentials Parameters used for Ar-C+ and C-Ar+ Interactions

Morse Potentials			
State	D_e (eV)	r_m (A°)	β (A^{o-1})
$^2\Pi$	0.938	1.995	1.714
$^4\Sigma$	2.130	1.782	2.101
$^4\Pi$	0.354	2.548	1.564

Exponential Potentials		
State	V_0 (eV)	B (A^{o-1})
$^2\Sigma^+$	27543.057	4.982
$^2\Sigma^-$	105.690	2.280

C – O+ and O – C+ Interactions

We have adjusted data from Krupenie [44] by exponential potentials. The potential curves for $X^2\Sigma^+$, $A^2\Pi$, $B^2\Sigma$ and $C^2\Delta$ states have been assigned to O-C+ interaction and those for $D^2\Pi$ to C-O+ interaction.

Table 15. Potentials Parameters used for C-O+ and O-C+ Interactions

State	V_0 (eV)	B (A^{o-1})
$X^2\Sigma^+$	-235.446	2.650
$D^2\Pi$	-128.802	2.498
$A^2\Pi$	-190.891	2.282
$B^2\Sigma$	-338.508	2.853
$C^2\Delta$	-537.137	2.421

C – C+ Interaction

Viscosity-type collision integrals:

It correspond to elastic interaction (l=2). Collision integrals calculations are performed in the same manner as neutral-neutral interactions. From Verhaegen [46] Spectroscopic data we assumed that this interaction can follow 4 curves corresponding to $X^4\Sigma^-_g$, $^4\Sigma^-_u$ $^2\Pi_u$ et $^2\Pi_g$ states. Data for fundamental state $X^4\Sigma^-_g$ have been fitted to a Morse potential and the other to exponential potentials:

Table 16. Potentials Parameters used for C-C+

Morse Potential			
State	D_e (eV)	r_m (A°)	β (A^{o-1})
$X^4\Sigma^-_g$	5.320	2.246	1.412

Exponential Potentials		
State	V_0 (eV)	B (A^{o-1})
$X^2\Sigma^-$	-292.560	2.874
$^2\Pi_u$	-121.640	1.884
$^2\Pi_g$	-106.150	2.035

Diffusion-type collision integrals:

It correspond to the case of inelastic collision with resonant charge exchange (l=1). Following the method described by Aubreton [2], we have made the gerade-ungerade difference of each couple of potentials and resolved the transcendental equation. The following values of A_n and B_n constants necessary for collision integrals derivation have been found:

$A_\Sigma = 21.137$ A° $B_\Sigma = 0.919$ A°

$A_\Pi = 28.000$ A° $B_\Pi = 1.218$ A°

3.2. *Charged - Charged Interactions*

Charged-charged interactions have been taken into account by Coulomb potential screened at Debye length. For Debye length calculation, following Devoto [48] and Murphy [65], we have chosen to take into account only electrons (neglecting ions). The corresponding collision integrals have been obtained from Mason[47] and Devoto [48] tables.

3.3. *Electron - Neutral Interactions*

In this case, collision integrals have been derived by numerical integration of momentum transfer cross-section found in the literature.

We have used data from Itikawa [49] and [50] for e-CH4, e-CO, e-CO2 interactions and those from Thomas [51] for e-C interaction.

Cross-section from Gee [52], who have extensively study electron-hydrocarbon interactions have been used for e-C2H2 and e-C2H4 interactions.

For interaction involving free radicals like CH, CH2, CH3 we have used the "methane-type" cross-section of Kleban [53].

In the case of interactions concerning species of Ar/N/O/H systems, we have directly used data from Aubreton [2], Mexmain [3], Carter [29] and Devoto [54] who have study respectively Ar/H, Ar/O, N/H and N/O systems.

By lack of data, the e-OH interactions have been assimilated to e-O2. In the same manner we have assimilated e-C2H to e-C2H2 interaction, e-C2 and e-C3 to e-C interaction, e-C2N and e-C2N2 to e-CN interaction, e-HCN and e-NH to e-NH3 interaction and finally e-HCO to e-CO interaction.

4. Results

Figure 1-a and 1-b present the chemical composition evolution of a mixture containing 50% of air and 50% of methane respectively between 500 and 5000 K and 5000 and 15000 K.

We have considered a standard dry air formed of 78.09% of nitrogen, 20.95% of oxygen, 0.93% of argon and 0.03% of carbon dioxide. These figures illustrate the evolution of molar fractions of the 35 species taken into account.

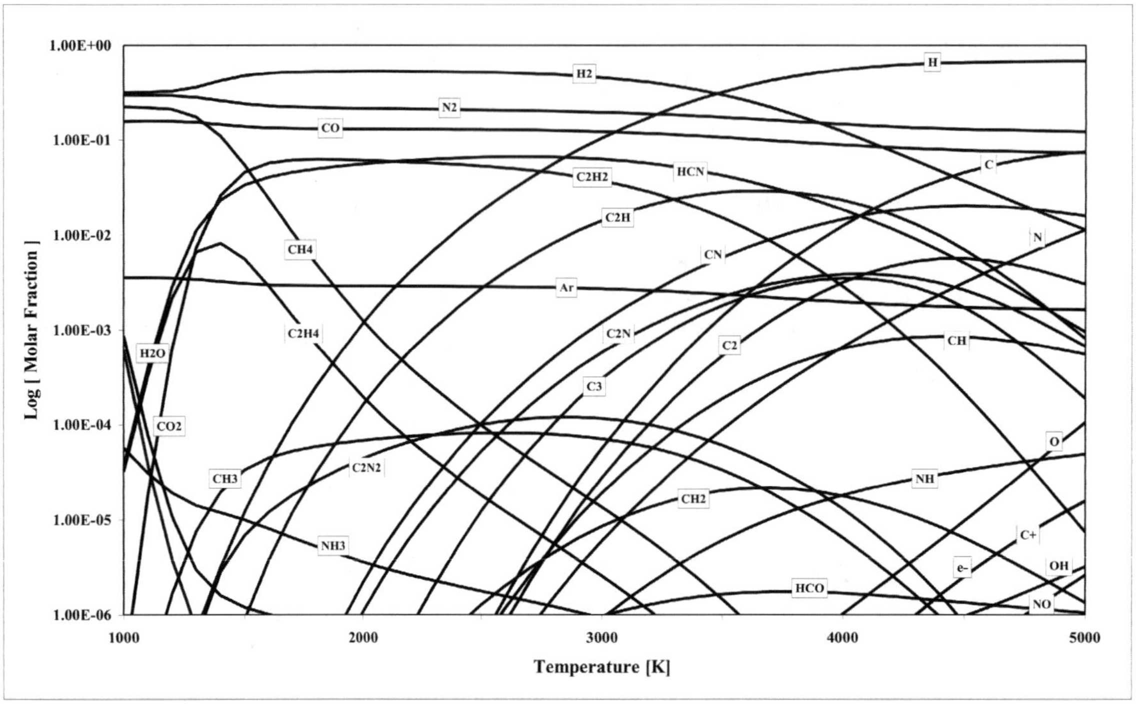

Figure 1a. Chemical composition of Air/CH4 mixture between 1000 and 1500K

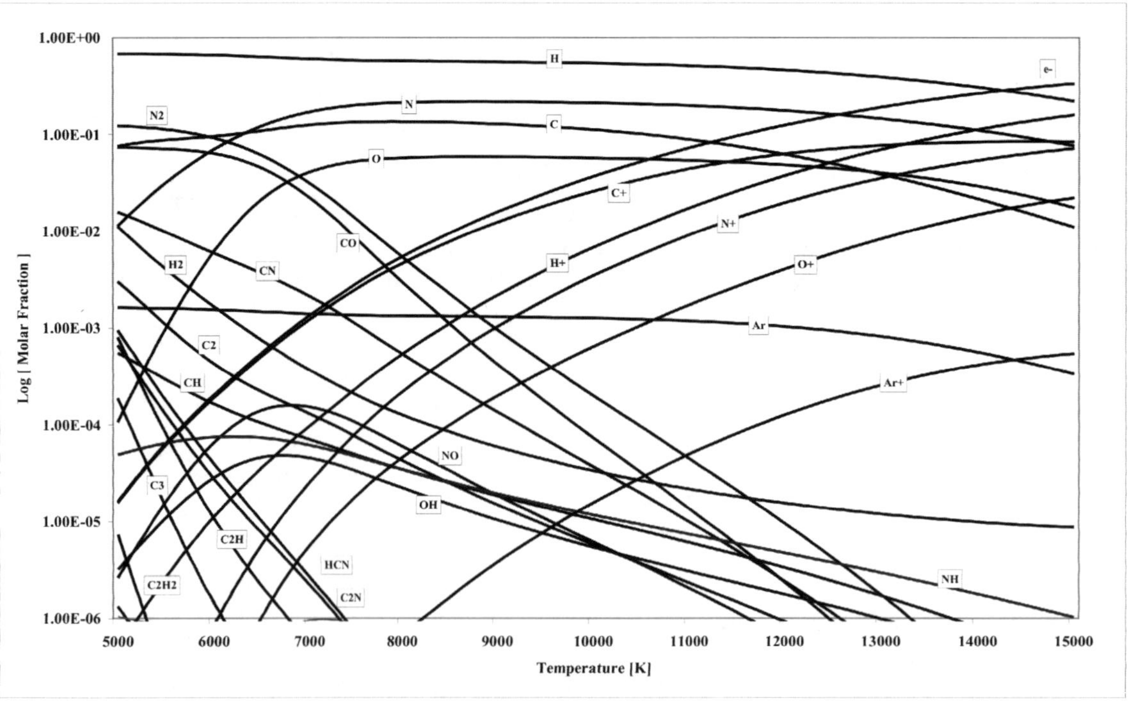

Figure 1b. Chemical Composition of Air/CH4 mixture between 5000 and 15000K

We have reported in figure 2 the thermal conductivity of methane with the contribution of different components. It can be seen the predominance of the reaction component from 1000 to 5500 K and from 11000 to 15000 K and those of heavy particles translation from 5500 to 11000 K. Internal thermal conductivity is appreciable only at low temperature. It reached a maximum of 0.400 W/m/K at 3300 K. Electron translation component become more and more important as the ionization increase. The 1st peak of thermal conductivity at 1400 K is due to CH4 molecules dissociation and the 2nd at 3800 K to the dissociation of H2 molecules formed in the system.

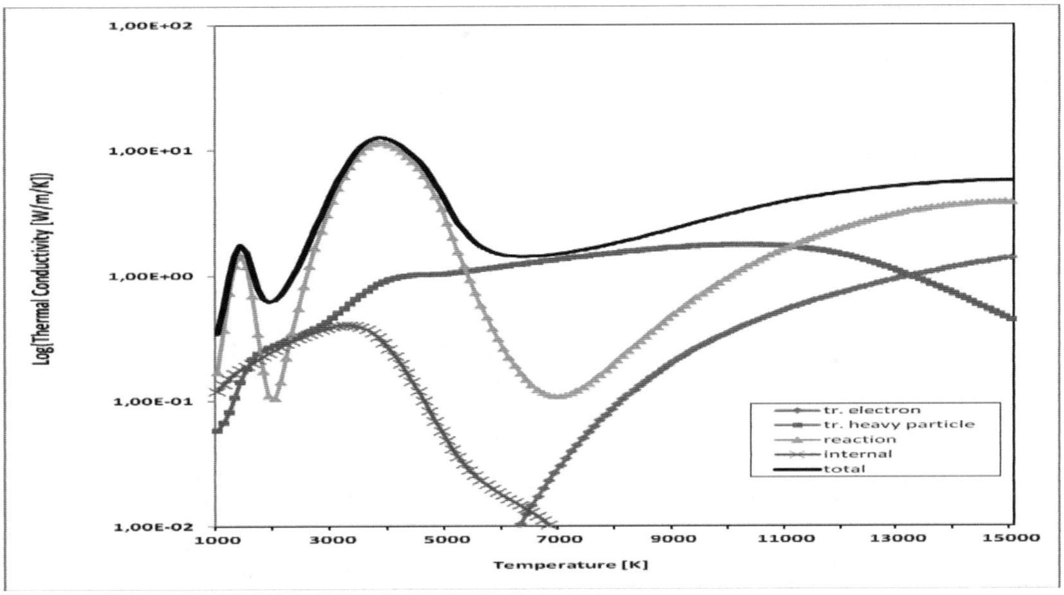

Figure 2. Thermal Conductivity of CH4 between 1000 and 15000K

Viscosity of methane is shown in figure 3. The maximum denoting the boundary between plasma dominated by neutral particles and plasma dominated by charged particles is at 9200 K.

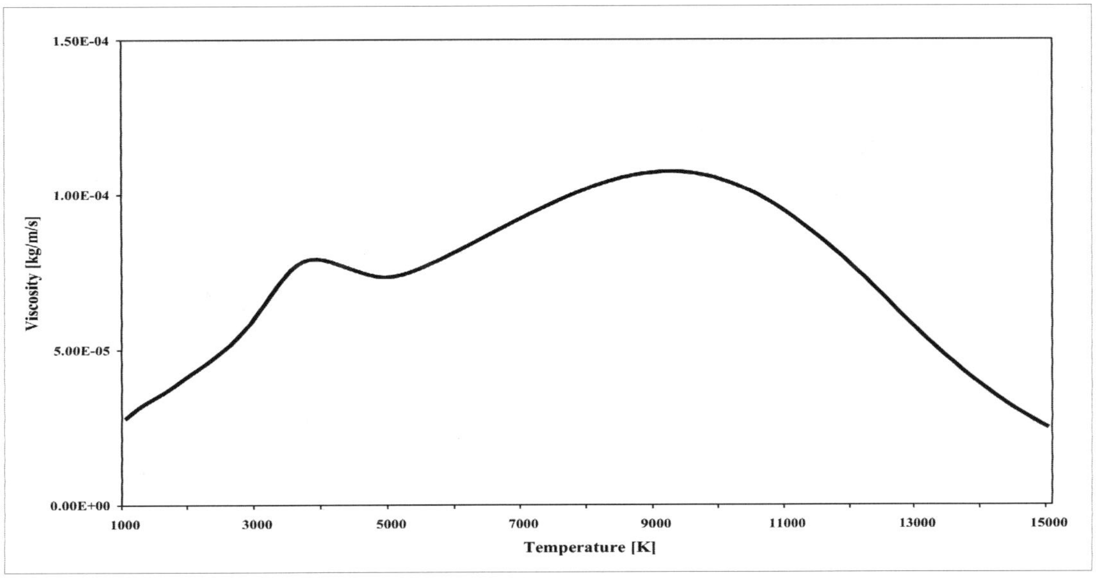

Figure 3. Viscosity of CH4 between 1000 and 15000K

Figure 4 show the evolution of electrical conductivity which, as electron translation thermal conductivity is directly dependant on the ionization degree.

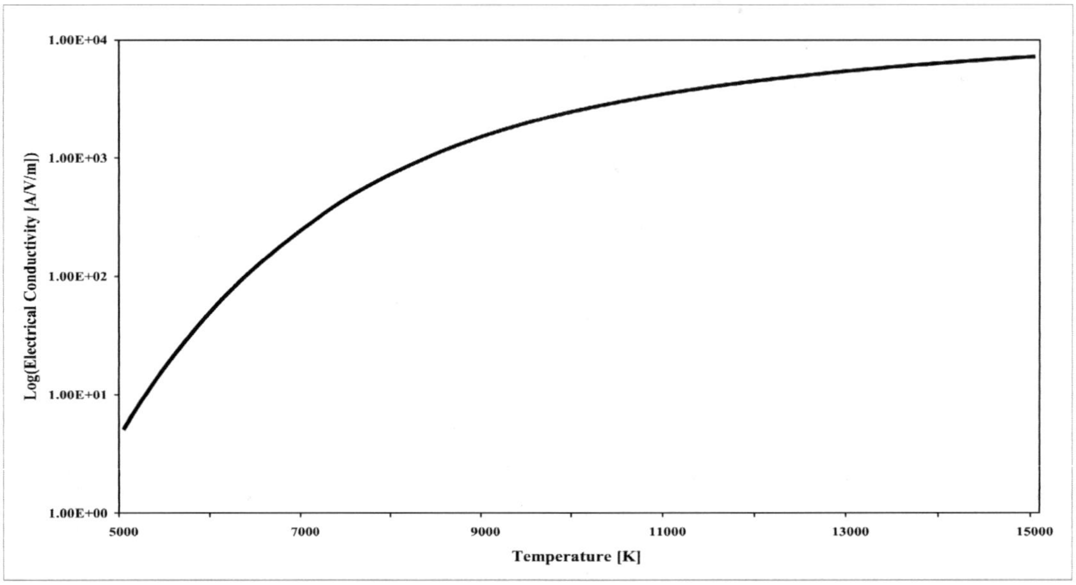

Figure 4. Electrical Conductivity of CH4 between 1000 and 15000K

4.1. *Transport Coefficients of Argon, Hydrogen, Oxygen and Nitrogen*

Figures 5, 6 and 7 give respectively thermal conductivity, viscosity and electrical conductivity of simple gases: argon, hydrogen, oxygen and nitrogen. We have reported in the same figures data from other authors.

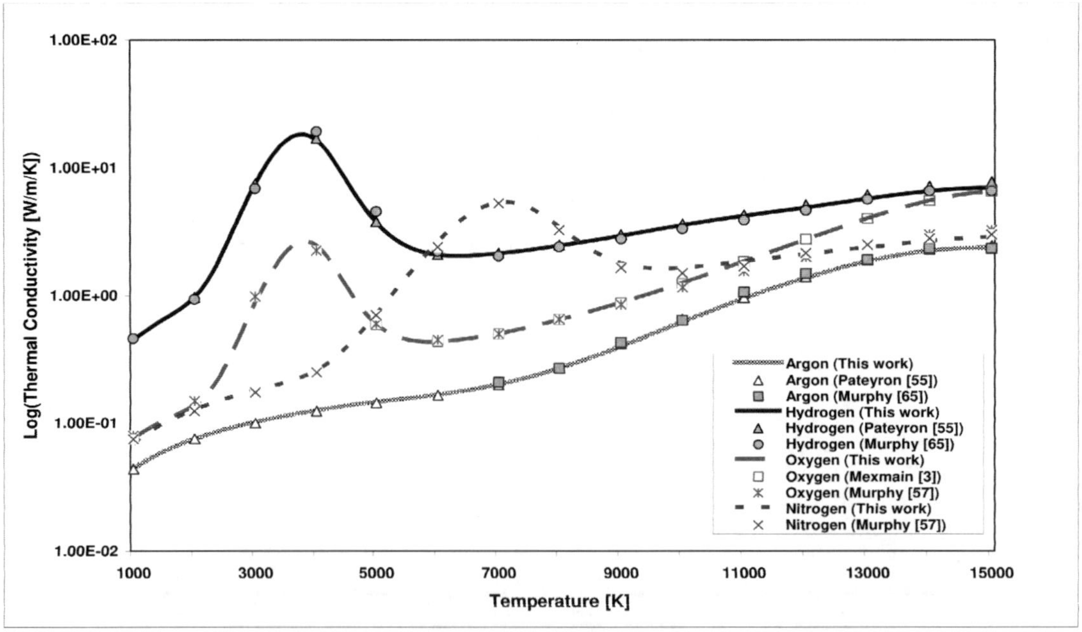

Figure 5. Thermal Conductivity of Argon, Hydrogen, Oxygen and Nitrogen

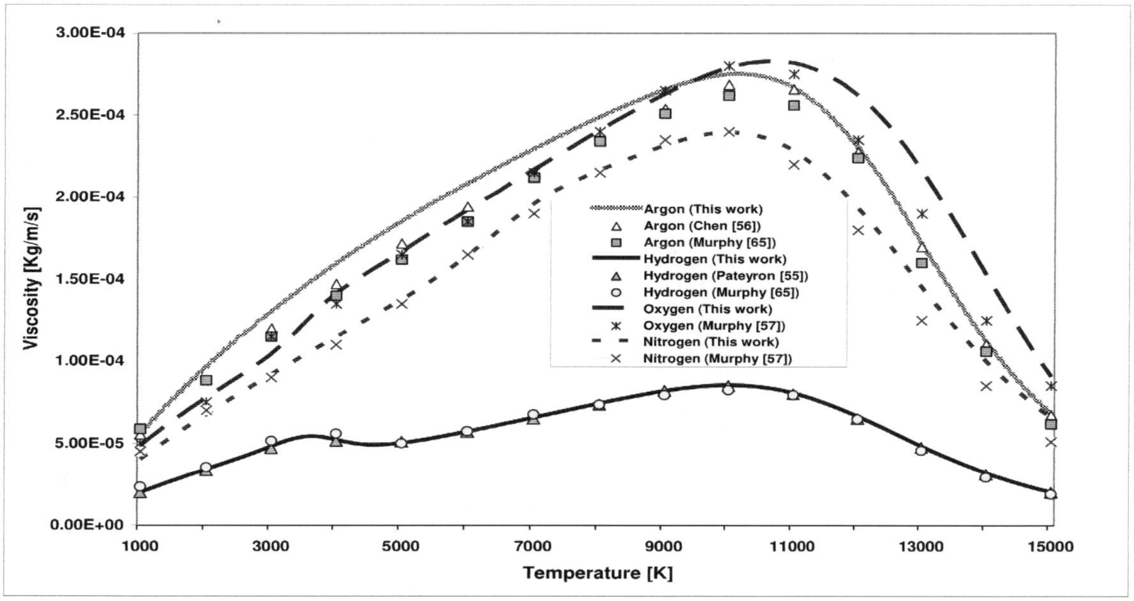

Figure 6. Viscosity of Argon, Hydrogen, Oxygen and Nitrogen

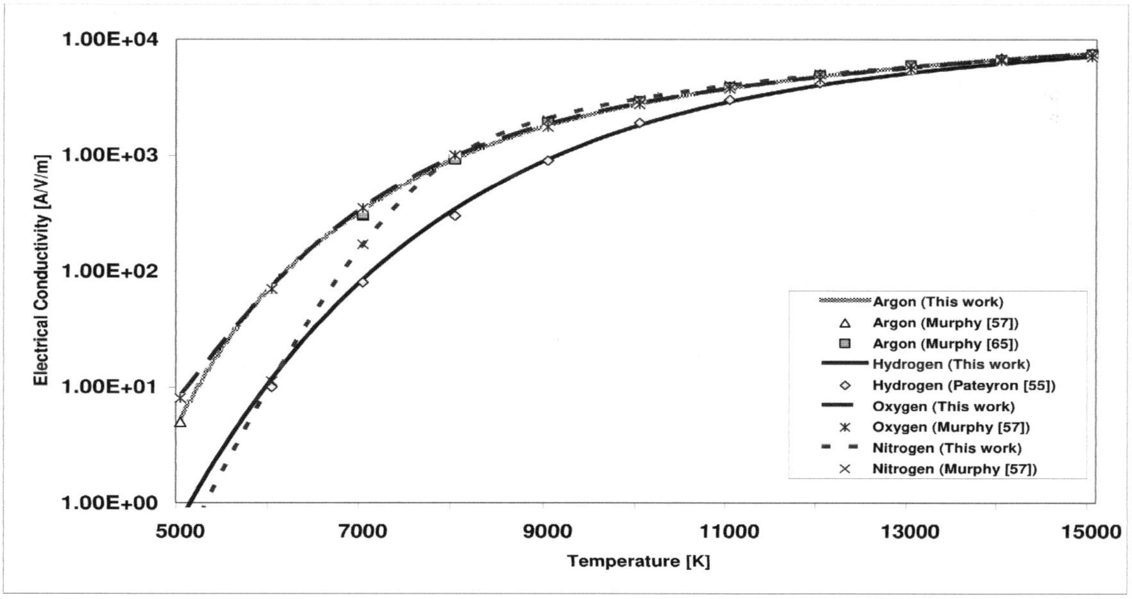

Figure 7. Electrical Conductivity of Argon, Hydrogen, Oxygen and Nitrogen

Our results for argon are in good agreement with those from Pateyron [55], since we have used almost the same potentials (namely those of Aubreton [2]). They are also in good agreement with those from Murphy [57] and Murphy [65] except in the case of viscosity. In this case we have reported on figure 6, recommended value of Chen [56] for argon viscosity and those from Murphy [65]. We denote an excellent agreement with our results only above 10000 K, in the decreasing region of viscosity. Differences exist in the increasing region but it never reaches 10%.

In the case of hydrogen, our results for the three transport coefficients are in very good agreement with those from Pateyron [55] as well as those from Murphy [65].

For oxygen, we denote a good agreement with results from Murphy [57] up to 10000 K. The values diverge above this temperature. This is to put in the count of O-O+ interaction. As Mexmain [3] we have used for this interaction, data from Beebe [58] while Murphy [57] have taken those from Rutherford [59]. In the contrary, for nitrogen, our results agree closely with the values from Murphy [57].

4.2. *Transport Coefficients of Steam, Carbon Dioxide, Air and Ammonia*

We have reported in figure 8, 9 and 10, the transport coefficients of steam, carbon dioxide, air and ammonia.

In the case of Air, we denote a reasonable agreement between our results and experimental values of thermal and electrical conductivity of Asinovsky [60]. Our results of viscosity are compared with those of Vargaftik [61] (up to 4000 K) and Nicolet [62]. There is a good agreement in the increasing region, but above 10000 K, values from Nicolet [62] are larger.

We have compared our results of viscosity and thermal conductivity with published data of Vargaftik[61] for steam and carbon dioxide. Except the stage in the viscosity of steam around 3000 K, where H2O molecule dissociation takes place, there is a fairly good agreement. Unfortunately data from Vargaftik [61] are limited at 4000 K for carbon dioxide and 6000 K for steam. In the temperature range above 9000 K, our results agree closely with those from Sourd [66]. . The 1st peak at 3500 K in the carbon dioxide thermal conductivity corresponds to CO2 molecules dissociation while the 2nd at 7200 K is due to CO molecule dissociation.

In the case of ammonia, we have compared ours results with those from Elchinger [63]. A fairly good agreement can be noted for thermal and electrical conductivity. The viscosity of Elchinger [63] is very low especially between 5000 and 12000 K. It should be noted that this author have greatly used simple potentials like polarization model or rigid sphere model for the derivation of N/H system collision integrals.

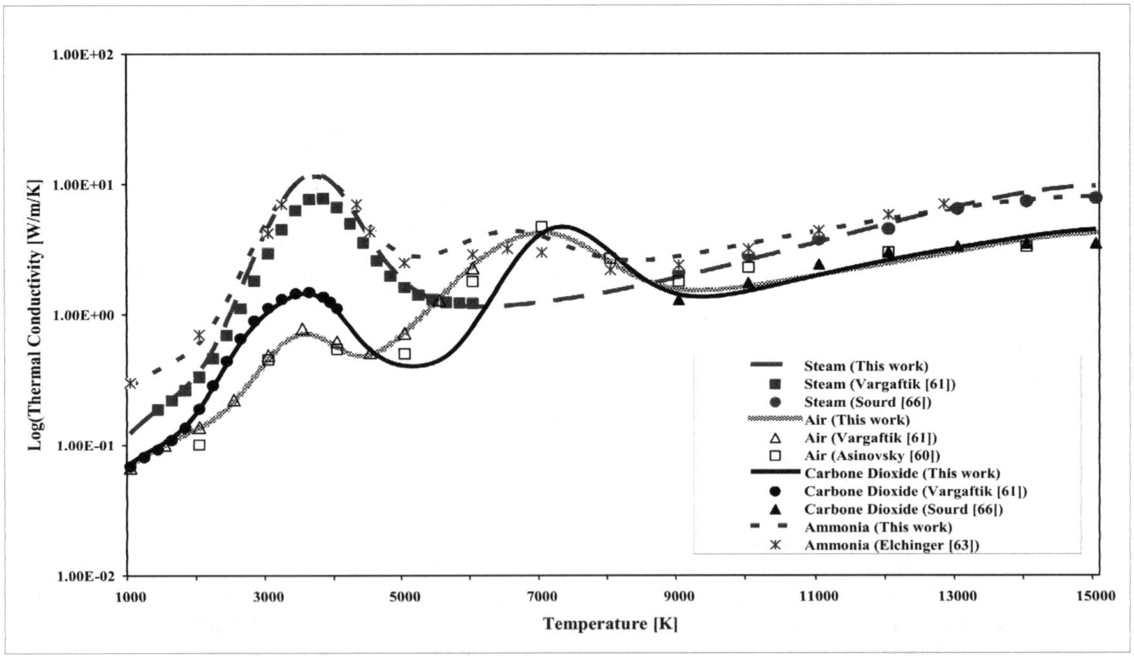

Figure 8. Thermal Conductivity of Steam, Carbon Dioxide, Air and Ammonia

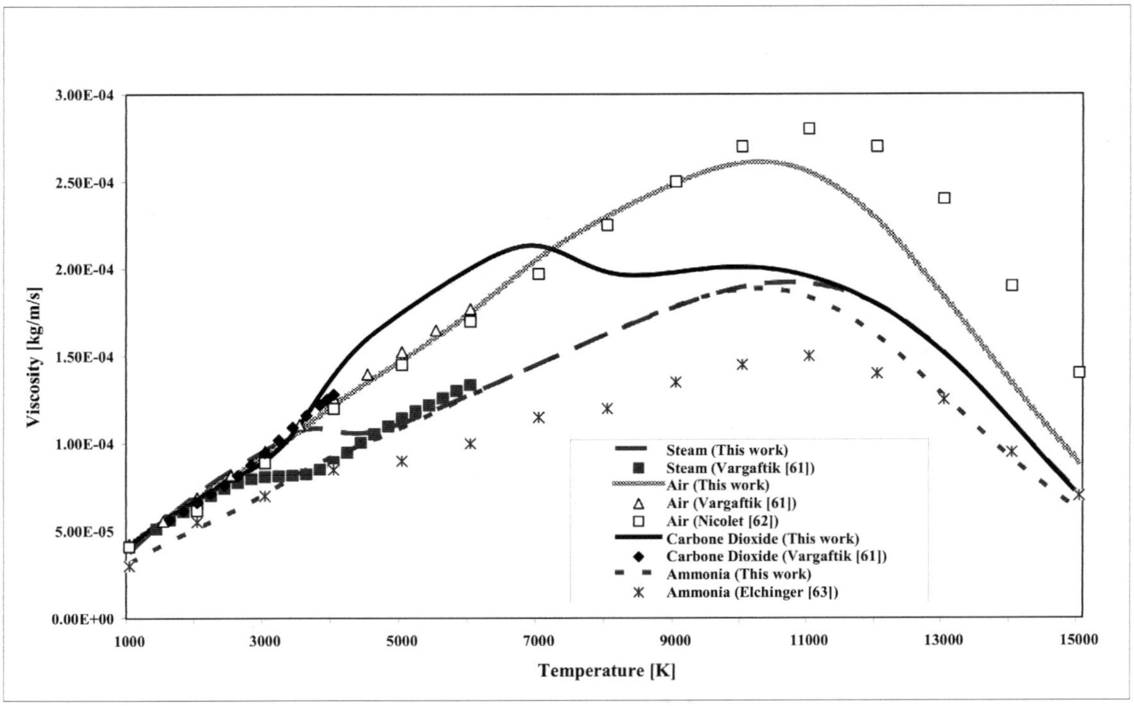

Figure 9. Viscosity of Steam, Carbon Dioxide, Air and Ammonia

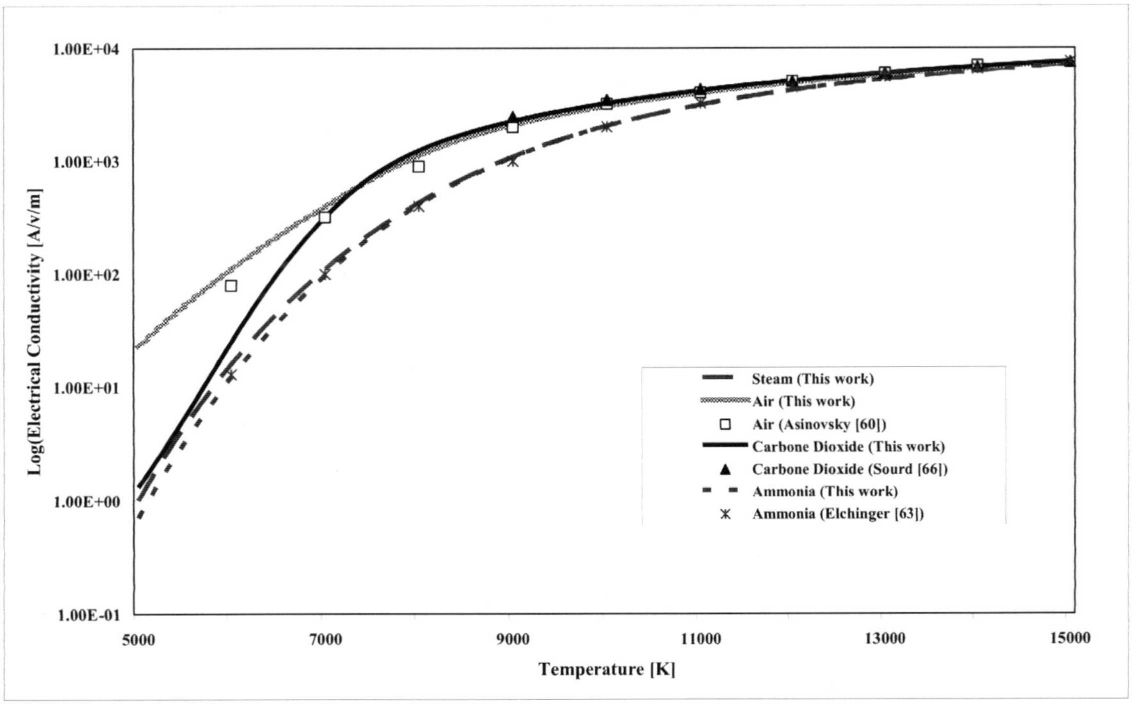

Figure 10. Electrical Conductivity of Steam, Carbon Dioxide, Air and Ammonia

4.3. *Transport Coefficients of some mixtures*

4.3.1. *Nitrogen-Oxygen-Methane mixtures.* Figures 11, 12 and 13 present the results for different mixtures of nitrogen, oxygen and methane.

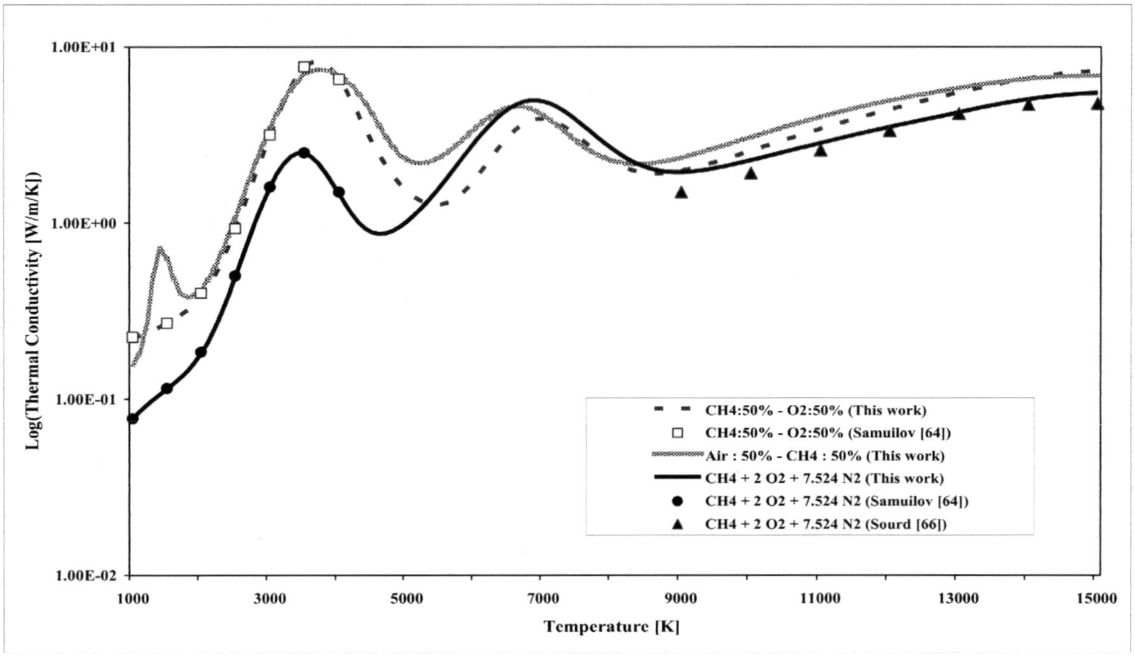

Figure 11. Thermal Conductivity of Nitrogen-Oxygen-Methane mixtures

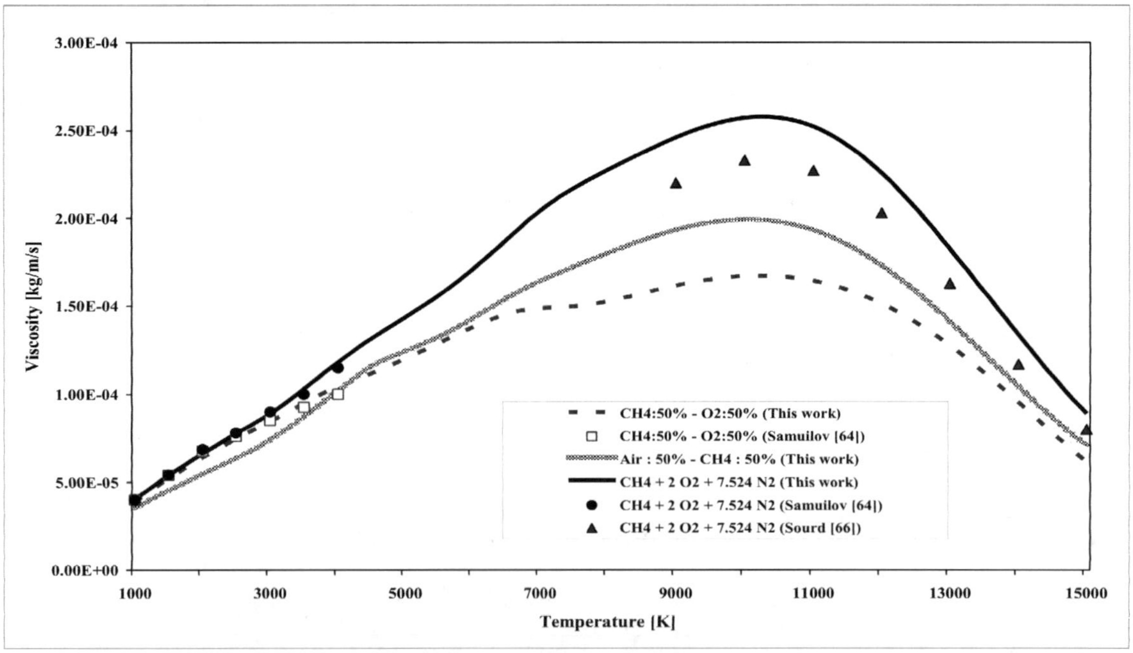

Figure 12. Viscosity of Nitrogen-Oxygen-Methane mixtures

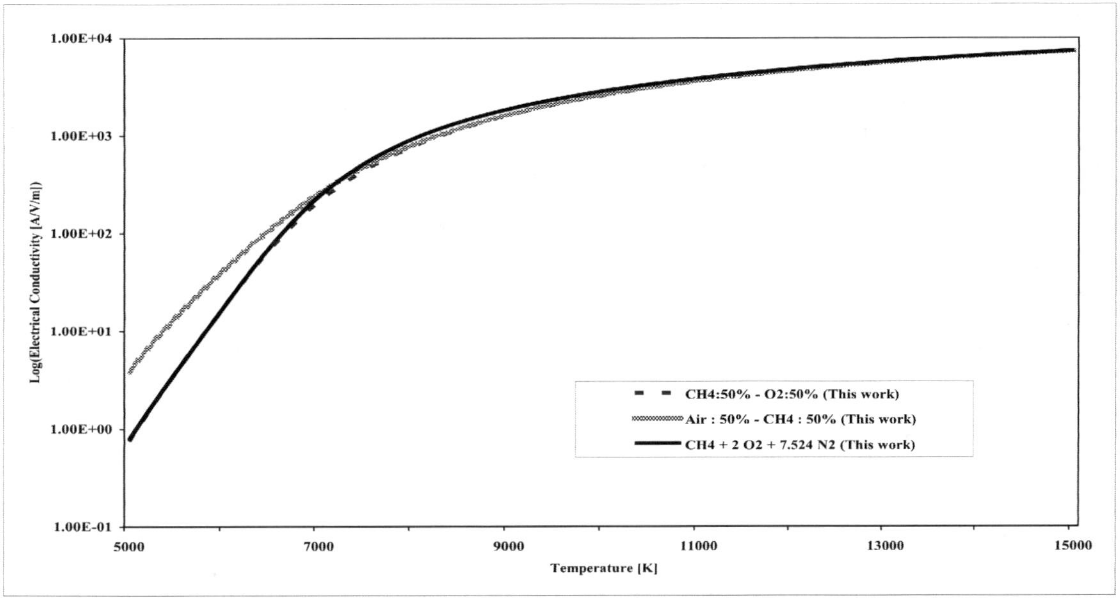

Figure 13. Electrical Conductivity of Nitrogen-Oxygen-Methane mixtures

Thus it can be seen the transport coefficients for the mixture of 50% of air and 50% of methane which chemical composition is shown in figure 1. It should be noted that for this mixture and especially in the case of mixture of 50% of oxygen and 50% of methane, the system is strongly under-stoichiometry. Actually the quantity of oxygen necessary to burn this methane is 2 times for 1st mixture and nearly 10 times for the 2nd.

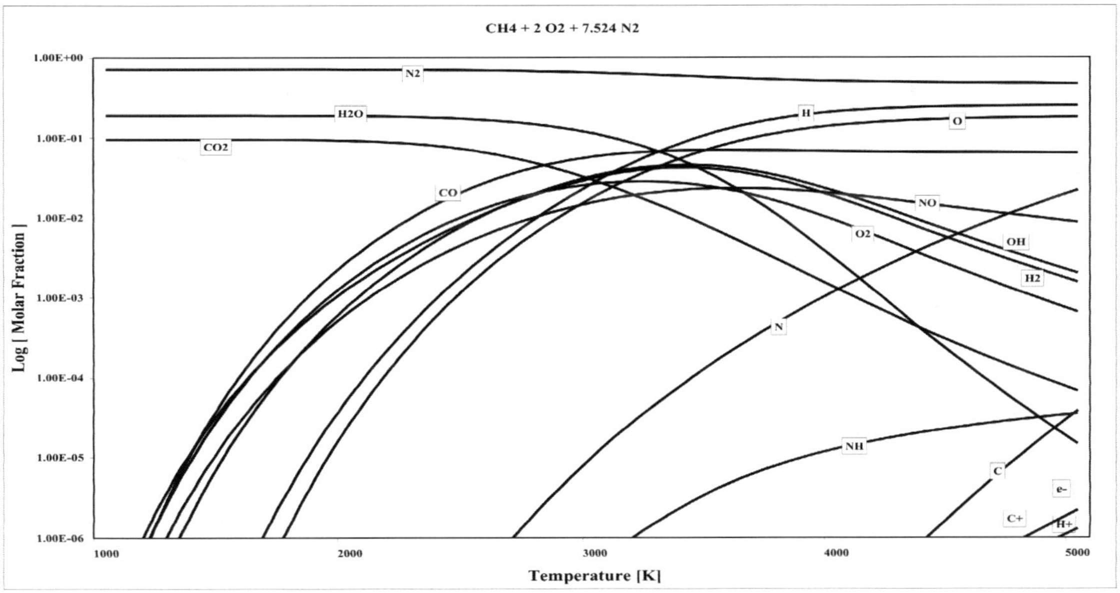

Figure 14. Chemical Composition of stoichiometric air-methane mixture

Figure 14 show the chemical composition evolution corresponding to this stoichiometric mixture. Compared with figure 1 the situation is radically different as it can be seen in the transport coefficients results.

So, while the thermal conductivity of under-stoichiometric mixture air-methane present 3 peaks corresponding respectively to CH4 molecules dissociation at 1400 K, H2 molecules dissociation at 3800 K and N2 and CO molecule dissociation at 6900 K, those of stoichiometric mixture present only 2 peaks corresponding to H2O molecules dissociation at 3500K and N2 dissociation at 7000 K. The relative augmentation of oxygen concentration and diminution of carbon concentration in the case of stoichiometric mixture increase the viscosity had the effect of increasing the viscosity and delaying the electrical conductivity.

We have reported in the same figures results from Samuilov [64] for Methane-oxygen mixture (50%-50%) and the results corresponding to a stoichiometric air- methane mixture (CH4 + 2 O2 + 7.524 N2) of Samuilov [64] and Sourd [66]. There is a good agreement with values given by Samuilov [64] up to 4000 K for thermal conductivity and viscosity. In the temperature range from 9000 to 15000K, our results for thermal conductivity are also in good agreement with those from Sourd [66], but in the case of viscosity our results are greater. It should be noted that Sourd [66] has taken into account multiply-charged ions (C++, O++, N++), which densities are not negligible in the temperature range 9000-20 000K.

4.3.2. *Argon-Steam mixtures.* Thermal plasma of Argon-Steam mixtures has many applications in Industries such as : Zircon or Aluminium oxide projection, Liquid wastes treatment, phosphoric acid synthesis …

Figure 15, 16 and 17 show the transport coefficients of different composition of argon-steam mixture. It can be seen the strong influence of steam on the thermal conductivity: around 2700 K, only 1% of steam in a argon plasma had the effect of multiplying by 2.3 the thermal conductivity. In the contrary this had no effect on the viscosity for which it is necessary to reach proportions of 50% order to reduce it sensibly.

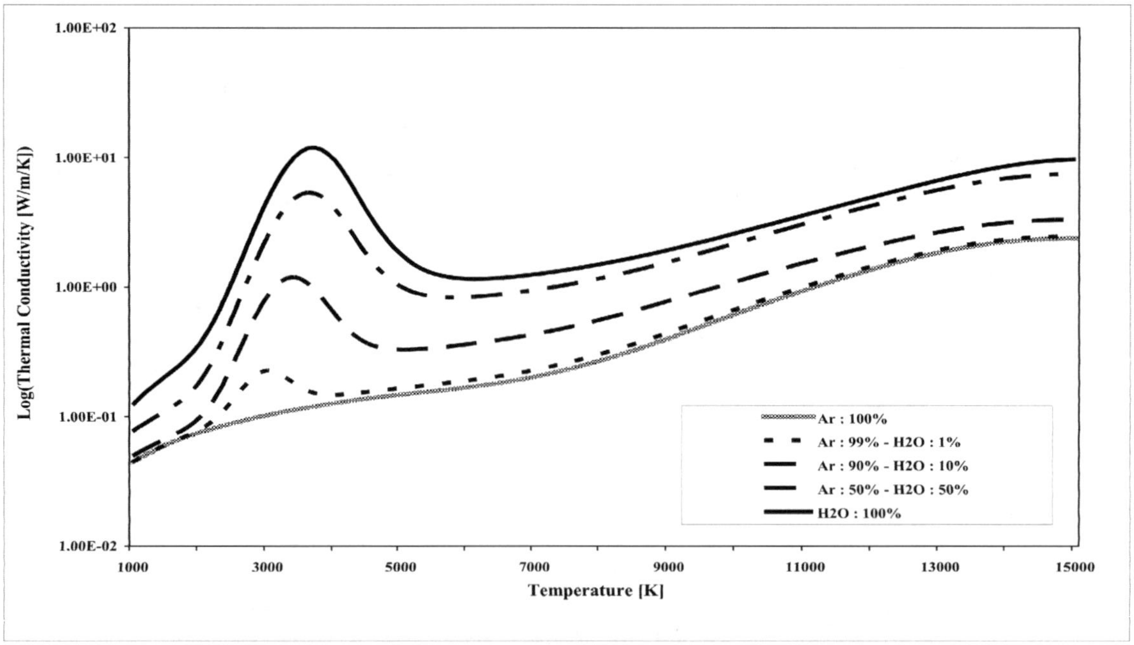

Figure 15. Thermal Conductivity of Argon-Steam mixtures

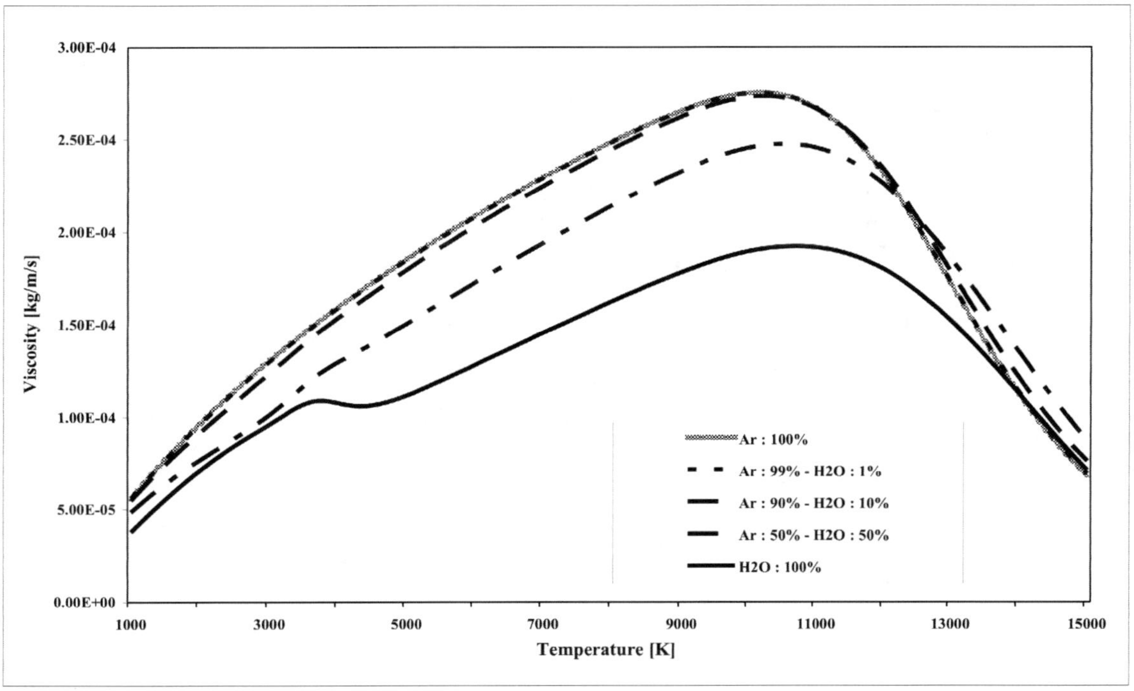

Figure 16. Viscosity of Argon-Steam mixtures

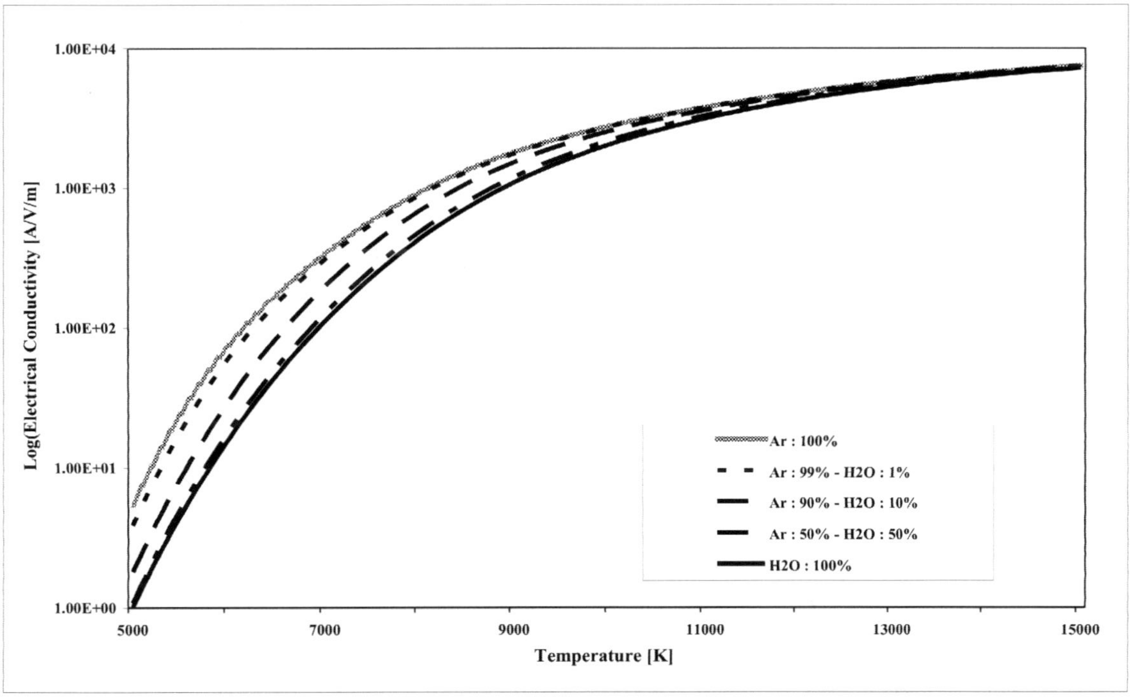

Figure 17. Electrical Conductivity of Argon-Steam mixtures

5. Conclusion

We have characterized the interactions between the major species of Ar/C/H/O/N systems allowing the calculation of transport coefficients of gases like argon, hydrogen, oxygen, nitrogen, methane … and their mixtures. The calculations are performed using standard approximation based on the Chapman-Enskog method, assuming the existence of local thermodynamic equilibrium.

We have taken special care with the important interactions atom-atom and atom-ion which can follow several potential curves. Interactions involving some species like free radicals or molecule-ion interactions have been taken into account with simple models. We have found that, this has not great influence on the results since the molar fractions of these species are low or when the molar fraction of 1st colliding partner is high those of the 2nd is low and vice versa.

Ours results for the transport properties of simple gases like argon, hydrogen, oxygen, nitrogen and their mixture agree with previously published data of the literature. Adding the carbon element to the system has permitted us to take into account hydrocarbon compounds. Thus, we have presented the transport coefficients of methane, oxygen-methane and air-methane mixtures. To our knowledge, it is the 1st published data involving hydrocarbon in the temperature range from 1000K up to 15000K.

1st International Symposium on Electrical Arc and Thermal Plasmas in Africa (ISAPA) IOP Publishing
IOP Conf. Series: Materials Science and Engineering **29** (2012) 012003 doi:10.1088/1757-899X/29/1/012003

Table 17.

	Ar	C	H	O	N	H2	O2	N2		CO		OH		NO	C2	C3	CH						CN		2	NH	·H3	HCO	HCN
Ar	MP	PP		MI		PM			LJ	LJ	LJ	LJ	LJ	LJ	RS	RS	DP	DP	DP	LJ	LJ	LJ	LJ	LJ	LJ	LJ	LJ	LJ	LJ
C		MI	MI	MI	MI		RS	RS	RS	RS	RS	RS	RS	RS		RS		RS	RS	RS	RS	RS	RS	RS	RS	RS	RS	RS	RS
H			MI	MI	MI				LJ	LJ	LJ	LJ	LJ	LJ		RS	Exp	MP	MP	LJ	LJ	LJ	LJ	LJ	LJ	LJ	LJ	LJ	LJ
O				MI			MI		PM	PM	LJ	LJ	LJ	LJ	RS	RS	LJ	LJ	LJ	LJ	LJ	LJ	LJ	LJ	LJ	LJ	LJ	LJ	LJ
N					MI				LJ	LJ	LJ	LJ	LJ		RS	RS	RS	RS	RS	LJ	LJ	LJ	LJ	LJ	LJ	LJ	LJ	LJ	LJ
H2									LJ	LJ	LJ	LJ	LJ	LJ	DP	DP	DP	DP	DP	LJ	LJ	LJ	LJ	LJ	LJ	LJ	LJ	LJ	LJ
O2												LJ	LJ		DP	DP	DP	DP	DP	LJ	LJ	LJ	LJ	LJ	LJ	LJ	LJ	LJ	LJ
N2									LJ	LJ		LJ	LJ	LJ	DP	DP	DP	DP	DP	LJ	LJ	LJ	LJ	LJ	LJ	LJ	LJ	LJ	LJ
										LJ	LJ	LJ	LJ	LJ	DP	DP	DP	DP	DP	LJ	LJ	LJ	LJ	LJ	LJ	LJ	LJ	LJ	LJ
CO											LJ	LJ		LJ	DP	DP	DP	DP	DP	LJ	LJ	LJ	LJ	LJ	LJ	LJ	LJ	LJ	LJ
												LJ	LJ	LJ	DP	DP	DP	DP	DP	LJ	LJ	LJ	LJ	LJ	LJ	LJ	LJ	LJ	LJ
OH												LJ	LJ	LJ	DP	DP	DP	DP	DP	LJ	LJ	LJ	LJ	LJ	LJ	LJ	LJ	LJ	LJ
H2O														LJ	DP	DP	DP	DP	DP	LJ	LJ	LJ	LJ	LJ	LJ	LJ	LJ	LJ	LJ
NO															DP	DP	DP	DP	DP	LJ	LJ	LJ	LJ	LJ	LJ	LJ	LJ	LJ	LJ
C2															DP	DP	DP	DP	DP	DP	DP	DP	DP	DP	DP	DP	DP	DP	DP
C3																DP	DP	DP	DP	DP	DP	DP	DP	DP	DP	DP	DP	DP	DP
CH																	DP	DP	DP	DP	DP	DP	DP	DP	DP	DP	DP	DP	DP
																		DP	DP	DP	DP	DP	DP	DP	DP	DP	DP	DP	DP
																			DP	DP	DP	DP	DP	DP	DP	DP	DP	DP	DP
																				LJ	LJ	LJ	LJ	LJ	LJ	LJ	LJ	LJ	LJ
																					LJ	LJ	LJ	LJ	LJ	LJ	LJ	LJ	LJ
																						LJ	LJ	LJ	LJ	LJ	LJ	LJ	LJ
CN																							LJ	LJ	LJ	LJ	LJ	LJ	LJ
																								LJ	LJ	LJ	LJ	LJ	LJ
																									LJ	LJ	LJ	LJ	LJ
NH																										LJ	LJ	LJ	LJ
																											LJ	LJ	LJ
																												LJ	LJ
																													LJ

References

[1] Boulos M I, Fauchais P and Pfender E 1994 *Thermal Plasmas - Fundamentals and Applications* Vol **1** (Plenum Press New York and London)
[2] Aubreton J 1983 *Revue de physique appliquée* **18** 51-66
[3] Mexmain J M 1983 *Thèse de Doctorat 3eme cycle* (Université de Limoges France)
[4] Bacri J and Raffanel S 1987 *Plasma Chem. and Plasma Process.* **7** 53
[5] Chapman S and Cowling T G 1970 *The Mathematical Theory of Non-Uniform Gases* (Cambridge University Press)
[6] Devoto R S 1967 *Phys. Fluids* **10-2** 354
[7] Hirschfelder J O, Curtiss C F and Bird R B 1954 *Molecular Theory of Gases and Liquids* (John Wiley and Sons Inc. New York)
[8] Kihara T, Taylor M H and Hirschfelder J O 1960 *Phys. Fluids* **3-5** 715
[9] Brokaw R S 1961 *Phys. Fluids* **4** 944
[10] Monchick L 1959 *Phys. Fluids* **2-6** 695
[11] Smith F J and Munn R J 1964 *J. Chem. Phys.* **41-11** 3560
[12] Samuilov E V and Tsitelauri N N 1969 *Teplofizika Visokikh Temperatur* **7-1** 168
[13] Mason E A 1954 *J. Chem. Phys.* **22-2** 169
[14] Neufeld P D, Jensen A R and Aziz R A 1972 *J. Chem. Phys.* **57-3** 1100
[15] Monchick L and Mason E A 1961 *J. Chem. Phys.* **35-5** 1676
[16] Hirschfelder J O 1958 J. Chem. Phys. **28** 197
[17] Aubreton J 1983 *Thèse de Doctorat d'Etat* (Université de Limoges France)
[18] Margeneau H and Kestner N R 1969 *The Theory of Intermolecular Forces* (Pergamon Press Oxford)
[19] Hulburt H and Hirschfelder J O 1941 *J. Chem. Phys.* **9** 61
[20] Kramer H Land Herschback D R 1980 *J. Chem. Phys.* **72** 5969
[21] Kong C L 1973 *J. Chem. Phys.* **59** 1953
[22] Smith F T 1972 *Phys. Rev. A* **5** 1708
[23] Debbagh-Nour E G 1984 *Thèse de Doctorat 3eme cycle* (Université de Toulouse France)
[24] Mason E A, Vanderslice J T and Yos J M 1959 *Phys. Fluids* **6** 688
[25] Cubley J S and Mason E A 1975 *Phys. Fluids* **18** 1109
[26] Herzberg G 1950 *Spectra of Diatomics Molecules* (D. Van Nostrand, New York)
[27] Alexander S A, McDonald C and Matsen F A 1983 *Int. J. Quant. Chem. Symp.* **17** 407
[28] Moiseyev N and Bar-Adon R 1984 *J. Chem. Phys.* **80-5** 1917
[29] Carter S, Mills L M and Murrel J N 1980 *J. Molec. Spectrosc.* **81** 110
[30] Capitelli M and Devoto R S 1973 *Phys. Fluids* **16-11** 1835
[31] Saxon R P and Liu B 1977 *J. Chem Phys.* **12** 5432
[32] Luiti G and Pirani F 1987 *J. Chem. Phys.* **87-9** 5266
[33] Holmer B K and Child M S 1983 *Faraday Discuss. Chem. Soc.* **75** 131
[34] Reid R C, Prautnitz J M and Sherwood T K 1977 *The Properties of Gases and Liquids* (McGraw Hill Book Company, 3rd edition)
[35] Kee R J, Warnatz J and Miller J A 1988 A Fortran Computer Code Package for the Evaluation of Gase Phase Viscosities, Conductivities and Diffusion Coefficients *Sandia Report SAND* 83-8209 UC-32
[36] Khayar A and Bonamy J 1982 *J. Quant. Spectrosc. Radial. Transp.* **28-3** 199
[37] Amdur I, Longmire M S and Mason E A 1961 *J. Chem. Phys.* **35** 895
[38] Eaker C W and Parr C A 1976 *J. Chem. Phys.* **64-4** 1322
[39] Van-Dishoek E F, Langhoff S R and Dalgarno A 1983 *J. Chem Phys.* **78-7** 4552
[40] Graff M M, Moseley J T, Durup J and Roueff E 1983 *J. Chem. Phys.* **78-5** 2355
[41] Kusunoki I, Sakai S, Kato S and Morokuma K 1983 *J. Chem. Phys.* **72**-12 6813
[42] Chambaud G and Launay J M 1980 *J. Phys. B* **13** 4205

[43] Rosen B 1970 *Données Spectroscopiques relatives aux molécules diatomiques* (Pergamon Press Oxford)
[44] Krupenie P H and Weissman S 1965 *J. Chem. Phy.* **43-5** 1529
[45] Hillier I H, Guest M F, Ding A, Karlau J and Weise J 1979 *J. Chem. Phys.* **70-02** 864
[46] Verhaegen G 1968 *J. Chem. Phys.* **49-10** 4696
[47] Mason E A, Munn R J and Smith F J 1967 *Phys. Fluids* **10-8** 1827
[48] Devoto R S 1973 *Phys. Fluids* **16-5** 616
[49] Itikawa Y 1974 *At. Data Nucl. Data Tables* **14** 1
[50] Itikawa Y 1978 *At. Data Nucl. Data Tables* **21** 69
[51] Thomas L D and Nesbet R K 1975 *Phys. Rev. A* **12-6** 2378
[52] Gee N and Freeman G R 1983 *J. Chem. Phys.* **78-4** 1951
[53] Kleban P and Davis H T 1978 *J. Chem. Phys.* **68** 2999
[54] Devoto R S 1976 *Phys. Fluids* **1-19** 22
[55] Pateyron B, Elchinger M F, Delluc G and Fauchais P 1992 *Plasma Chem. Plasma Process.* **12-4** 421
[56] Chen W L T, Heberlein J and Pfender E 1996 *Plasma Chem. Plasma Process.* **16-4** 635
[57] Murphy A B and Arundell C J 1994 *Plasma Chem. Plasma Process.* **14-4** 451
[58] Beebe N H, Thulstrup E W and Andersen A 1976 *J. Chem. Phys.* **64** 2080
[59] Rutherford J A and Vroom D A 1974 *J. Chem. Phys.* **61** 2514
[60] Asinovsky E I, Kirilin V A, Pakhomon E P and Shabahov V 1971 *Proc. of IEEE* **59** 592
[61] Vargaftik N B 1975 *Tables of Thermophysical Properties of Liquids and Gases* (Hemisphere Publishing Corp. 2nd Edition)
[62] Nicolet WE, Shepard C E, Clark K J, Balakrishnan A, Kesserling J P, Suchsland K E and Reese J J 1975 *Report AEDC-TR-75-47*
[63] Elchinger M F, Pateyron B, Delluc G and Fauchais P 1990 *J. Phys. Colloque C5* **n°18 T.51** 536
[64] Samuilov E V and Tsitelauri N N 1970 *Teplofisika Visokikh Temperature* **6-8** 1174
[65] Murphy A B 2000 *Plasma Chem. Plasma Process.* **20-3** 279
[66] Sourd B, Aubreton J, Elchinger M-F, Labrot M and Michon U 2006 *J. Phys. D : Appl. Phys.* **39** 1105

IOP Conf. Series: Materials Science and Engineering **29** (2012) 012004 doi:10.1088/1757-899X/29/1/012004

Calculation of air-water vapor mixtures thermal plasmas transport coefficients

A K Kagoné[1], Z Koalaga[1] and F Zougmoré[1]

[1]LAME-UO, Laboratoire de Matériaux et Environnement, Département de Physique, UFR/SEA, Université de Ouagadougou, 03 BP 7021 Ouagadougou 03, Burkina Faso

Email: kagone@yahoo.fr, koalaga@univ-ouaga.bf

Abstract. Knowledge of transport coefficients of air-water vapor mixtures thermal plasmas is important to estimate the performances of electrical arc cutting in this gas by a circuit breaker. In this paper, air-water vapor mixtures thermal plasmas dynamical viscosity, thermal and electrical conductivities coefficients are calculated in a temperature range from 5000K to 30000K. The calculations are carried out by supposing local thermodynamic equilibrium at pressure of 1.0 atm, 5.0 atm and 10 atm. The calculation results show the influence of the water vapor initial proportion but also that of the plasmas pressure on these transport coefficients. The plasma of the mixture constituted by 20% of air and 80% of water vapor presents at first sight good characteristics for the electric current cut. It has the highest thermal conductivity and the lowest electrical conductivity among plasmas of studied mixtures.

1. Introduction

The electrical current breaking in some circuit breakers is obtained by separating electrodes in the air and the compressed air. This electrodes separation generates an electrical arc which interacts with surrounding gas and creates plasma of this gas. Circuit breakers made object of numerous theoretical and experimental studies [1, 2, 3, 4, 5]. Other studies on the electrical arc have shown the influence of the insulator vapors [6, 7, 8, 9], metals [4, 10, 11] and hydrogen [6, 9] on the engendered plasmas characteristics. In this study, we are interested by hydrogen influence on plasma characteristics for electrical current breaking. This hydrogen can be brought by a mixture of water vapor and air. In our knowledge, there are not yet works on the air-water vapor mixture plasma transport coefficients.

Indeed, a bibliographical research shows that only the transport properties of water and argon mixture plasma [12, 13] and those of pure water plasma [14, 15] were already studied. So we undertake in this study the calculation of air-water vapor mixtures thermal plasmas transport coefficients. The mixture can be naturally formed because of air humidity. But the water vapor partial pressure in the atmospheric air even in zones with great humidity led to a maximum percentage of water vapor around 7% at temperature of 40°C. As this value is not sufficient to bring enough hydrogen for better performances, we choose to study air-water vapor mixtures with high water vapor percentages: 80% air-20 % water vapor, 50% air-50 % water vapor and 20% air-80 % water vapor. The proportions of air and water vapor of chosen mixtures are given in volume. These three (03) mixtures are chosen to be able to determine the water vapor influence on the plasmas transport coefficients. The increase of water vapor in the mixture could be done during the breaking of the electrical current by water injection.

Supposing local thermodynamic equilibrium, calculations are made in a temperature range from 5000K to 30000K at pressure of 1.0, 5.0 and 10 atm for the three mixtures. We suppose that the air is constituted by 80 % of nitrogen and 20 % of oxygen. The three values of the pressure are also considered to estimate the impact of the pressure on the plasmas transport coefficients. Given the temperature range considered here, the various chemical species taken into account in the plasma composition are electrons, diatomic molecules, neutral atoms and their corresponding ions charged one or two times: e^-, H, N, O, H_2, N_2, O_2, OH, HN, NO, H^+, O^+, N^+, OH^+, HN^+, NO^+, H_2^+, N_2^+, O_2^+, N^{2+} et O^{2+}.

2. Transport coefficients

The plasma transport coefficients calculation is based on the resolution of Boltzmann intégro-differential equation by the method of Chapmann-Enskog [16] which is an approximate method.

Devoto [17] developed theoretical expressions which allow calculating of the transport coefficients of gas mixtures with υ components until the 4th order approximation, except viscosity (2nd order). However calculations become complicated for gases composed of many particles, starting from the 3th order approximation. The same author shows that one can bring simplifications to the expressions of the transport coefficients of υ component mixture if one of chemical species has a very low mass compared to those of the other species: that is the case of the plasma of our study. During a binary collision between an electron and a heavy particle, only the momentum and the speed of the heavy particle are unchanged. One also admits that the variation of the heavy species function of disturbance is weak, compared with that of the electrons. Under these conditions, the transport properties of the electrons and heavy particles can then be treated independently and defined with good approximations by means of expressions which will be presented.

The orders of approximation used to determine the transport coefficients will be specified for each one of them.

2.1. Thermal conductivity coefficient

The thermal conductivity is the add of four terms: translational term of electrons, translational term of heavy species, internal term due to internal degree of freedom and reactional term due to chemical reactions. It is given by the following relation:

$$\lambda = \lambda_{tr}^{e} + \lambda_{tr}^{h} + \lambda_{\text{int}} + \lambda_{R} \tag{1}$$

λ_{tr}^{e} and λ_{tr}^{h} are respectively electrons and heavy particles translation thermal conductivities [18, 19]. λ_{int} is the internal thermal conductivity bound to the freedom degrees existence due to the vibrations and the rotations of molecules [20]. λ_R is the thermal conductivity due to the ionization, recombination and dissociation reactions [21, 22].

The formula of the electrons thermal conductivity of translation λ_{tr}^{e} is given by Devoto [17] in the 3th order approximation in the method of Chapmann-Enskog.

$$\lambda_{tr}^{e} = \frac{75}{8} n_e^{\,2} k . \left(\frac{2\pi kT}{m_e} \right)^{1/2} \frac{1}{q^{11} - \left(q^{12}\right)^2 / q^{22}} \tag{2}$$

The expression of heavy particles thermal component of translation λ_{tr}^{h} is independent of the electrons properties. Only the interactions between heavy particles are to be taken into account. In the 2th order approximation of Chapmann-Enskog method, Muckenfuss and Curtiss [19] proposed a simplified expression of this coefficient as follow:

$$\lambda_{tr}^{h} = 4\begin{vmatrix} L_{11} & \cdots & L_{1\upsilon} & x_1 \\ \vdots & & \vdots & \vdots \\ L_{\upsilon 1} & \cdots & L_{\upsilon\upsilon} & x_\upsilon \\ x_1 & & x_\upsilon & 0 \end{vmatrix} \Bigg/ \begin{vmatrix} L_{11} & \cdots & L_{\upsilon} \\ \vdots & & \vdots \\ L_{\upsilon\upsilon} & \cdots & L_{\upsilon\upsilon} \end{vmatrix} \quad (3)$$

where x_i is the molar fraction of the particle i. The term $L_{i\,j}$ is a function of the heavy particles collision integrals, masses molars and the composition [16].

Because of the existence of chemical reactions in the plasma, the energy transfer can be made under the enthalpy chemical shape. The reaction thermal conductivity reports no elastics processes. It can be calculated by an expression given by Meador and Staton [22] in the case of partially ionized gases. In the 1[rst] order approximation they write:

$$\lambda_R = \frac{-1}{RT^2}\begin{vmatrix} 0 & \Delta H_1 & \cdots & \Delta H_\mu \\ \Delta H_1 & A_{11} & \cdots & A_{1\mu} \\ \vdots & \vdots & & \vdots \\ \Delta H_\mu & A_{\mu 1} & \cdots & A_{\mu\mu} \end{vmatrix} \Bigg/ \begin{vmatrix} A_{11} & \cdots & A_{1\mu} \\ \vdots & & \vdots \\ A_{\mu 1} & \cdots & A_{\mu\mu} \end{vmatrix} \quad (4)$$

where ΔH_i represents the variation of enthalpy due to the ith chemical reaction. R is the constant of perfect gases.

The energy transfer in internal movements (rotation and vibration) of molecules is expressed by the internal conductivity. By extension of the Eucken theory, formulated for a pure gas, Yun et al. [23] propose an analytical expression, with 1[rst] order approximation for internal thermal conductivity:

$$\lambda_{\mathrm{int}} = \sum_{i=1}^{\upsilon} \frac{(\lambda_{\mathrm{int}})_i}{1 + \sum\limits_{\substack{j=1 \\ j\neq i}}^{\upsilon} \frac{D_{ii}}{D_{ij}} \cdot \frac{x_j}{x_i}} \quad (5)$$

$(\lambda_{\mathrm{int}})_i$ is the internal thermal conductivity of the ith component.

D_{ij} are defined by the relation:

$$D_{ij} = \frac{3}{16\sqrt{\pi}} \cdot \frac{(kT)^{3/2}}{P.\overline{\Omega}_{ij}^{(1,1)}} \cdot \left(\frac{M_i + M_j}{2M_i.M_j}\right)^{1/2} \quad (6)$$

2.2. Electrical conductivity coefficient

By neglecting the contribution of ions to the current transport, Devoto [17] developed with the 3th order approximation a theoretical simplified expression (7) of the electrical conductivity of a partially ionized gas, supposed in local thermodynamical equilibrium. He gives:

$$\sigma = \frac{3}{2} e^2 n_e^2 \left(\frac{2\pi}{m_e kT} \right)^{1/2} \begin{vmatrix} q^{11} & q^{12} \\ q^{12} & q^{22} \end{vmatrix} \Bigg/ \begin{vmatrix} q^{00} & q^{01} & q^{02} \\ q^{01} & q^{11} & q^{12} \\ q^{02} & q^{12} & q^{22} \end{vmatrix} \tag{7}$$

n_e is the electrons numerical density and m_e the electron mass. q^{ij} depends on numerical densities and on average effective sections of collision of particles.

2.3. *Dynamical viscosity coefficient*

The dynamical viscosity coefficient characterizes the transport of momentum in plasma. It is thus practically independent of the electrons properties because of their low mass compared to those of the heavy particles. A good approximation at 1[rst] order [16] of this coefficient is obtained by means of the simple expression established by Buddenberg and Wilke [24] which we use in this paper:

$$\eta = \sum_{i=2}^{\upsilon} \frac{x_i^2}{\frac{x_i^2}{\eta_i} + 1,385 \sum_{\substack{j=2 \\ j \neq i}}^{\upsilon} x_i x_j \frac{RT}{P.M_i.D_{ij}}} \tag{8}$$

where m_i and x_i are respectively the mass and the molar fraction of the species i; η_i is the viscosity coefficient of a pure gas defined by Hirschfelder like:

$$\eta_i = \frac{5}{16\sqrt{\pi}} \frac{(M_i kT)^{1/2}}{\overline{\Omega}_{i,i}^{(2,2)}} \tag{9}$$

2.4. *Collision integrals*

The knowledge of the collision integral [25, 26, 27] is determining for the transport coefficients calculation. It is thus important to choose the types of interaction which are appropriate with the nature of the particles populating studied plasma. The collision integrals calculation is made by considering four categories of interaction: Neutral – neutral interaction; neutral – ion interaction; electron – neutral interaction and charged – charged interaction. Different methods are usually used for the collision integrals determination: numerical integration, the method of published results smoothing and the reduced temperatures method. In this study we privilege the method of published results smoothing. In general, the published collision integrals are given for $\Omega_{ij}^{(1,1)}$ and $\Omega_{ij}^{(2,2)}$. Thus to obtain the others we use the formula of recurrence established by Hirschfelder [16]:

$$\Omega_{ij}^{(l,s+1)} = \Omega_{ij}^{(l,s)} + \frac{T}{s+2} \frac{\partial \Omega_{ij}^{(l,s)}}{\partial T} \tag{10}$$

2.4.1. *Neutral-neutral interaction.* Some neutral-neutral interactions have been already calculated by authors.

N-N Interaction: We use the values published by Capitelli et al [28] from the Morse potential.

N-N_2 and N_2-N_2Interaction: These interactions are represented by a repulsive exponential potential. The corresponding collision integrals are proposed by Capitelli et al [28].

H-H, H-H_2 and H_2-H_2 Interaction: We retain the Vanderslice et al results [29] but for the collision integral $\Omega^{(2,2)}$ of the H-H interaction, we chose the values proposed by Belov [30]. This author considers a temperature interval covering our domain of study.

O-O interaction: We use the collision integral published by P. André et al. [31].

For O-O_2 and O_2-O_2 Interaction: We use the values obtained by Mexmain [32] who made a detailed study of these interactions.

By the reduced temperatures method, we use the reduced collision integrals published by Hirschfelder [16] according to the reduced temperature T^*. For other interactions between not polar neutral species, the used potential is the one of Lennard-Jones. The reduced temperature is given by:

$$T^* = \frac{k}{\varepsilon} T \tag{11}$$

$$\frac{\varepsilon}{k} = 0{,}77.\frac{3}{2} T_b \tag{12}$$

where T_b is the boiling point of the particle and ε is the maximum of attractive energy for an interaction between similar species.

2.4.2. Electron-neutral interaction. We use in the present study the integral of: e-H, e-H_2, e-O, e-O_2, e-N, e-N_2, e-NH, e-NO, and e-OH published by P. André et al. [33].

2.4.3. Neutral-ion interaction. We use the collision integrals calculated by Capitelli et al [28] for the $H - H^+$ and $N - N^+$ interactions and those calculated by Eymard [34] for the $N_2 - N_2^+$ interaction. For the $O - O^+$ and $O_2 - O_2^+$ interactions we use the values calculated by Mexmain [32]. Other neutral - ion interactions were considered purely elastic and the corresponding integrals of collision are determined from Maxwell's potential.

2.4.4. *Charged-charged interaction.* In our study, the collision integrals corresponding to the interactions charged-charged are obtained by means of the expressions (13) and (14) established by Devoto [17]. He obtained these formulas from the transport sections efficacies of Liboff who uses the screened Coulomb potential by introducing a distance L included between the average value of the shorter approach distance $\Delta = \frac{Z_i Z_j e^2}{2kT}$ and the Debye length λ_D ($\Delta \le L \le \lambda_D$), such as:

- for distances lower than L, the interactions are controls by the potential of Coulomb;
- for distances higher than L one uses the screened Coulomb potential.

$$\overline{Q}^{(1,s)} = \frac{4\pi}{s(s+1)} b_0^2 \left[\ln\left(\frac{2\lambda_D}{b_0}\right) - \frac{1}{2} - 2\gamma + \psi(s) \right] \tag{13}$$

$$\overline{Q}^{(2,s)} = \frac{12\pi}{s(s+1)} b_0^2 \left[\ln\left(\frac{2\lambda_D}{b_0}\right) - \frac{1}{2} - 2\gamma + \psi(s) \right] \tag{14}$$

where $\psi(s) = \sum_{n=1}^{s-1} \left(\frac{1}{n}\right)$, $\psi(1) = 0$, γ is the Euler constant and $b_0 = \Delta$.

In the literature the authors are much divided on the calculation of Debye length [35], [31]. Certain authors calculate it by taking only into account the presence of the electrons and others authors by taking into account the presence of electrons and ions. In this study, we chose to calculate the Debye length by taking into account the presence of the electrons and ions.

$$\lambda_D = \left(\frac{\varepsilon_0 k}{e^2} \sum_{i=1}^{\upsilon} \frac{T}{n_i Z_i^2} \right)^{\frac{1}{2}} \tag{15}$$

3. Results and analyses

In this part we present and analyze the results of the transport coefficients calculation for the considered mixtures plasmas. As transport coefficients, we have retained dynamical viscosity, electrical and thermal conductivities. In order to be able to make comparisons, the transport coefficients of dry air plasma and water vapor plasma are also calculated. For writing lightening reasons, we will use the notation indicated in the table 1 for studied plasmas.

Table 1. Plasmas notation

Plasma name	Notation
Dry air	Air
Mixture of 80% air and 20% water vapor	Mix 1
Mixture of 50% air and 50% water vapor	Mix 2
Mixture of 20% air and 80% water vapor	Mix 3
Mixture of simple water vapor	Water

First at all, we have tested our program with a well-know plasma such as dry air plasma. On figure 1, we give the dynamical viscosity evolution according to temperature at atmospheric pressure. We also report on this figure M Capitelli computation results. For temperature higher than 14000K, we obtained same results. But for lower temperature, some differences are observed. The maximum variation is less than 15%. This variation at low temperature could be due to the equilibrium composition calculations as we used the same collision integrals for the major part of the particles interactions.

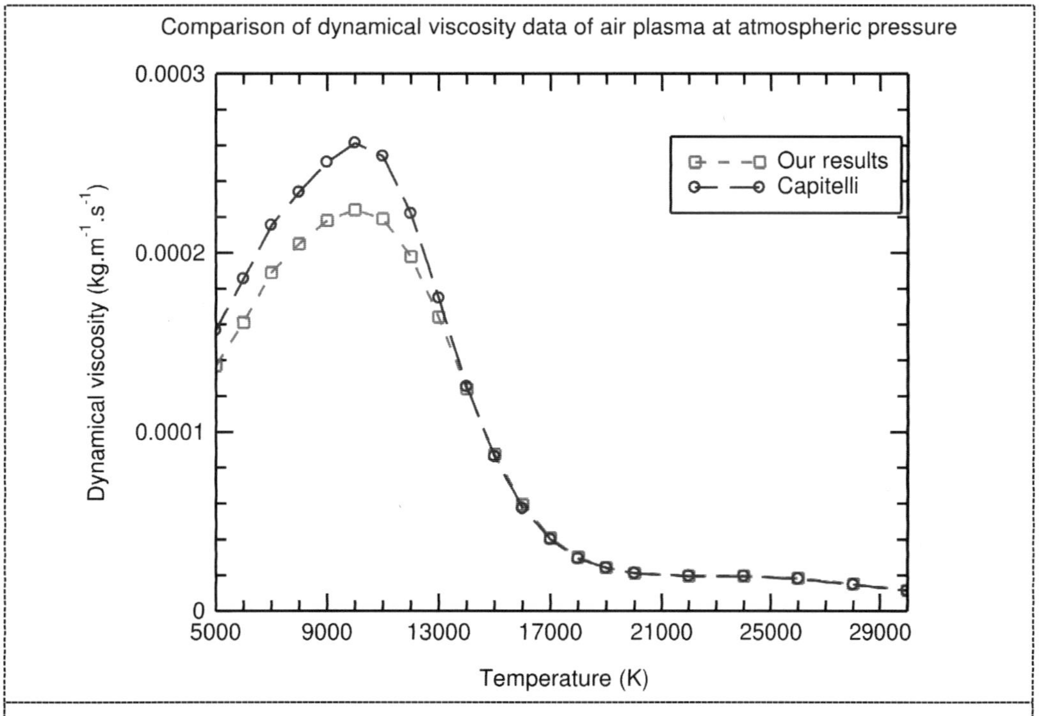

Figure 1. Comparison of dynamical viscosity data of air plasma at atmospheric pressure..

We have also undertaken the same testing approach for thermal and electrical conductivities of dry air plasma at atmospheric pressure. The results are given on figure 2 and figure 3. On these figures, are also reported computation results of M Capitelli and P Fauchais. For thermal conductivity our results are in concordance with those of P Fauchais on all the range of temperature. On the other hand, there is some difference between ours and those of Capitelli for temperatures ranging between 10000K and 20000K. For electrical conductivity, our results are similar to those of P Fauchais and M Capitelli respectively on 5000K-12000K and 12000K-30000k intervals of temperature. But they are slightly higher than those of P Fauchais on 12000K-30000K interval of temperature.

To get an idea on the contribution of each total thermal conductivity component, we have plotted on figure 4 their evolutions. We have chosen the case of Mix 2 plasma. Thus this reveals that at the low temperatures (small ionization level), total thermal conductivity is due primarily to the various chemical reactions and the translations of the heavy particles. Precisely, the components relating to the heavy particles translation and the reactions become dominating between 5000 K and 20000 K. When the temperature exceeds 20000K, thermal conductivity is due in major part to the component of electrons translation. Plasma is almost completely ionized in this zone of temperatures. The component λ_{int} appears only at the low temperatures and its contribution is much negligible on all the temperature range. For studied mixtures plasmas, It should be noted that the contributions on total thermal conductivity of these various components are similar.

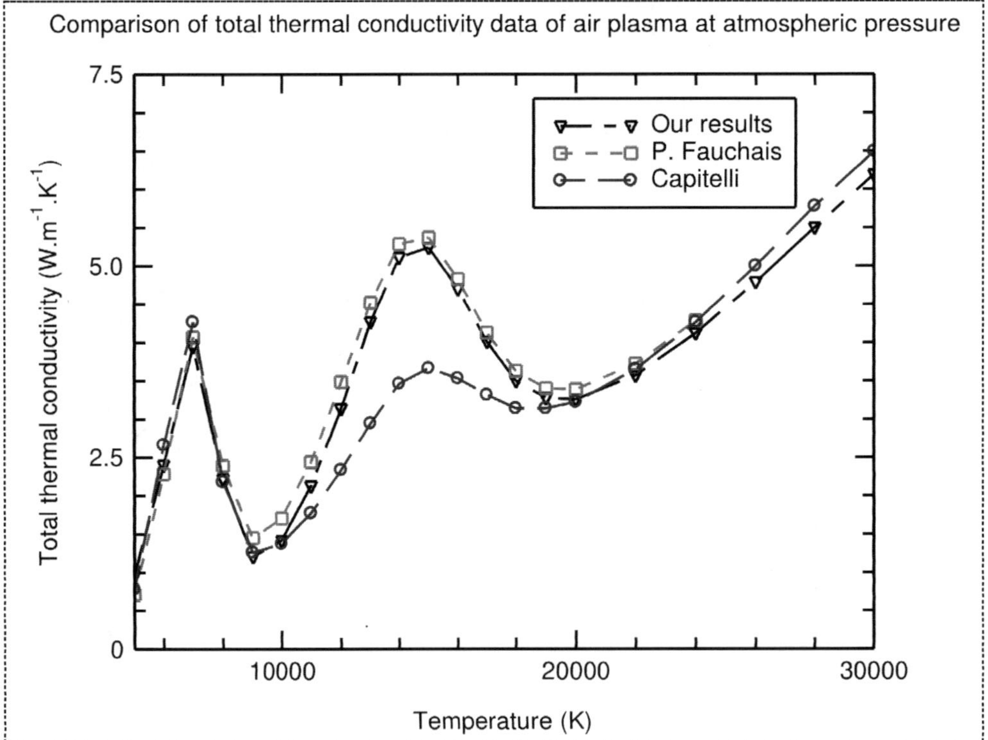

Figure 2. Comparison of total thermal conductivity data of air plasma at atmospheric pressure.

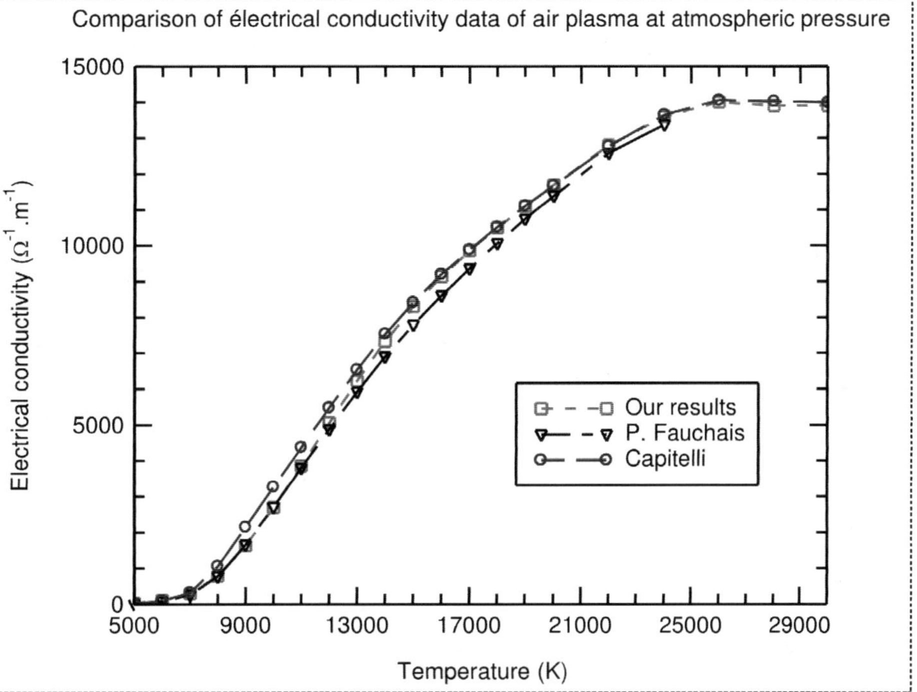

Figure 3. Comparison of electrical conductivity data of air plasma at atmospheric pressure.

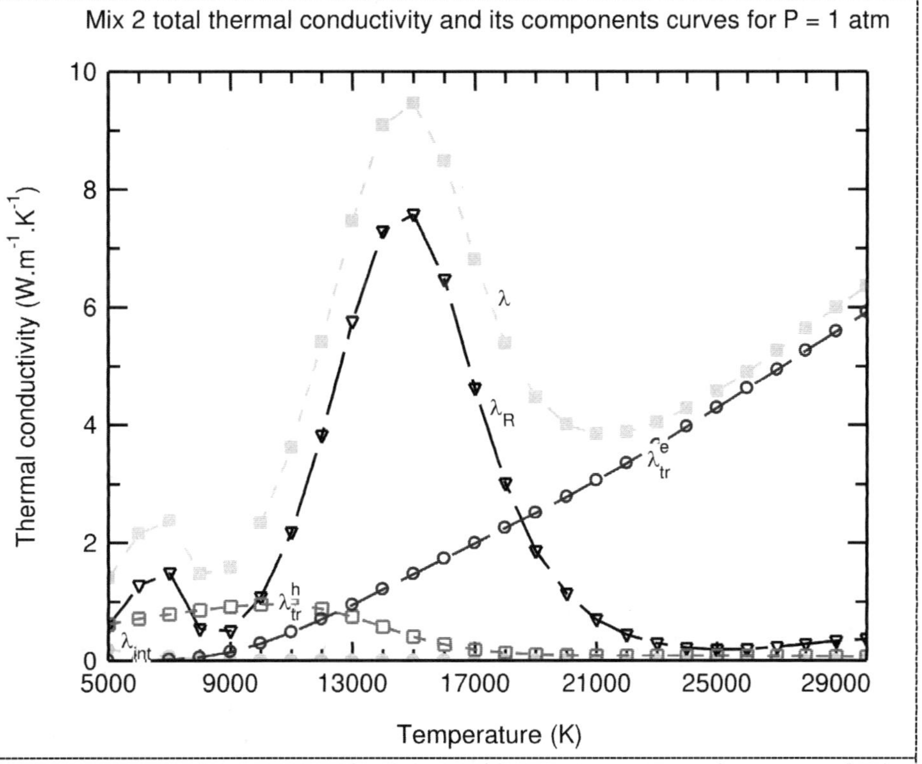

Figure 4. The total thermal conductivity and its components results curves of Mix 2 (P = 1 atm).

Figures 5, 6 and 7 respectively represent each one the evolution of dynamical viscosity, thermal conductivity and the electrical conductivity of the three studied mixtures, the dry air and the pure water vapor plasmas. With these figures the influence of the water vapor on the plasma transport coefficients appears clearly.

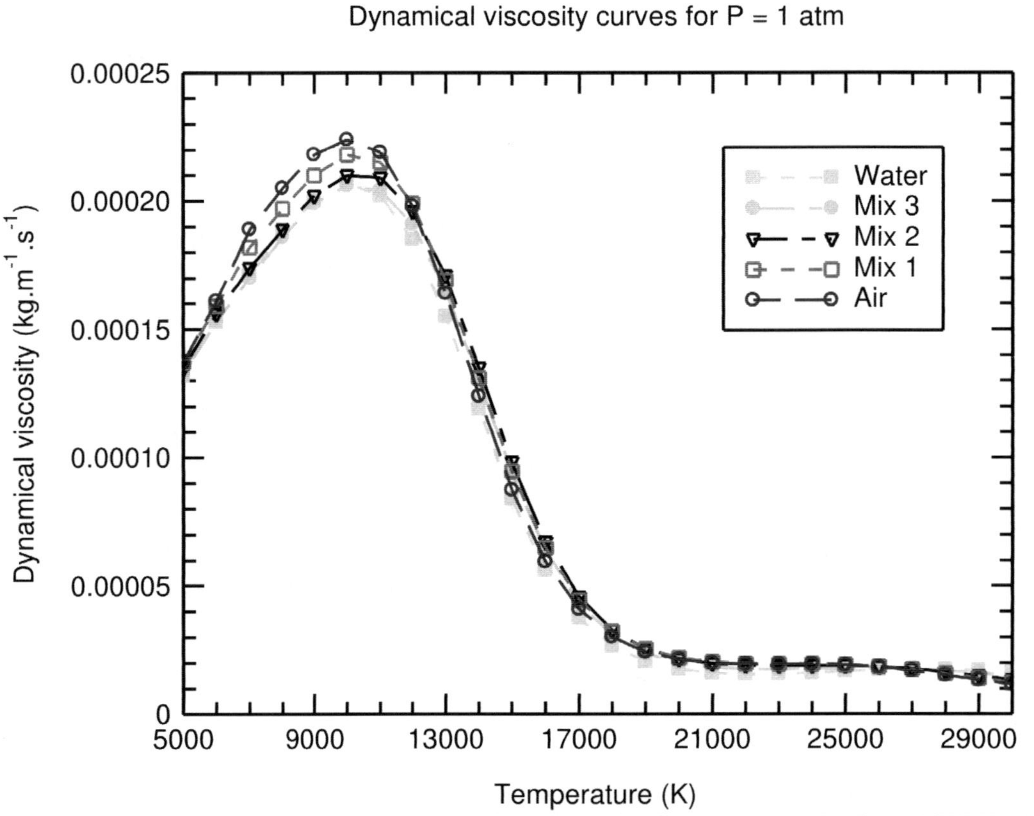

Figure 5. The dynamical viscosity results curves of the five plasmas (P = 1 atm).

The dynamical viscosity (η) of studied plasmas is given on figure 5. For all these plasmas, the dynamical viscosity maximum is between 9000K and 11000K. In the interval of temperatures 7000K at 13000K, the plasma dynamical viscosity decreases with the water vapor proportion in the mixture. The dry air plasma has a dynamical viscosity slightly higher than those of other plasmas. That can be explained by the great proportion of nitrogen in the air. Indeed nitrogen molecule (N_2) has the highest dissociation energy. For temperatures higher than 13000K, we observe a fast decrease of viscosity for all various plasmas because of the progressive disappearance of molecules (dissociation) and atoms (ionization).

Figure 6 gives the total thermal conductivity (λ) evolution of studied plasmas according to the temperature. The various peaks observed correspond to the molecules dissociation and the atoms ionization reactions. For all curves, these peaks appear at temperatures lower than 23000K where the reactions of molecules dissociation and atoms ionization take place. The peaks observed at around 7000K correspond to the nitrogen molecule (N_2) dissociation. We noted that at around 7000K the air plasma has the highest value of λ due to the fact that air holds the strongest proportion of nitrogen. For the other diatomic species the dissociation peaks will appear at temperatures lower than 5000K, so out of our range of temperature. But the shape of the curves let to think that the water vapor plasma will have the highest peak at very low temperatures. The influence of H, O and N atoms ionization appears in the temperatures interval from 13000K to16000K. In this temperatures interval Mix 3 plasma (plasma of 20% air and 80% vapor of water) has the highest peak of thermal conductivity. In general, except for Mix 3, the thermal conductivity of the mixture plasma increases with the water vapor proportion at medium temperatures (10000 K to 21000 K) and decreases with this one at low temperatures. For temperatures higher than 20000K, thermal conductivity is especially due to the electrons translation thermal conductivity (λ_{tr}^{e}). Indeed at these temperatures, all plasmas are almost completely ionized what explains the similarity of their thermal conductivities.

In addition, the studies undertaken by various authors Hertz [36], Vacquié [37] showed that thermal conductivity has a very significant role on the time-constant of arc τ. Consequently the shape of the thermal conductivity curve gives information on the performance of gas used for the electric arcs. The importance of thermal conductivity particularly appears through its reactional component λ_R. From the study of several gases, Hertz [36] proposes an empirical relation linking the arc time-constant τ to the maximum value of the reaction thermal conductivity coefficient of gas: $\tau.\lambda_R^{max} = C^{te}$. This relation shows that the reaction thermal conductivity has a paramount role in the speed of electric arcs extinction. The more significant the maximum λ_R is, and the shorter the arc lifespan is. In addition, some others studies [6, 9] have shown that more the percentage of atoms H is raised and more the form of λ(T) of the plasma is interesting.

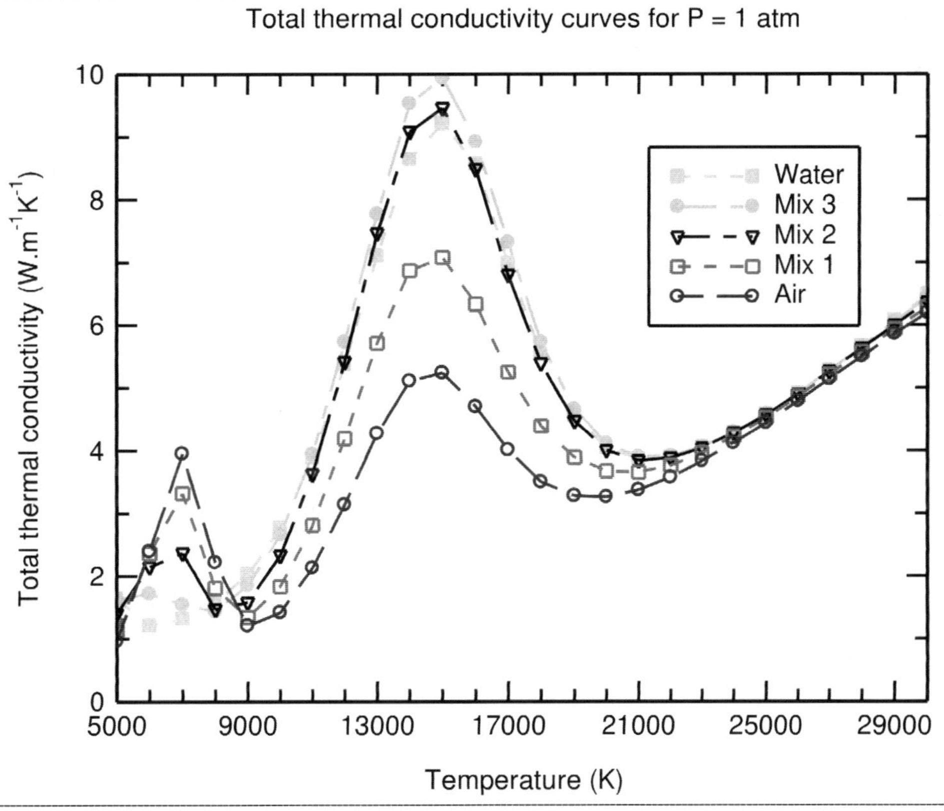

Figure 6. The total thermal conductivity results curves of the five plasmas (P = 1 atm).

On figure 7 appear the evolutions of electrical conductivities (σ) of studied plasmas. Globally, we can say that the various plasmas specially the 3 mixtures of air- water vapor have a similar electrical conductivity for temperature below 24000 K. However, the electrical conductivity of the plasma of dry air is slightly higher than those of all the other plasmas studied. That is explained by the low value of the energy of ionization of the molecules NO present in great quantity in the plasma of air. When the temperature exceeded 24000 K, the second ionizations of atoms involve some differences in plasmas electrical conductivities.

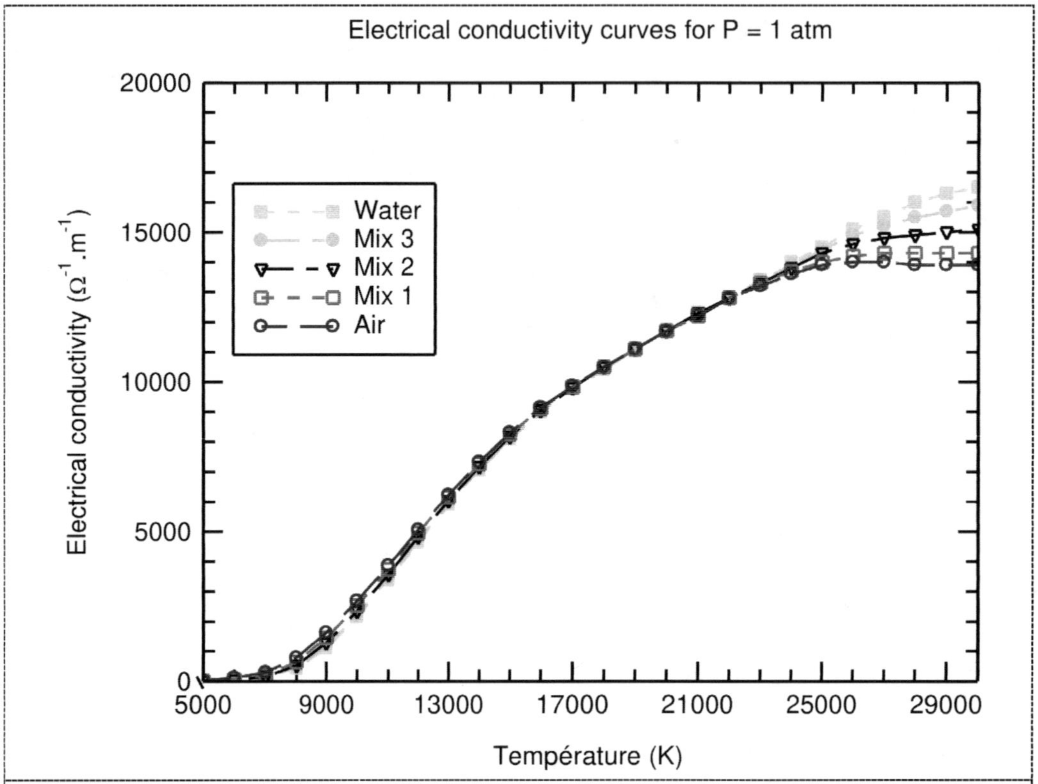

Figure 7. The electrical conductivity results curves of the five plasmas (P = 1 atm).

Figures 8, 9 and 10 respectively present the evolution according to the temperature of thermal conductivity, electrical conductivity and the dynamical viscosity of Mix 2 plasma for the three values of the pressure. It is advisable to announce that we have only presented the results obtained of Mix 2. The results of the other mixtures are perfectly similar to those of Mix 2. The peaks of thermal conductivity move towards the high temperatures and increase slightly when the pressure increases. This translation of the peaks is due to the displacement of chemical reactions.

The value of the electrical conductivity of plasma drops when the temperature increases for temperature lower than 13000 K. That is related to a significant reduction in the ionization rate of plasma in this zone of temperature caused by the pressure increase. For temperature higher than 13000K the electrical conductivity increases when the pressure increases.

The dynamical viscosity of studied plasmas is independent of the pressure at low temperature ($T < 9000$ K). For high temperature, we observe a displacement towards the high temperatures of the maximum of the dynamical viscosity coefficient when the pressure increases. This displacement is attached to that of chemical equilibrium with the pressure. Indeed the heavy particles dissociations take place at high temperatures when the pressure increases. For a given temperature, we observe also an increase of the dynamical viscosity coefficient value with the pressure increase.

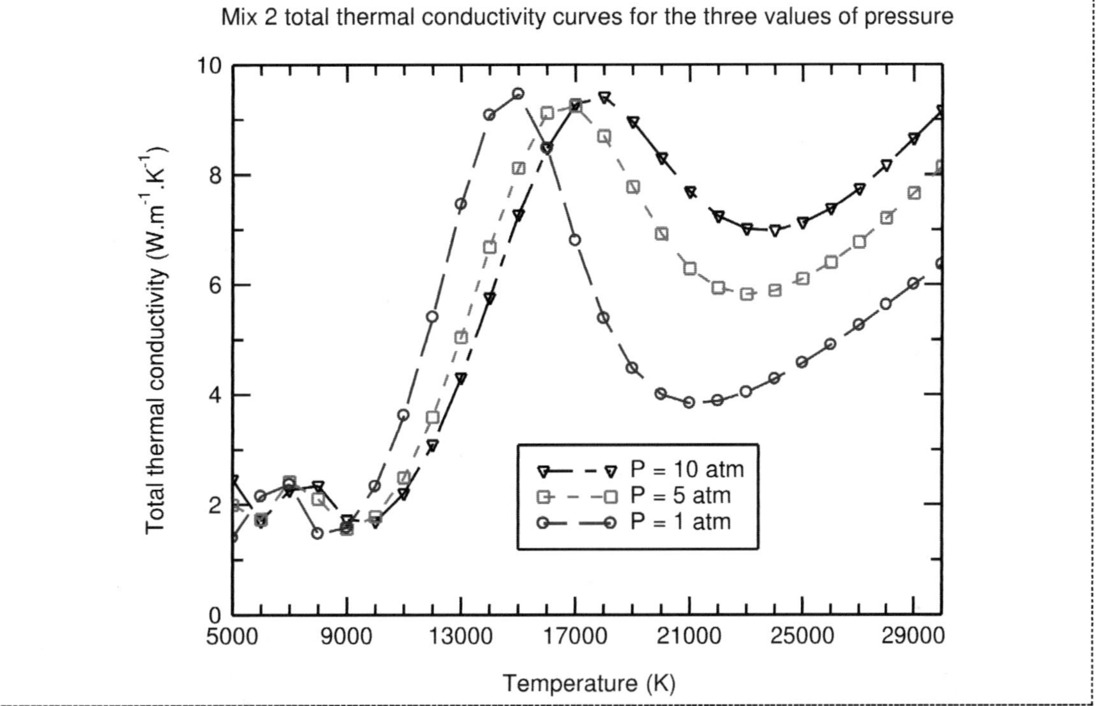

Figure 8. The total thermal conductivity results curves of Mix 2 for the three values of pressure.

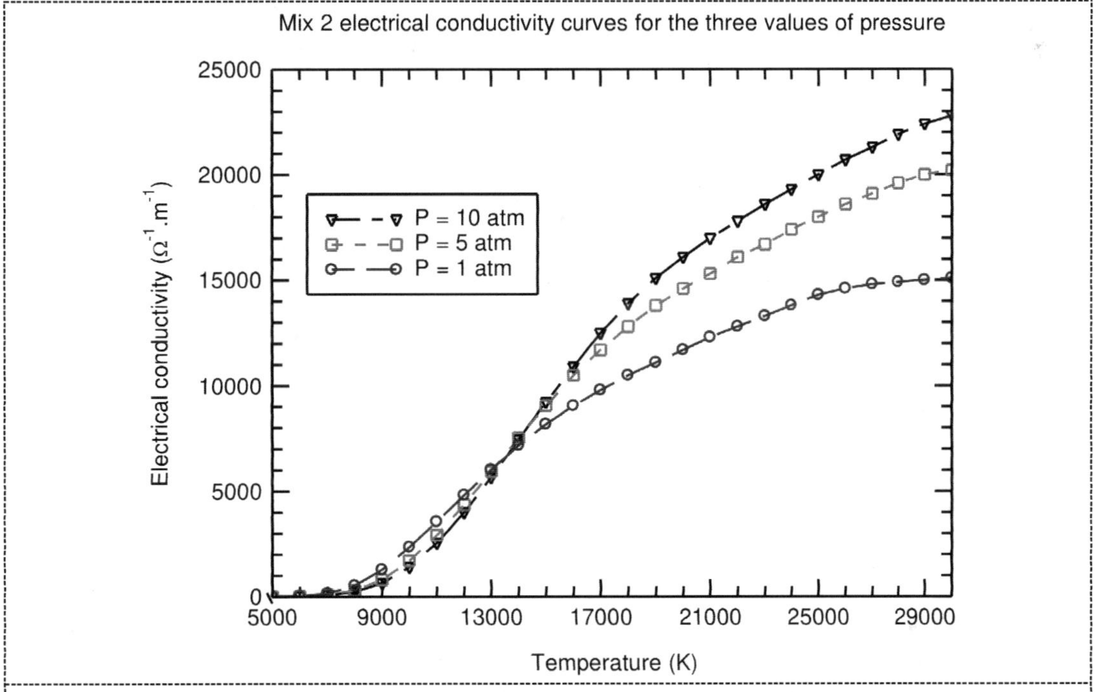

Figure 9. The electrical conductivity results curves of Mix 2 for the three values of pressure

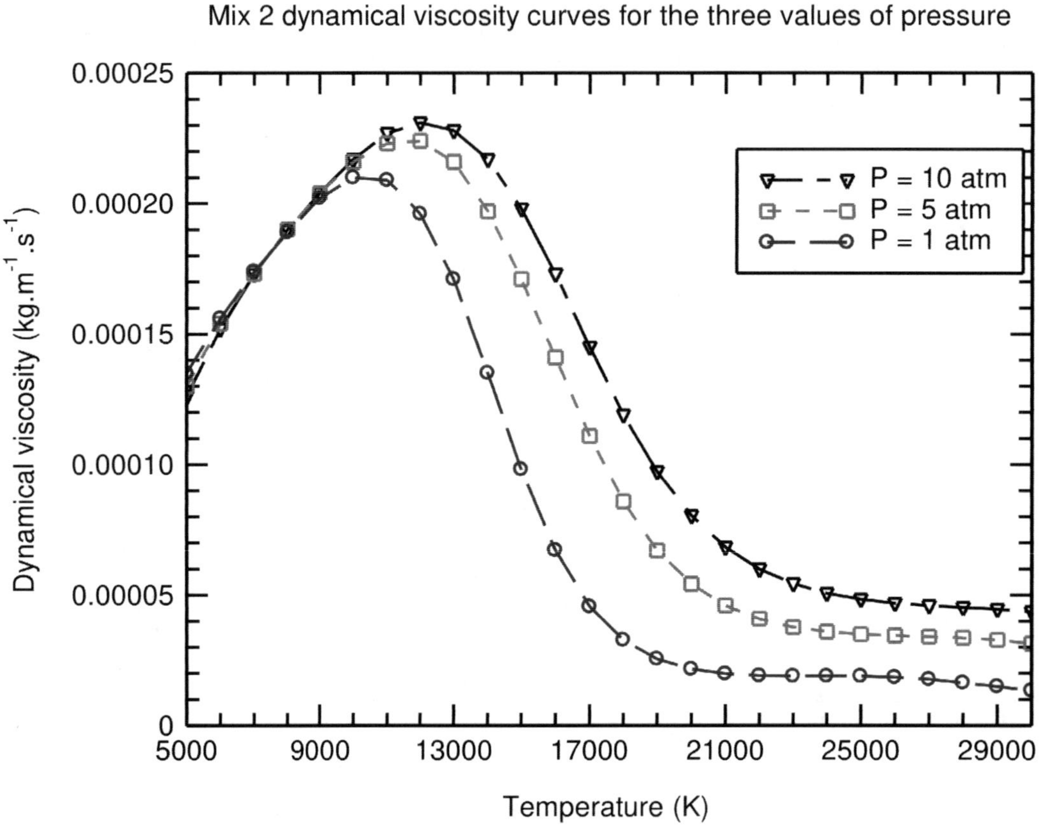

Figure 10. The dynamical viscosity results curves of Mix 2 for the three values of pressure.

4. Conclusion

In this paper, the calculation of thermal plasmas transport coefficients of three air and water vapor mixtures is carried out. Calculation is made in the range of temperature going from 5000 K to 30000 K and for three values of pressure based on the assumption of local thermodynamic equilibrium.

The results of calculation show that the evolutions of dynamical viscosities of mixtures plasmas considered here are almost similar on all the range of temperature. The maximum are between 9000 K and 11000 K. However a little increase in viscosity is noted when the water vapor percentage increases in the interval of temperature [7000 K - 13000 K].

For all studied plasmas, electrical conductivities have similar evolutions according to the temperature at low temperature. But at high temperatures (T > 24000 K), electrical conductivity increases with the percentage of the water vapor in the mixture.

The peaks of the plasmas thermal conductivities of the various mixtures appear at temperatures lower than 23000 K. From the value of temperature of 25000 K, all plasmas have practically the same value of thermal conductivity. It also arises that the thermal conductivity of plasma increases when the percentage of water vapor in the mixture increases. Mix 3 has a total thermal conductivity higher than those of the air and the two other mixtures. This is in good agreement with the results of the studies [38] which showed that thermal conductivity increases when the quantity of hydrogen (H) in gas increases. However we obtained a thermal conductivity peak of Mix3 higher than that of the water plasma. According to us, this seems to be related to the oxygen of which the quantity in water is higher than that in Mix 3. According to the empirical relation [36] who shows that more the maximum of the reaction thermal conductivity λ_R is high and the shorter the arc lifespan is; characteristics of Mix 3 for the electrical arc cut could be better than those of the pure water vapor which however contains the greatest quantity of hydrogen (H). But, it is clear that these results are not sufficient to be able to come to a conclusion about this subject. Although some other parameters (electrodes oxidation, dielectrical rigidity, etc.) will have to be taken into account; these results can already be used in the modeling of the discharges of electrical arcs in the air-water vapor mixture.

The influence of the pressure is also studied. It results that the pressure increase involves a translation of the characteristics of plasma towards high temperatures.

References

[1] André P, Brunet L, Bussière W, Caillard J, Lombard J M and Picard J P 2004 Transport coefficients of plasmas consisting of insulator vapours: Application to PE, POM, PMMA PA66 and PC *Eur. Phys. J. Appl. Phys.* **25** 169–182
[2] André P 1997 The influence of graphite on the composition and thermodynamic properties of plasma formed in ablated vapour of PMMA, PA6-6, PETP, POM and PE used in circuit-breakers *J. Phys. D: Appl. Phys.* **30** 475–493
[3] Abbaoui M, Koalaga Z and Lefort A 1992 Propriétés thermodynamiques et de transport des Plasmas issus de la vaporisation des isolants PTFE et PE *J. Phys. III France* **2** 455-472
[4] Andanson P, Lefort A and Roche J 1979 Erosion des contacts électriques sous l'effet d'un arc électrique de forte intensité *J. Phys D : Appl. Phys.* **12**
[5] André P 1997 Etude d'un plasma de SF6 hors d'équilibre thermique *J. Phys. III Fance* **7** 1339-1359
[6] Koalaga Z 1991 Contribution à l'étude expérimentale et théorique des plasmas d'arcs électriques laminés *Thèse de doctorat d'université* (Clermont Fd France)
[7] Cheminat B and Andanson P 1986 Etude expérimentale d'une décharge d'arc électrique contaminée par des vapeurs d'isolants *Revue Phys. Appl.* **21** 187-193
[8] Cheminat B 1989 Influence de l'ablation des parois sur les caractéristiques d'un arc électrique laminé *Revue Phys. Appl.* **24** 277-284
[9] Abbaoui M, Koalaga Z and Lefort A 1993 Composition et coefficients de transports des plasmas de matériaux plastiques (polymères) *Can. J. Phys.* **71** 1291
[10] Ouajji H, Cheminat B and Andanson P 1986 Composition et conductivité d'un plasma air-cuivre *J. Phys. D : Appl. Phys.* **19** 1903-1916
[11] Andanson P and Leffort A 1984 Calcul de l'érosion par vaporisation au niveau de la tache cathodique *J. Phys.D :Appl. Phys.* **17** 2377-2386
[12] Petr . K 2008 Thermophysical properties of H_2O-Ar plasmas at temperatures 400-50000K and pressure 0,1MPa *Plasma Chem Plasma Process* **28** 107-122
[13] Hrabovsky´ M, Kopecky´ V, Sember V, Kavka T, Chumak O and Konra´d M 2006 *IEEE Trans Plasma Sci* **34** 1566
[14] Aubreton J, Elchinger M F and Vinson J M 2008 Transport coefficients in water plasma: Part I: Equilibrium *Plasma Plasma Chem Plasma Process* **29** 149-171
[15] Hrabovsky M and Krenek P 1993 *Proceedings of the 11th international symposium on plasma Chemistry* Loughborough p 315
[16] Hirschfelder J O, Curtis C F and Bird R B 1954 *Molecular theory of gases liquids* John Wiley and Sons Inc.New-York
[17] Devoto R S 1967 *Phys. Fluids* **10** 354 – 64
[18] Devoto R S 1967 *Phys. Fluids* **10** 2105-12
[19] Muckenfuss C and Curtis C F 1958 *J. Chem. Phys.* **29** 1273
[20] Vanderslice J T, Welsman S, Mason E R and Fallon R J Phys 1962 *Fluids* **5** 155
[21] Butler J N and Brokaw R S J 1960 *Chem. Phys.* **32** 1005
[22] Meador W E and Staton L D 1965 *Phys. Fluids* **8** 1694
[23] Yun K S, Weissman S and Mason E A 1962 *Phys. Fluids.* **5** 672
[24] Ruddenberg J W and Wilke C R 1949 *Ind . Eng. Chem.* **41** 1345
[25] Rat V, Andre´ P, Aubreton J, Elchinger MF, Fauchais P and Lefort A 2001 *Phys Rev E* **64**

026409

[26] Rat V, Aubreton J, Elchinger MF, Fauchais P and Murphy AB 2002 *Phys Rev* **66** 056407

[27] Rat V, Andre´ P, Aubreton J, Elchinger M F, Fauchais P and Lefort A 2002 *plasma Chem and Plasma Process* **22** 475

[28] Capitelli M and Devoto R S 1973 *The Phys. of Fluids* **16** Il 1835 – 40

[29] Vanderslice J T, Weissman S, Mason E A and Fallon R J 1962 *The Phys. of Fluids* **5** 155.

[30] Belov V A and Teplofiz Vys 1967 *Temp* **5** 37

[31] André P, Bussière W and Rochette D 2007 Transport Coefficients of Ag-SiO2 Plasmas *PCPP* **27** Issue 4 381-403

[32] Mexmain J M 1983 *Thèse de 3ème cycle* N°83-31 Université de Limoges (France)

[33] André P, Brunet L, Buissière W, Caillard J, Lombard J M and Picard J P 2004 Transport coefficients of plasmas consisting of insulator vapours: Application to PE, POM, PMMA PA66 and PC *Eur. Phys. J. Appl. Phys.* **25** 169–182

[34] Eymard M 1984 *Thèse de 3ème cycle* N°3101 Université de Toulouse (France)

[35] Murphy A B 2000 Transport coefficients of hydrogen and argon-hydrogen plasmas *Plasma Chemistry and Plasma Processing* **20** N°3 279

[36] Hertz W 1971 *Zeit. Fur Phys.* **245** 105-125

[37] Vacquié S Rôle de la conductivité thermique sur la constante de temps d'arc Université Paul Sabatier Toulouse

[38] Koalaga Z, Abbaoui M and Lefort A 1993 Calcul des propriétés thermodynamiques des Plasmas d'isolants $C_\alpha H_\beta O_\gamma N_\theta$ *J. Phys. D : Appl. Phys.* **26** 393-403

Thermodynamics equilibrium and non equilibrium of plasma flows

M Dudeck[1], P André[2], A Kaminska[3], M Lino da Silva[4]

[1]Institut Jean Le Rond d'Alembert, Université Pierre et Marie Curie
4 place Jussieu, 75252 Paris cedex, France
[2]Laboratoire des Arcs Electriques et Plasmas Thermiques, Université Blaise Pascal
24 avenue des Landais, 63177 Aubière cedex, France
[3]Institute of Electric Power Engineering, Poznan University of Technology
60-965 Poznan, Poland
[4]Instituto de Plasmas e Fusao Nuclear, Laboratorio Associado, Instituto Superior Tecnico Av. Rovisco Pais, 1049-001, Lisboa, Portugal

E-mail: michel.dudeck@upmc.fr, pascal.andre@univ-bpclermont.fr

Abstract. A plasma flow is described using the properties of the entropy function. The plasma flow is a multi-temperature mixture with reactive processes (translation: T for heavy particles, T_e for electrons and T_{vj} for the vibrational temperature of the species j), diffusion and non equilibrium chemical reactions. The entropy rate production is determined for a plasma at rest and for a plasma flow. Equilibrium conditions are deduced and phenomenological equations are presented for non equilibrium plasma flow conditions.

1. Introduction

Generally, plasmas in space and plasmas in laboratories present a strong non equilibrium of the populations of the excited levels and ionisation levels related to Boltzmann equilibrium. In this case, only an equivalent temperature can be defined but not the kinetic temperature of the species (from energy distribution functions or the ground state and the first excited levels for the vibrational temperatures). In many case a macroscopic description can be considered as a satisfying approach with a set of temperatures associated to each species (kinetic temperatures T_j, rotational temperatures T_{rj}, vibrational temperatures T_{vj}). In our description we assume an thermal equilibrium for the kinetic temperature of heavy particles j, then $T_j = T$ and a temperature for electrons $T_e \neq T$. Vibrational temperatures T_{Vj} are introduced for the molecular species.

In the section 2, the entropy production is calculated for a plasma at rest with two kinetic temperatures ($T_e \neq T$) and with chemical reaction. For a light thermal non equilibrium, the exchange of energy between electrons and heavy particles is assumed to be described by a linear law. This rate exhibits the usual expression for the chemical term with the affinity function and the degree of advancement conducting to equilibrium laws and linear equations for near equilibrium processes.

IOP Conf. Series: Materials Science and Engineering **29** (2012) 012005 doi:10.1088/1757-899X/29/1/012005

In the section 3, the entropy rate production is determined for a plasma flow with two temperatures ($T_e \neq T$) with the use of the local thermodynamical equilibrium assumption (L.T.E) without chemical reaction and without diffusion effect. Different cases are studied: heat transfer without diffusion (section 3.1), diffusion without heat transfer (section 3.2), heat transfer and diffusion (section 3.3) and vibrational relaxation (section 3.4). From the entropy rate production, phenomenological equations are established from the entropy rate production.

In the section 4, two examples of calculation of non equilibrium plasma are presented: the first one concerns the composition of a plasma at equilibrium with two kinetic temperatures and the second one presents the calculation of a plasma flow in a arc-jet with a thermal non equilibrium and excited internal levels.

2. Energy exchanges and chemical reaction in a plasma at rest

The entropy rate production is determined by the use of a thermodynamic description [1, 2]. For an uniform plasma at rest in a volume *V*, the internal energy of electrons U_e (*Uj* for the heavy species *j*) is assumed to be dependent of extensive variables as entropy S_e for electrons (S_j for each heavy species), of volume *V*, and of masses M_e, M_j as $U_e = U_e(S_e, V, M_e), U_j = U_j(S_j, V, M_j)$. For a small variation of all the extensive variables (S_e, S_j, V, M_e, M_j) the Gibbs relations are $\delta U_e = T_e \delta S_e - p_e \delta V + g_e \delta M_e + \ldots$, $\delta U_j = T\,\delta S_j - p_j \delta V + g_j \delta M_j + \ldots$ where p_e, p_j are the partial pressures and g_e, g_j the potentials. With the property of additivity of the entropy of each subsystem $S = S_e + \sum_j S_j$, one obtains

$$\delta S = \frac{1}{T_e}(\delta U_e + p_e \delta V) + \frac{1}{T}(\sum_j \delta U_j + \sum_j p_j \delta V) - (\frac{g_e}{T_e}\delta M_e + \frac{\sum_j g_j}{T}\delta M_j)$$

With the equations of internal energies $\delta U_e = -p_e \delta V + \delta Q + \dot{E}_{e-j}\delta t$ and $\delta U_j = -p_j \delta V + \delta Q_j + \dot{E}_{j-e}\delta t$ where $\dot{E}_{e-j}$ is the energy exchange in the volume *V* between the electrons and the species *j* per unit of time with $\dot{E}_{e-j} + \sum_j \dot{E}_{j-e} = 0$, the previous relation becomes

$$\overset{\circ}{\overline{\delta S}} = \frac{\overset{\circ}{\overline{\delta Q_e}}}{T_e} + \frac{\sum_j \overset{\circ}{\overline{\delta Q_j}}}{T} + \dot{E}_{e-j}(\frac{1}{T_e} - \frac{1}{T}) + \frac{1}{T}\tilde{A}.\dot{\xi}$$

with $\overset{\circ}{\overline{\delta S}} = \frac{\delta S}{\delta t}$, $\overset{\circ}{\overline{\delta Q_e}} = \frac{\delta Q_e}{\delta t}$, $\overset{\circ}{\overline{\delta Q_j}} = \frac{\delta Q_j}{\delta t}$, $\dot{\xi} = \frac{\delta \xi}{\delta t}$. The variations of mass $\delta M_e, \delta M_j$ are expressed with the degree ξ defined as $\delta M_e = a_e \mathcal{M}_e \delta\xi$ and $\delta M_j = a_j \mathcal{M}_j \delta\xi$ where a_e, a_j are the stoichiometric coefficients of the chemical reaction (it can be easily extend to the case of multi-chemical reactions with a degree ξ_r associated to each process *r*). Using $\overset{\circ}{\overline{\delta_e S}} = \frac{\overset{\circ}{\overline{\delta Q_e}}}{T_e} + \frac{\sum_j \overset{\circ}{\overline{\delta Q_j}}}{T}$ related to the exchange of energy, the entropy production rate is deduced as

$$\overset{\circ}{\overline{\delta_i S}} = \dot{E}_{e-j}(\frac{1}{T_e} - \frac{1}{T}) + \frac{\tilde{A}.\dot{\xi}}{T} \tag{1}$$

The chemical affinity $\tilde{A}$ in presence of a thermal non equilibrium (T, T_e) is defined with the molar potentials (μ_j, μ_e) as $\tilde{A} = -(\sum_j \mu_j a_j + \frac{T}{T_e}\mu_e a_e)$. In the case of a thermal equilibrium ($T_e = T$) the chemical affinity $\tilde{A}$ is reduced to the current expression $\tilde{A} = A = -(\sum_j \mu_j a_j + \mu_e a_e)$ and the entropy production rate becomes $\overset{\circ}{\overline{\delta_i S}} = \frac{A.\dot{\xi}}{T}$. For an evolution near the equilibrium conditions, the linear phenomenological equations are linked to the generalized flux $(\dot{E}_{e-j}, \dot{\xi})$ and the generalized strengths $(T_e - T, \tilde{A})$ as

$$\begin{bmatrix} \dot{E}_{e-j} \\ \dot{\xi} \end{bmatrix} = \begin{bmatrix} L_1 & 0 \\ 0 & L_2 \end{bmatrix} \cdot \begin{bmatrix} T - T_e \\ \tilde{A} \end{bmatrix} \tag{2}$$

where the two phenomenological coefficients L_1 and L_2 are positive (positive entropy source). The parameter L_1 is related to electron-heavy particles elastic and non elastic collisions. The elastic collisions effect can be expressed as $L_1 = 3k_B n_e \sum_j \frac{m_e}{m_j}\nu_{ej}$ where k_B is the Boltzmann constant, n_e the plasma density and ν_{ej} is the elastic collision electron frequencies with neutral species and ions.

The second phenomenological equation introduces the positive coefficient L_2. This coefficient is not often used because one can introduce the specific chemical rates of the reactive processes whatever the thermodynamics situation (near the equilibrium or far from the equilibrium).

3. Energy, mass exchanges and chemical reaction in a plasma flow

In the section 2, we analysed the properties of an uniform plasma at rest with two temperatures (T, Te) and now we extend the analysis to a plasma flow in thermal non equilibrium (T, T_e). From Gibbs, Euler and Gibbs-Duhem equations, we will deduce the expression of the entropy variation ds and with the assumption of a quasi-local equilibrium condition, we determine the expression of the local entropy production rate. From this expression, equilibrium conditions are deduced and phenomenological relations are obtained and discussed. The Gibbs, Euler and Gibbs-Duhem equations for the electrons and the heavy species are

$$\bar{u}_e = T_e\bar{s}_e - p_e v + g_e Y_e \quad 0 = \bar{s}_e dT_e - v dp_e + Y_e dg_e \quad \text{and} \quad d\bar{u}_e = T_e d\bar{s}_e - p_e dv + g_e dY_e$$
$$\bar{u}_j = T\bar{s}_j - p_j v + g_j Y_j \quad 0 = \bar{s}_j dT - v dp_j + Y_j dg_j \quad \text{and} \quad d\bar{u}_j = T d\bar{s}_j - p_j dv + g_j dY_j$$

where $\bar{u}_e = \frac{U_e}{M}, \bar{u}_j = \frac{U_j}{M}$ are the specific internal energies calculated with the mass M of the mixture, $\bar{s}_e = \frac{S_e}{M}, \bar{s}_j = \frac{S_j}{M}$ are the specific entropies, $Y_e = \frac{M_e}{M}, Y_j = \frac{M_j}{M}$ are the mass fractions and v the specific volume (the small letters stand for the specific amounts). By these relations and with the expression of the entropy $s = \frac{S}{M} = \frac{S_e}{M} + \sum_j \frac{S_j}{M} = \bar{s}_e + \sum_j \bar{s}_j$

$$ds = \frac{d\bar{u}_e}{T_e} + \sum_j \frac{d\bar{u}_j}{T} + \left(\frac{p_e}{T_e} + \frac{\sum_j p_j}{T}\right)dv - \left(\frac{g_e}{T_e}dY_e + \frac{1}{T}\sum_j g_j dY_j\right)$$

If a local thermodynamical equilibrium (L.T.E.) is assumed following the macroscopic (centre-of-mass) plasma flow the balance of entropy becomes

$$\rho\frac{ds}{dt} = \frac{1}{T_e}\rho\frac{d(Y_e u_e)}{dt} + \frac{1}{T}\sum_j \rho\frac{d(Y_j u_j)}{dt} + \left(\frac{p_e}{T_e} + \frac{\sum_j p_j}{T}\right)div\vec{V} - \left(\frac{g_e}{T_e}\rho\frac{dY_e}{dt} + \frac{\sum_j g_j}{T}\rho\frac{dY_j}{dt}\right) \quad (3)$$

with $\frac{1}{v}\frac{dv}{dt} = div\vec{V}$ and where the internal energies (u_e, u_j) are $u_e = \frac{U_e}{M_e}, u_j = \frac{U_j}{M_j}$. This equation is changed by the use of the laws of conservation of mass and the energy balance equations.

3.1. *Heat transfer without diffusion, chemical reaction and viscous effect*

Without diffusion and chemical reaction the previous balance of entropy (3) is reduced to

$$\rho\frac{ds}{dt} = \frac{1}{T_e}\rho\frac{d(Y_e u_e)}{dt} + \frac{1}{T}\sum_j \rho\frac{d(Y_j u_j)}{dt} + \left(\frac{p_e}{T_e} + \frac{\sum_j p_j}{T}\right)div\vec{V}$$

The energy balance relation is $\frac{\partial(\rho_{e,j}u_{e,j})}{\partial t} = -div(\vec{q}_{e,j} + \rho_{e,j}u_{e,j}\vec{V}_{e,j}) - \overline{\overline{P}}_{e,j}.\overline{\overline{grad\vec{V}}}_{e,j} + \dot{e}_{j,e}$

where $\vec{V}_{e,j}$ is the velocity of the species *e, j*, $\vec{q}_{e,j}$ are the heat flux, and $\dot{e}_{j,e}$ are the rate of energy exchange per unit of volume. Then, the energy balance becomes in absence of diffusion and viscosity

$$\rho\frac{d(Y_{e,j}u_{e,j})}{dt} = -div\vec{q}_{e,j} - p_{e,j}div\vec{V}_{e,j} + \dot{e}_{j,e}$$

and the balance of entropy is reduced to the following expression (the quantity $-\frac{p_e}{T_e}\vec{\Delta}_e - \sum_j \frac{p_j}{T}\vec{\Delta}_j$ where $\vec{\Delta}_{e,j} = \vec{V}_{e,j} - \vec{V}$ is neglected)

$$\rho\frac{ds}{dt} = -div\left(\frac{\vec{q}_e}{T_e} + \frac{\sum_j \vec{q}_j}{T}\right) + \vec{q}_e.\overrightarrow{grad}\frac{1}{T_e} + \sum_j \vec{q}_j.\overrightarrow{grad}\frac{1}{T} + \dot{e}_e\left(\frac{1}{T_e} - \frac{1}{T}\right) \quad (4)$$

The entropy flux is $\vec{J}_S = \frac{\vec{q}_e}{T_e} + \frac{\sum_j \vec{q}_j}{T}$ which is reduced to $\vec{J}_S = \frac{\vec{q}}{T}$ with $\vec{q} = \vec{q}_e + \sum_j \vec{q}_j$ for a thermal equilibrium $(T_e = T)$. The total entropy flux is $\vec{J}_s = \frac{\vec{q}_e}{T_e} + \frac{\sum_j \vec{q}_j}{T} + \rho s\vec{V}$. The rate of entropy production is

$$\dot{w}_s = \vec{q}_e.\overrightarrow{grad}\frac{1}{T_e} + \sum_j \vec{q}_j.\overrightarrow{grad}\frac{1}{T} + \dot{e}_e\left(\frac{1}{T_e} - \frac{1}{T}\right) \quad (5)$$

The linear relations associated to a near equilibrium evolution are

$$\vec{q}_e = -\lambda_e \overrightarrow{gradT_e}, \quad \vec{q} = \sum_j (\vec{q}_j^*) = -\lambda \overrightarrow{gradT}$$
$$\dot{e}_e = L(T - T_e) \tag{6}$$

where λ_e, λ are the heat conduction coefficients for electrons and heavy species respectively. The third phenomenological relation for the energy exchange by elastic collisions is equivalent to the relation established in section 1 but now related to the unit of volume of the plasma flow. All the coefficients are positive $(\lambda_e > 0,\ \lambda > 0,\ L > 0)$ as consequence of the second principle of thermodynamics. For a plasma flow in thermal equilibrium ($T = T_e$), one obtains the usual Fourier's law for the heat transfer.

3.2. *Diffusion effects without heat transfer, chemical reaction and viscous effect*

The diffusion effects are now introduced in presence of two different but uniform kinetic temperatures (T, T_e), without chemical reaction, viscous effect and heat flux.

The mass conservation is $\rho \dfrac{dY_{e,j}}{dt} = -div\vec{J}_{e,j}$ where $\vec{J}_{e,j}$ are $\vec{J}_{e,j} = \rho_{e,j}(\vec{V}_{e,j} - \vec{V})$ and the balance of energy is written as

$$\frac{\partial(\rho_{e,j} u_{e,j})}{\partial t} = -div(\vec{q}_{e,j} + \rho_{e,j} u_{e,j} \vec{V}_{e,j}) - p_{e,j} div\vec{V}_{e,j} + \dot{e}_{j,e}$$

$$\rho \frac{d(Y_{e,j} u_{e,j})}{dt} = -div(\vec{q}_{e,j} + u_{e,j} \vec{J}_{e,j}) - p_{e,j} div\vec{V}_{e,j} + \dot{e}_{j,e}.$$

In order to be coherent with the balance of total energy, one assumes that $(\vec{V}_{e,j} \approx \vec{V})$ only in the balance energy, not in the mass conservation, $-\sum p_{e,j} div\vec{V}_{e,j} \approx -(\sum p_{e,j}) div\vec{V} = -p\,div\vec{V}$ and without heat transfer the previous balance equation becomes $\rho \dfrac{d(Y_{e,j} u_{e,j})}{dt} = -p_{e,j} div\vec{V} + \dot{e}_{j,e}$.

The equation of entropy is

$$\rho \frac{ds}{dt} = -div[-\frac{g_e}{T_e}\vec{J}_e - \sum_j \frac{g_j}{T}\vec{J}_j] - \frac{1}{T_e}\vec{J}_e.\overrightarrow{gradg_e} - \frac{1}{T}\sum_j \vec{J}_j.\overrightarrow{gradg_j} + \dot{e}_e(\frac{1}{T_e} - \frac{1}{T}) \tag{7}$$

The flux of entropy is $\vec{J}_S = -\dfrac{g_e}{T_e}\vec{J}_e - \sum_j \dfrac{g_j}{T}\vec{J}_j$ and its total flux is $\vec{J}_S + \rho s\vec{V}$.

The entropy production rate is

$$\dot{w}_S = -\frac{1}{T_e}(\frac{\partial g_e}{\partial Y_e})_{Te,p}\vec{J}_e.\overrightarrow{gradY_e} - \frac{1}{T}\sum_j(\frac{\partial g_e}{\partial Y_j})_{Tj,p}\vec{J}_j.\overrightarrow{gradY_j} + +\dot{e}_e(\frac{1}{T_e}-\frac{1}{T}) \quad (8)$$

and the linear (and direct) relations associated to a near equilibrium evolution are

$$\begin{aligned} \vec{J}_e &= -\mathcal{D}_e\overrightarrow{gradY_e}, \quad \vec{J}_j = -\mathcal{D}_j\overrightarrow{gradY_j} \\ \dot{e}_e &= L(T-T_e) \end{aligned} \quad (9)$$

The thermal non equilibrium $(T \neq T_e)$ does not introduce new correlation but one can include the diffusion coefficients as $D_{e,j} = D_{e,j}(T_{e,j,}p,Y_{e,j}) > 0$ in agreement with a positive production entropy rate and with $(\frac{\partial g_e}{\partial Y_e})_{Te,p} > 0, (\frac{\partial g_e}{\partial Y_j})_{Tj,p} > 0$. In this description, the diffusion of the electrons is only correlated to the gradient of the electron mass fraction. An ambipolar electron diffusion is generally used in plasma flow with $D_a = \frac{3kT_e}{4\rho Q_{in}}$ where Q_{in} is the ion-neutral cross section [4]. Crossed effect can be introduced with $\mathcal{D}_{ij}$ coefficients.

3.3. *Heat transfer and diffusion effects without chemical reaction and viscous effect*

Let us now consider a plasma flow with heat transfer and diffusion effects but without chemical reaction and viscous effect. The plasma flow is described by two kinetic temperatures (T_e, T). The balance of energy is written as $\rho\frac{d(Y_{e,j}u_{e,j})}{dt} = -div\vec{q}_{e,j} - p_{e,j}div\vec{V} + \dot{e}_{j,e}$ and the entropy equation is

$$\begin{aligned} \rho\frac{ds}{dt} &= -div[\frac{\vec{q}_e - g_e\vec{J}_e}{T_e} - \sum_j\frac{\vec{q}_j - g_j\vec{J}_j}{T}] + \vec{q}_e.\overrightarrow{grad}\frac{1}{T_e} + (\sum_j\vec{q}_j).\overrightarrow{grad}\frac{1}{T} \\ &- \vec{J}_e.\overrightarrow{grad}\frac{g_e}{T_e} - \sum_j\vec{J}_j.\overrightarrow{grad}\frac{g_j}{T} \end{aligned} \quad (10)$$

The flux of entropy is $\vec{J}_S = \frac{\vec{q}_e - g_e\vec{J}_e}{T_e} - \sum_j\frac{\vec{q}_j - g_j\vec{J}_j}{T}$ and the production rate is

$$\dot{w}_S = \vec{q}_e.\overrightarrow{grad}\frac{1}{T_e} + (\sum_j\vec{q}_j).\overrightarrow{grad}\frac{1}{T} - \vec{J}_e.\overrightarrow{grad}\frac{g_e(T_e,p,Y_e)}{T_e} - \sum_j\vec{J}_j.\overrightarrow{grad}\frac{g_j(T,p,Y_j)}{T}$$

Then after modifications of the potentials [2],

$$\begin{aligned} \dot{w}_S &= -\frac{1}{T_e^2}\vec{q}_e^*.\overrightarrow{gradT_e} - \frac{1}{T^2}(\sum_j\vec{q}_j^*).\overrightarrow{gradT} - \frac{1}{T_e}(\frac{\partial g_e}{\partial Y_e})_T\vec{J}_e.\overrightarrow{gradY_e} - \sum_j\frac{1}{T}(\frac{\partial g_j}{\partial Y_j})_T\vec{J}_j.\overrightarrow{gradY_j} \\ &+ \dot{e}_e(\frac{1}{T_e}-\frac{1}{T}) \end{aligned} \quad (11)$$

with the heat fluxes $\vec{q}_e^* = \vec{q}_e + h_e \vec{J}_e$, $\vec{q}_j^* = \vec{q}_j + h_j \vec{J}_j$ where the specific enthalpies are defined as $h_e = g_e + T_e s_e$ and $h_j = g_j + Ts_j$. The linear and direct phenomenological equations for heat transfer and diffusion are written as

$$\begin{aligned}
&\vec{q}_e^* = -\lambda_e \overrightarrow{gradT_e}, \quad \vec{q} = \sum_j (\vec{q}_j^*) = -\lambda \overrightarrow{gradT} \\
&\vec{J}_e = -\mathcal{D}_e \overrightarrow{gradY_e}, \quad \vec{J}_j = -\mathcal{D}_j \overrightarrow{gradY_j} \\
&\dot{e}_e = L(T - T_e)
\end{aligned} \tag{12}$$

Crossed effects between heat transfer and diffusion could be added without difficulty. As previous remark, all the phenomenological coefficients are positive.

3.4. *Vibrational relaxation*

Now, let us consider a gas flow with two different kinetic temperatures for heavy species and electrons $(T \neq T_e)$ without diffusion effect and reactions but with a vibrational temperature T_{vj} associated at each molecule with the relation $d\overline{u}_{vj} = T_{vj} d\overline{s}_j$. Then, the equation of entropy requires other terms with the vibrational energy u_{vj},

$$\rho\frac{ds}{dt} = \frac{1}{T_e}\rho\frac{d(Y_e u_e)}{dt} + \frac{1}{T}\sum_j \rho\frac{d(Y_j u_j)}{dt} + \sum_j \frac{1}{T_{vj}}\rho\frac{d(Y_j u_{vj})}{dt} + \left(\frac{p_e}{T_e} + \frac{\sum_j p_j}{T}\right) div\vec{V}$$
$$-\left(\frac{g_e}{T_e}\rho\frac{dY_e}{dt} + \frac{\sum_j g_j}{T}\rho\frac{dY_j}{dt}\right)$$

The balance equation for the vibrational energy of *j* is

$$\rho\frac{d(Y_j u_{vj})}{dt} = -div(\vec{q}_{vj} + u_{vj}\vec{J}_j) + \dot{Q}_{e-j}^{V-Tr} + \sum_{i\neq j} (\dot{Q}_{i-j}^{V-Tr} + \dot{Q}_{i-j}^{V-V})$$

where $\vec{q}_{vj}$ is the vibrational heat flow, $\dot{Q}_{e-j}^{V-Tr}, \dot{Q}_{i-j}^{V-Tr}$ are the vibration exchange terms respectively with electron and other heavy species vibration (Vibration-Translation : V-T), $\dot{Q}_{i-j}^{V-V}$ is the vibration exchange term between the species *i* and *j* (Vibration-Vibration : V-V),

The balance equation for energy requires new terms to introduce the energy vibration exchanges.

Electron: $\rho\frac{d(Y_e u_e)}{dt} = -div\vec{q}_e - p_e div\vec{V} + \dot{e}_e + \sum_j \dot{Q}_{j-e}^{V-Tr}$

Species *j*: $\rho\frac{d(Y_j u_j)}{dt} = -div\vec{q}_j - p_j div\vec{V} + \dot{e}_j + \dot{Q}_{e-j}^{V-Tr} + \sum_{i\neq j} (\dot{Q}_{j-i}^{V-Tr} + \dot{Q}_{j-i}^{V-V})$

The balance equation of entropy is now written as

$$\rho\frac{ds}{dt}=-div[\frac{\vec{q}_e-g_e\vec{J}_e}{T_e}+\sum_j(\frac{\vec{q}_j-g_j\vec{J}_j}{T}+\frac{\vec{q}_{vj}}{T_{vj}})]$$
$$+\vec{q}_e^*\,\overrightarrow{grad}\frac{1}{T_e}+\sum_j(\vec{q}_j^*)\overrightarrow{grad}\frac{1}{T}+\sum_j(\vec{q}_{vj})\overrightarrow{grad}\frac{1}{T_{vj}}$$
$$-\frac{1}{T_e}(\frac{\partial g_e}{\partial Y_e})_T\vec{\jmath}_e.\overrightarrow{gradY_e}-\sum_j\frac{1}{T}(\frac{\partial g_j}{\partial Y_j})_T\vec{\jmath}_j.\overrightarrow{gradY_j} \qquad (13)$$
$$+\dot{e}_e(\frac{1}{T_e}-\frac{1}{T})+\sum_j\dot{Q}_{e-j}^{V-Tr}(\frac{1}{T_e}-\frac{1}{T_{vj}})+\sum_j\dot{Q}_{j,i}^{V-Tr}(\frac{1}{T}-\frac{1}{T_{vj}})+\sum_{i,j}\dot{Q}_{j,i}^{V-V}(\frac{1}{T_{vi}}-\frac{1}{T_{vj}})$$

The flux of entropy is $\vec{J}_S=\frac{\vec{q}_e^*}{T_e}+\sum_j(\frac{\vec{q}_j^*}{T}+\frac{\vec{q}_{vj}^*}{T_{vj}})$ and the linear and direct phenomenological equations for a near equilibrium evolution are

$$\vec{q}_e^*=-\lambda_e\overrightarrow{gradT_e},\quad \vec{q}=\sum_j(\vec{q}_j^*)=-\lambda\overrightarrow{gradT},\quad \vec{q}_{vj}=-\lambda_{vj}\overrightarrow{gradT_{vj}}$$
$$\vec{\jmath}_e=-\mathcal{D}_e\overrightarrow{gradY_e},\quad \vec{\jmath}_j=-\mathcal{D}_j\overrightarrow{gradY_j} \qquad (14)$$
$$\dot{e}_e=L\,(T-T_e)$$
$$\dot{Q}_{e-j}^{V-Tr}=L_{e-j}^{V-Tr}(T_{vj}-T_e),\quad \dot{Q}_{j,i}^{V-Tr}=L_{j-i}^{V-Tr}(T_{vj}-T),\quad \dot{Q}_{j,i}^{V-V}=L_{j-i}^{V-V}(T_{vi}-T_{vj})$$

The vibrational heat transfer of j is proportional to the gradient of the vibrational temperature of j and the exchanges of energy related to the vibrational transfers are proportional to the differences of temperatures. According to the sign of the entropy production, the coefficients $\lambda_{vj}, L_{e-j}^{V-Tr}, L_{j-i}^{V-Tr}, L_{j-i}^{V-V}$ are positive. The expressions of the energy exchanges $(\dot{Q}_{e-j}^{V-Tr}, \dot{Q}_{j,i}^{V-Tr}, \dot{Q}_{j,i}^{V-V})$ are to be compared with the Landau-Teller law for the vibrational energies relaxations [5].

4. Examples of non equilibrium plasma

One can find numerous descriptions of non-equilibrium plasmas in litterature. Here we will present two sample cases. In the first case, the composition of a plasma has been calculated [6, 7] with two froozen kinetic temperatures (T, T_e). The figures 1a presents the time evolution of the concentrations of an air plasma and the figures 1b shows for the same plasma, the concentrations at quasi-equilibrium [6]. The gas composition is determined by the use of the minimization of the Gibbs free energy.

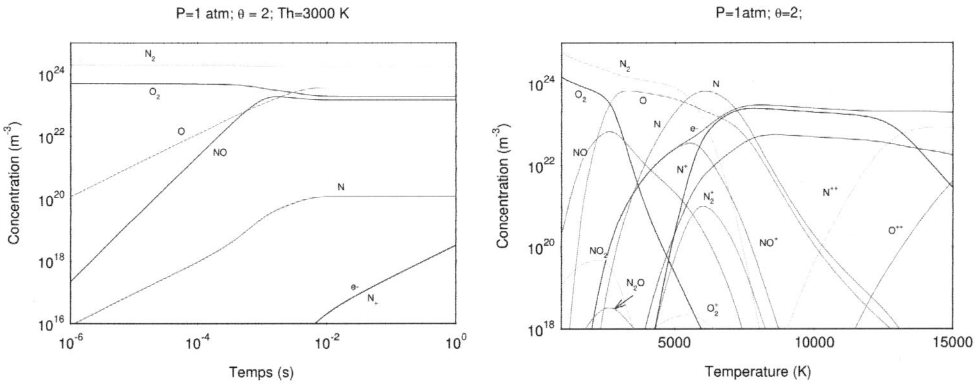

a) *b)*

Figure 1. Air (N_2-O_2) plasma without flow at a pressure of 1atm and with $\theta = Te/T = 2$ from [6].
a) Time evolution of the concentrations b) Concentrations at equilibrium

The second case presents a more complex configuration. Here, we simulate a nozzle flow as obtained in the low-pressure SR5 plasma arc-jet facility available at the ICARE/CNRS laboratory in Orléans, France. In this facility, a convergent-divergent nozzle is placed in a low pressure chamber (about 2-10Pa background pressure). Different mixture gases are injected in the convergent section of the nozzle, and an electrical arc between a cathode and the nozzle throat (which acts as an anode) heats the flow which is then supersonically expanded in the divergent section. The schematics of the SR5 nozzle are presented in Fig. 2.

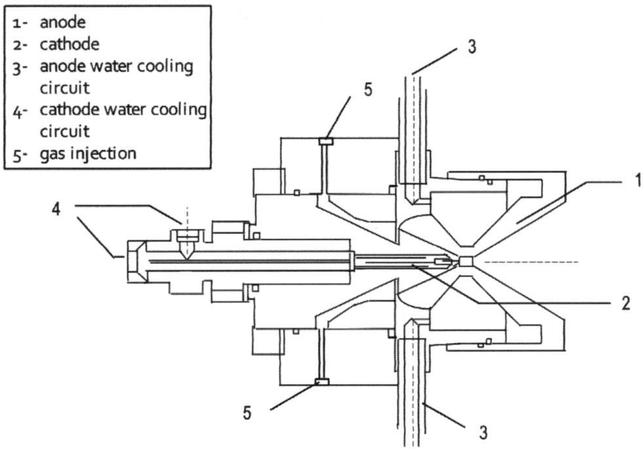

Figure 2. SR5 Nozzle schematics.

In a first approach, the internal nozzle flow has been simulated using a 1D axisymetric two-fluid model (with the motion of electrons and heavy species considered separately. The details of this approach are described in [8, 9].

Figure 3 from [8, 9] shows the evolutions of the two kinetic temperatures (T, T_e) of a non equilibrium plasma flow in an axisymmetric arc-jet running at reduced pressure with argon. The figure 2.a presents the evolution along the axis of the throat of the nozzle and the figure 2.b in the divergent part of the nozzle. The 2D fluid model introduces heat flux for electrons and heavy species, elastic collisions, on equilibrium ionisation and diffusion effects. The results are presented for different values of the arc current (from 60A to 140A).

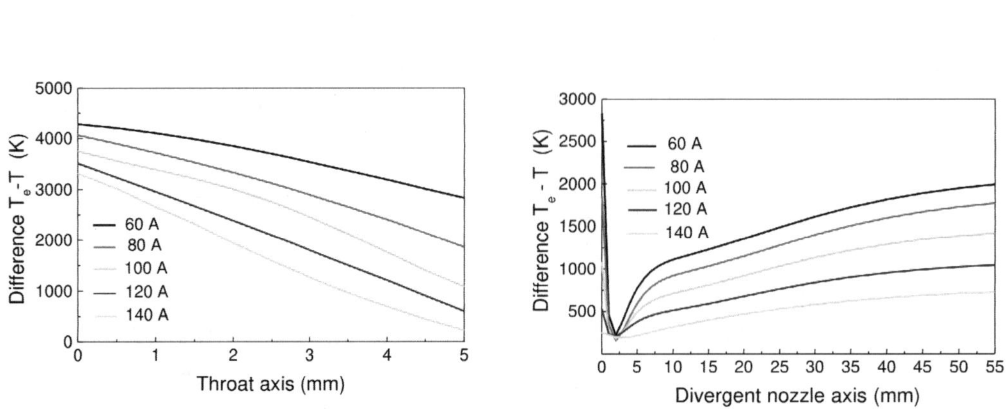

a) *b)*

Figure 3. Difference T_e - T along the axis of an arc-jet with argon [8, 9].
a) Along the cylindrical throat axis b) Along the axis of the divergent part of the nozzle

In a second approach, a much more simplified quasi-1D fluid model [10] has been used to determine the rotational and the vibrational temperatures in the same arc-jet but for a mixture N_2-CH_4 representative of Titan. The simplified hydrodynamic model allows a very detailed description of the exchange processes between the flow species (temperatures for CN, CH, N_2, NH, H_2 and C_2 are reported in Fig. 4) internal modes (electronic excitation, vibrational, and rotation) and the chemical reactions (dissociation, ionization, recombination) without large computational overheads. Energy exchange between the translational, vibrational and rotational modes is modelled considering the Landau-Teller equation. Figure 4 shows the simulation results for a sample flow of 0.45g/s with an arc energy input of 5.4kW

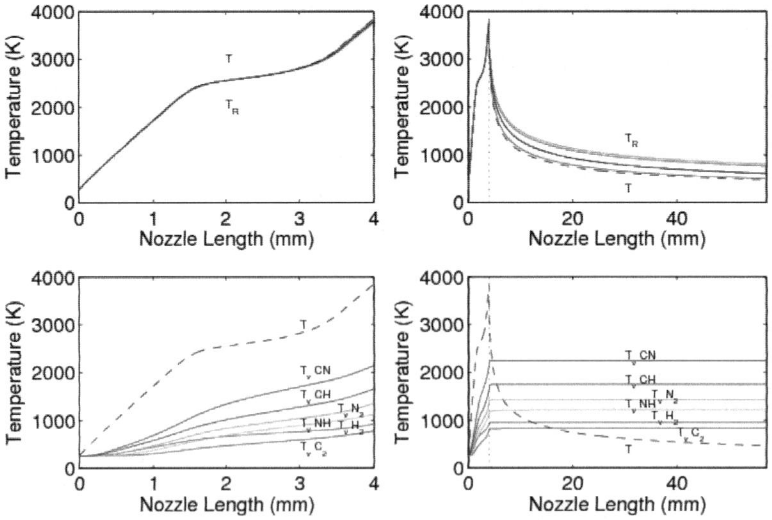

Figure 4. Rotational (top) and vibrational (bottom) temperatures of the molecular species in the nozzle throat (left) and diverging section (right), for an arc-jet with a N_2-CH_4 mixture.

In the nozzle throat region, we can see the translational temperature steadily increasing due to the energy input, up to the point where endothermic chemical reactions are activated, at which point a plateau in the heating is reached until all triatomic species (CH_4, CH_3, and CH_2) are fully dissociated, at which point, heating of the gas resumes. As expected, rotation-translation R-T exchange processes are very efficient, leading to an equilibrium between Translational T and rotational Tr temperatures. Vibration-translation V-T processes are slightly less efficient, and the vibrational temperatures increase more slowly than the translational temperature, leading to an increasing nonequilibrium between these two modes.

In the nozzle divergent region, the freezing of the V-T exchanges due to the rapidly expanding flow induces the appearance of a strong non equilibrium in the divergent part of the nozzle (figure 3). This effect is not introduced in the thermodynamic model described in this paper but it is easy to add this effect in the entropy equation. As R-T processes are very efficient, rotational temperatures in the flow do not reach a freezing point, however they cannot accompany the translational temperatures, hence leading also to the establishment of a slight rotational-translational nonequilibrium at the nozzle exit.

5. Conclusion

The flow of a gas in non-equilibrium conditions have been studied for many years using an approach Thermodynamic of Irreversible Phenomena (TIP) based on the rate of entropy production (see [1, 2]). This approach led to the proposal of linear phenomenological laws when the evolution takes place in near-equilibrium conditions. This approach was extended to the case of a multi-temperature gas flow with two kinetic temperatures and temperatures of vibration. It led to the introduction of linear phenomenological equations associated with a thermal non equilibrium, gadient of températures, diffusion and reaction. As usual, this approach does not reveal new physical phenomena, but shows the consistency of the TIP approach extended to a medium in non thermal equilibrium. Also, if this approach introduces exchange coefficients whose sign is specified, it does not give their expressions which are a strong limitation to its applications, it is complementary to a microscopic approach.

References

[1] Napolitano L G 1971 *Thermodynamique des systèmes composites en équilibre ou hors équilibre, Mémoire des Sciences Physiques*, fasc. 71, Gauthier Villars

[2] De Groot S R and Mazur P 1969 *Non equilibrium thermodynamics*, North Holland Publishing Company

[3] André P, Brunet L, Duffour E and Lombard J M 2002 Composition, pressure and thermodynamic properties calculated in plasma formed in insulator vapours of PC and POM at fixed volume *Eur. Phys. J. AP* **17** 53-64

[4] Devoto R S 1966 *Phys. of Fluid* **9** 1230-1240

[5] Lino da Silva M 2004 Simulation des propriétés radiatives du plasma entourant un véhicule traversant une atmosphère planétaire à vitesse hypersonique *Thesis* (in French) University of Orléans France

[6] André P *to be published*

[7] André P et al 2002 Theorical study of column of discharge with liquid non-metalloc (tap water) electrodes in air at atmospheric pressure *J. of Physics D:Appl. Phys.* **35** 1846-1854

[8] Kaminska A, Lopez B, Izrar B and Dudeck M 2008 Modelling of an argon plasma jet generated by a dc arc *Plasma Source Sci. Technol.* **17** 035018

[9] Katsonis K, Bérenguer Ch, Kaminska A and Dudeck M 2011 Argon 4s and 4p excited states atomic data applied in arc-jet modeling *Int. J. of Aerospace Engineering, accepted for publication*

[10] Le Quang D, Lino da Silva M and Dudeck M 2009 Non equilibrium plasma flows in a low-pressure supersonic Arcjet facility *3rd European Conference of Space and Aerospace Science* Versailles France

IOP Conf. Series: Materials Science and Engineering **29** (2012) 012006 doi:10.1088/1757-899X/29/1/012006

Theory about cathode arc root: a review

A Lefort [1,2] and M Abbaoui [1,2]

[1]Clermont Université, Université d'Auvergne, LAEPT, BP 10448, F-63000 CLERMONT-FERRAND
[2]Clermont Université, Université Blaise Pascal, LAEPT, BP 10448, F-63000 CLERMONT-FERRAND

E-mail: Andre.Lefort@univ-bpclermont.fr, Mhammed.Abbaoui@univ-bpclermont.fr

Abstract. Works on root arc cathode back to the early 1900s with its emphasis on emission cathode. The 70s show a definite change of theoretical work with a more pronounced growth phase over the past decade because of the calculation means available. The following article discusses some of the main models implemented.

1. Introduction

The 70s are a milestone in the theoretical work of cathode arc root. This very small and disturbed part of the discharge is not easily accessible experimentally, it is between a metal electrode that can be considered as belonging to it and the ionized column will be considered as a boundary. The geometric transition between the two parts is done through a contracted shape from the column to the metal surface, the distance over which this transition will vary with the circumstances of $0.1mm$ to $1mm$. During his lifetime, the arc root is subjected to several procedures that will profoundly change the existing physical phenomena. The magnitudes of the evolution phases of the phenomenon are ranging between the nanosecond and microsecond. These very short times explain the difficulties encountered in optical, electrical, thermal or geometric measurements.

Except for theories explaining the creation of the cathode spot by the explosion of a molten bridge or by an emitting metal tip, most models are stationary and ignoring the time factor, the particle flows vary only with the position. The experimental results have mostly shown for the metal area carrying the flow of charges a circular limit. However, this limit is situated along a circular surface to which the authors give different forms: flat, spherical, concave, convex spherical, conical ... Some authors have modelled the enthalpy evolution of this area showing , for non-refractory materials, the appearance of liquid and vapour phases resulting from a positive balance energy at this level. The axial symmetry is almost always carried out, the theories proposed are limited to a representation axi symmetric or mostly one-dimensional. Although the dimensions of this contraction zone increase from the surface of the metal electrode to the junction with the column, the particle flows are almost always parallel to the axis of symmetry.

The phenomenon of sheath of a metal rod inserted into plasma has been naturally transposed to the arc root both for the spot and diffuse regime. So for all models on the metal surface there is a space-charge sheath (SCS) whose thickness varies from the cases according to the authors of $0.1\mu m$ to $10\mu m$, in this area the charged particles are moving without collision they correspond to the current carried by the arc root. In this zone, an almost constant value positive charge creates a very intense electric field ($10^6 - 10^{10} V/m$) which will cause the emission of an electron flow. The metal electrode may be considered refractory, in this case the flow of neutral atoms emitted is considered negligible, and otherwise the flow becomes dominant and feeds the contraction zone with metal atoms. The electron emission physics laws are well known to this day, current density obtained depends on the properties of the emitting material, the electric field present and the surface temperature, these laws are obtained by integration of the Schrödinger equation, however, a problem persists because the proximity to the emitting surface of positive ions will modify the shape of the potential barrier, eventually most authors used an approximate law where this phenomenon is neglected. We will review this physical phenomenon in the first part of this article.

When the electron flow and the flow of metal vapour emitted in the case of a non-refractory cathode, move away from the area previously presented SCS, they will collide with neutral particles and ions, the corresponding energies are modified as well as particle densities. This change is continuous, but almost all authors decompose the space between the SCS area and the column in several areas with specific parameters. The computing resources required for all physical quantities related to this area show that it will soon be possible to describe this area continuously. Although the radius of this zone varies with the distance from the metal emitting surface, no author has yet modelled the evolution of the radius and the particle flux are always oriented parallel to the axis of the discharge. A portion of the different models developed will be presented in chronological order in the second part.

2. The electron emission

A metal surface whose surface is heated to a temperature T emits from the relationship of Ridchardson-Dushman [1] the current density:

$$J_T = \frac{4\pi mek^2T^2}{h^3}\exp\left(-\frac{eU}{kT}\right) \tag{1}$$

with: m, e mass and charge of the electron, k and h Boltzmann and Planck constant and U work function of the material.

In the presence of a high applied electric field E near the metal surface, the potential barrier controlling the exit of electrons from the metal is lowered with the magnitude: $\Delta U = \left(\frac{eE}{4\pi\varepsilon_0}\right)^{1/2}$

and the above relation becomes:

$$J_{TF} = \frac{4\pi mek^2T^2}{h^3}\exp\left(-\frac{e(U-\Delta U)}{kT}\right) \tag{2}$$

ε_0 represents the permittivity of vacuum.

This relationship, YP Raizer [2], describes the thermionic (thermal electron) emission with Schottky effect.

By solving the Schrödinger equation by the WKB method, EL Murphy and RH Good [3] obtain a relationship that gives the current density emitted when the metal surface is heated to a temperature T and subjected to the electric field E independently of the values taken by these two physical parameters:

$$J_{MG} = e\int_{-W_a}^{\infty} D(W,E)N(W,T)dW \tag{3}$$

W_a is the actual energy potential of electrons in the metal, $D(E,W)$ is the electron tunnelling probability across the potential barrier and $N(W,T)dW$ is the number of electron having energy between W and $W+dW$.

Depending on the values of the electric field, equation 3 is simplified and leads for weak fields to equation 1 and for strong fields to equation 2. This equation is only the potential barrier modified by a uniform electric field, several authors [4, 5, 6] showed that the presence of a positively charged ion located a few nanometers from the metal surface gives new variations of this potential barrier. By integration of the Schrödinger equation, Gayet et al [6] obtain a relationship taking into account the density n_i and the charge z_i of ions and this changes the relation 3 by the factor β :

$$J_{MG+I} = \beta(T, n_i, z_i) J_{MG} \tag{4}$$

Figure 1 shows the current density emitted for a copper cathode with $n_i = 10^{23} m^{-3}$ and $E = 10^8 Vm^{-1}$ according to [5]. The influence of the presence of ions (curve JMG + I) near the metal surface gives an increase in the current density emitted and resulting of the two other calculations.

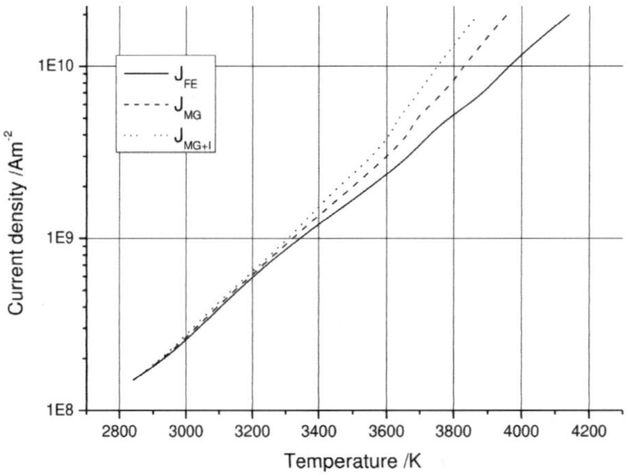

Figure 1. Comparison of emitted current densities versus surface temperature of the cathode. (from [5]).

3. Modelisation of the contraction zone

3.1. Simple models for the cathode spot

3.1.1. *A model on electrode surface.* Observing the shape of the end of the plasma column to the cathode of an electric arc generated a lot of approaches to describe this physical phenomenon. Logically different theories have begun with the emission of electrons from the metal surface, this finding demonstrates the need for a high value of the surface temperature T and also a high electric field E. These authors show that the energy required to cross the metal surface is provided by the metal ions located in the plasma situated in front of the surface of the electrode, this leads to define the two particle flux responsible of the passage of electric current, what is translated here by the variable s called electronic fraction which represents the ratio between the current density due to electrons and the total current density J. Thus their model is limited to four unknowns: J, T, E and s. As in many models to be developed subsequently the need to take the radius r as an additional variable is necessary because the four unknowns depend on r. The conventional material constants are assumed known, but other constants are necessary for writing the various equations it is the case of the cathode fall V_c, the values of M P Reece [8] are used.

Ions accumulate in front of the cathode surface creating intense electric field E that is easily calculated by the equation SS Mackeown [9] obtained by integrating the Poisson equation:

$$E^2 = \frac{4}{\varepsilon_0} J\left[(1-s)(\frac{M}{2e})^{1/2} - s(\frac{m}{2e})^{1/2}\right]V_C^{1/2} \tag{5}$$

M is the mass of a metal ion.
Conserving energy at the electrode surface results in equalizing the power supplied and the power lost:

$$P_i = P_v + P_c + P_r + P_e \tag{6}$$

For P_i authors [8] give the following value:

$$P_i = J_+(V_i + V_c - U) = (1-s)J(V_i + V_c - U) \tag{7}$$

V_i is the ionization potential and V_c the cathode fall. In newer models, it is considered the ion temperature T_i and enthalpy of vaporization W_{ev} calculated from the liquid metal [10, 11] :

$$P_i = (1-s)\times(\frac{5kT_i}{2e} + V_C + V_i - U + \frac{W_{ev}}{e})\times J \tag{8}$$

The hypothesis of an energy input Joule negligible compared to the energies involved in this model is made, it is in almost all the models developed thereafter. The power P_v expended by evaporation of neutrals is calculated by the Dushmann equation [12]:

$$\log P_v = C' - \frac{1}{2}\log T - \frac{B}{T} \tag{9}$$

C' and B are constants.
For the power P_e resulting from the emission of electron flow, it is obtained by the work of TH Lee [13] which gives values between 2 and $4eV$ per electron emitted for materials having a work function located in interval $3.5-4.5eV$ and temperatures from 1000 to $4000K$. This cooling effect of the cathode, first studied by Nottingham [14] is more complex than indicated by the studies cited above [13], several authors [15-17] have shown that the calculation based on the function of distribution of electrons in the metal is actually more complex. For the calculation of the energy exchanged with the cathode:

$$\varepsilon_N = \langle W \rangle - \langle W' \rangle \tag{10}$$

to simplify the calculations $\langle W' \rangle$ is taken equal to the Fermi level W_F^0 set to $T = 0K$. ($\langle W \rangle$ represents the average energy of electrons emitted by the cathode and $\langle W' \rangle$ the energy of the electrons used for the replacement of the electrons belonging to the metal). This calculation shows that the quantity ε_N calculated from temperature and electric field applied to the metal surface, and taking into account the physical properties of this metal can be positive or negative, change conditioned mainly by the electric field E occurs at a temperature T_i called temperature inversion, [15] gives the value (in K):

$$T_i \cong 5.32.10^{-5} \frac{E}{\sqrt{U}} \quad (11)$$

E in V/cm and U in eV.

In the relation 6, the power P_r emitted by radiation in this model is considered negligible compared to the values of other powers considered, this approximation is found in virtually all models of the cathode spot. To calculate the power P_c dissipated by conduction TH Lee and A. Greenwood used an approximate value of the temperature u_∞ rise obtained by extrapolating the analytical solution of the heat equation:

$$u_\infty = \frac{qa}{K\sqrt{\pi}} \quad (12)$$

with: q density of energy input, a radius of the spot and K thermal conductivity.

Finally the problem has five unknowns: J, s, E, a and T for only four equations: (3) (5) (6) and $I = \pi a^2 J$. The relationship (12) is too vague to give a valid solution. Even by imposing two additional boundary conditions on s and on the pressure p, results obtained give ranges of values for unknown quantities.

For example for a work function of $4.5eV$ for surface temperatures ranging from $3600K$ to $6800K$, the current density varies from 10^7 to $10^{12} Am^{-2}$, the electric field of 10^8 to $10^9 Vm^{-1}$ and the electronic fraction s from 0.8 to a value close to 1.

3.1.2. *The Existence Diagram [18].* G. Ecker noted that the equations available to describe the phenomena at the cathode spot are either inaccurate or insufficient, even in the case of a very simple model reduced to three parts: the electrode, the sheath and the plasma body. Using equations 3 and 5 and the estimated current density J_+ for ions [19]:

$$\frac{J_+}{J} \leq \frac{V_c}{V_c + V_i} \quad (13)$$

he plotted $T_{cc}(J)$ (see Figure 2) corresponding to lower values of T. The next limit curve that appears is also given by equations 3 and 5 but with the condition requiring equality between the ions back on to the electrode and those issued by vaporization, this leads to the curve $T_{nc}(J)$. The third limit curve $T_{ec}(J)$ is given in the same way with the simplification of equation 6 in which certain processes such as heat conduction, taking into account the latent heat of fusion are neglected. The curves thus define a surface that matches the conditions of existence of the spot. This model has been applied (eg [20]) and supplemented by its author [21] which will show the existence of several surfaces corresponding to the problem and propose a model with several areas to go from the surface of the electrode to the plasma arc column.

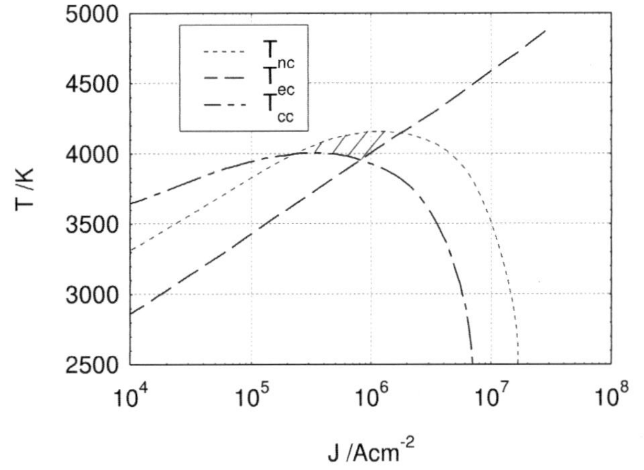

Figure 2. Example for a copper cathode for an arc in a vacuum. The area of existence is the curvilinear triangle (hatched area) created by the three curves.

3.2. Models with multiple zones

The models summarized in § 3.1 describes a cathode spot moving in stationary mode that is to say, not taking into account the phases of creation, evolution and extinction. Fursey's work & all [22, 23] show the initiation and maintenance of the cathode spot on micro tip place of existence of a high electric field favouring the electron emission. More advanced models [24-31] have been developed in this way leading to the identification of spots of type I corresponding to a model explosive and fast-moving spots and type II corresponding to a quasi-stationary state and moving slowly.

3.2.1. Model with three zones. As shown in the diagram in Figure 3 these authors [32] have chosen a spherical model applied to a cavity of radius r_0 dug into the metal electrode.

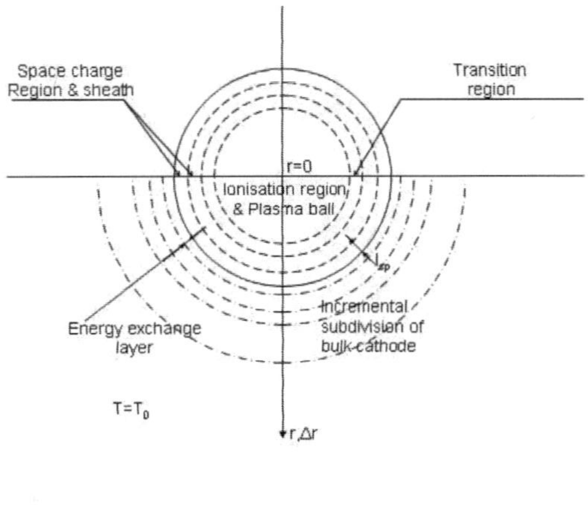

Figure 3. Cathode spot geometry and model zones

The plasma ball is connected to the metal through three zones as in the model defined by Ecker [21]: a space charge region lsp, an inertia sheath, and a transition region. The equations of Ecker's article is here used to characterize the "plasma region, except the equation for the emission of electron current where the Hantzsch's formula [33] more convenient for calculations was used:

$$J_e = K(AT^2 + BE^{9/8})\exp[-(CT^2 + DE^2)^{-1/2}] \tag{14}$$

the coefficients A, B, C, D depend on the metal emissive use. The Nottingham effect was taken into account in the heat exchange with the surface of the material.

The conservation equations for the complete system, associated with the initial conditions and boundary conditions give, by numeric resolution, the evolution with time of the modelled spot. Among the equations used is the one giving the thickness of the space charge region:

$$l_{sp} = \left(\frac{4\varepsilon_0}{9}\right)^{1/2}\left(\frac{2e}{m_i}\right)^{1/4} V_c^{3/4}\frac{1}{\sqrt{J_i}} \tag{15}$$

For the energy input by the ions, rather than the relationship 7, the following relationship:

$$P_i = J_i(V_c + V_i) + J_i(\Delta H_v / N_A.e) \tag{16}$$

takes into account the second term of the condensation energy ions (ΔH_v), N_A is Avogadro's number. Temperature to calculate ΔH_v is the boiling point of the metal under atmospheric pressure. Another input of energy, often neglected by the way, is taken into account, it is the Joule effect which the volume density ψ_J is given power by the classical relationship:

$$\psi_J = (J_e + J_i)^2 \rho \tag{17}$$

ρ is the electrical conductivity, it depends on temperature.

For the energy dissipation given by relations 16 and 17, the authors take into account the thermal radiation (Stefan's law) and the heat conduction including phase changes.

The vapour pressure p emitted leads to the rate of evaporation Γ of the material:

$$\Gamma = \frac{mv}{\sqrt{2\pi mkT}} \tag{18}$$

To solve numerically the equation corresponding to the Fourier heat flow, the cathode was divided into successive hemispheres (Figure 3). At each calculation step, with the Crank-Nicolson method, the physical quantities depending directly or indirectly from the temperature $T(r,t)$ were calculated: evaporation rate, heat current density and evaporation cooling, material coefficients, electron current density, ions current density, electric field and Joule heat power density. At time $t = 0$ a constant electric field E_0 is applied uniformly over the hemisphere of cathode radius r_0. By imposing limits on physical processes considered the authors obtain results that correspond to the « existence area » , and considering the changing values of the electric field, they obtain « spot initiation area » and « spot operation area ». A large quantity of results is presented as curves where two physical quantities evolve according to a third, considered as a constant, this to plot the curve. The advantage of this model is to introduce time and interactive variables considered, contributing to areas of stability (stationary case) and of instability (thermal runaway). The comparison with experimental results is satisfactory.

IOP Conf. Series: Materials Science and Engineering **29** (2012) 012006 doi:10.1088/1757-899X/29/1/012006

Another example of a three-zone model [34] has the same concept as above, but the difference comes from a planar geometry for the electrode surface (see Figure 4).

For calculations the three areas identified are: the cathode, the space charge region called the sheath and the ionization zone.

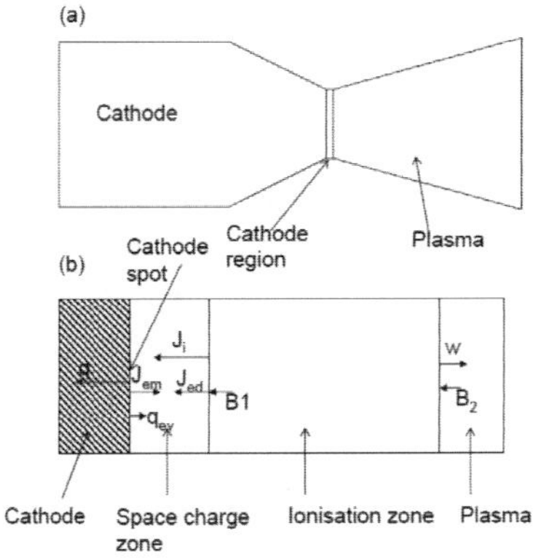

Figure 4. Schematic drawing of the cathode region: (a) the ard-cathode region and (b) the cathode region.

The assumed value of the sheath thickness is between the free path of electrons and ions, it is between 10^{-6} and $10^{-7}\,m$ in this area exist a strong electric field. Three streams of charged particles are considered, one corresponding to electrons emitted by the metal and the other two headed towards the electrode that is to say, the ion flux and flow of backscattered electrons. Electrons under the effect of cathode potential drop acquire energy which gives them a temperature T_e and a Maxwellian distribution, heavy particles emitted by the liquid portion of the electrode have the corresponding temperature T_c. In the ionization zone, the ions and electrons have the same density, but different temperatures, to describe this area the authors use the Steenbeck minimum principle [35], this gives a potential drop in the cathode area equal to:

$$V_t = V_c + V(T_e) + V_0 \tag{19}$$

$V(T_e) + V_0$ represents the potential drop in the ionization zone and V_0 is a constant determined on the boundary B2 depending on the total current and temperature T_e. On this boundary B2 between the ionization zone and the plasma column, the LTE is assumed to be reached.

To describe all the phenomena taken into account the equations mentioned in previous models are included with some modifications, and additional equations are added. On the boundary B1 densities of species and the electron temperature are linked by the Richley Tuma relationship [36]:

$$\frac{n_e n_i}{n_o} = \frac{2Z_i}{Z_0} \frac{(2\pi m_e k T_e)^{3/2}}{h^3} \exp\left(-\frac{\chi_i - \Delta\chi}{kT_e} \right) \tag{20}$$

the lowering ($\Delta\chi$) of the ionization energy is given in reference [37], Z_i and Z_0 are the partition functions of ions and neutral particles.

From the following data: total current, pressure and temperature, the model determines at the boundary B2, the physical quantities: cathode temperature, electron temperature, cathode spot size, thermionic electron current density, ion current density, electric field at the cathode spot, voltage distribution in the space charge zone heat transfer to the cathode and voltage drop in the sheath. For example for a current of $200A$, a tungsten electrode and argon arc at atmospheric pressure the following values are obtained:

- minimum voltage drop of $11.8V$ for an electron temperature of $1.44\times10^4 K$,
- electronic fraction of about 0.8,
- total current density : $7.4\times10^7 Am^{-2}$,
- cathode spot temperature: $3710K$,
- radius of the cathode spot: $9.0mm$,
- electric field: $1.7\times10^8 V/m$,
- evaporation rate: $2\times10^{-5} g/s$.

Contrary to the stationary model I Beilis et al [38] give a three-zone model with time-dependent results. The equations for this model are the result of several approaches [39-42]. For copper and chromium, the results for a cathode fall of $U_C = 15V$, a current of $I = 40A$ and an erosion of $G = 30\mu g/C$ showed a decrease with time of the important parameters of cathode spot (see table 1).

Table 1. Variations of Cathode Surface Temperature, Electric Field and Total Current Density for Two Time Values.

	Cr		Cu	
t (s)	10^{-8}	2×10^{-3}	4×10^{-8}	2×10^{-3}
T_c (K)	4350	3750	4600	3930
E (Vm^{-1})	3.8×10^9	3×10^8	3.8×10^9	1.3×10^9
J (Am^{-2})	3.5×10^{10}	0.5×10^8	4×10^{10}	3×10^9

3.2.2. *More complex models.* In the chapter entitled "Theories of the cathode spots" E Hantzsch [43] shows on a relatively simple diagram, reproduced here in Figure 5, all the problems to be solved to describe the cathode spot in a theoretical approach. In the case of the use of a refractory material (eg W) for the electrode, the liquid layer (2 in Figure 5) does not exist, but it still has metal particles (neutral and ionized) present the cathode spot. The area just after the metal (liquid or solid) is considered by some authors as an area where the flow of ballistic particles are independent of each other, and by others as an area of space charge as in the model presented previously [34]. The following areas will also differ according to many authors, two examples are now given.

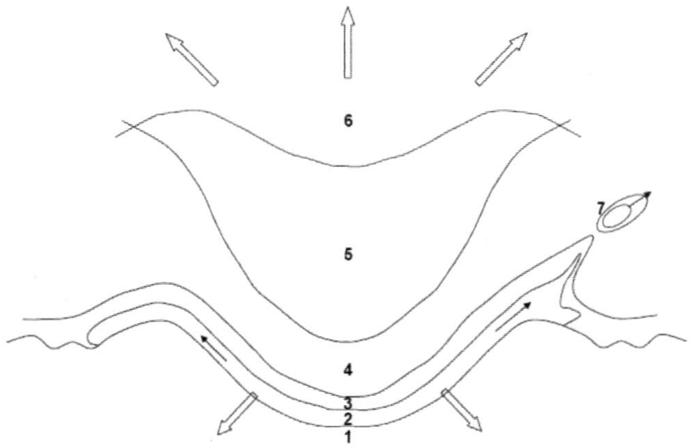

Figure 5. Schematic section of a typical developed cathode arc spot on clean surfaces :1)solid cathode metal metal below the spot (arrows indicate current and heat flow); 2) molten metal layer; 3) space charge layer; 4) ionization and thermalization layer of the spot plasma; 5) dense central spot plasma; 6) plasma expansion region (arrows indicate plasma flow); 7) ejection of molten droplets [43].

- *Model for a vacuum arc [42]*

As shown in Figure 6 this model has 4 areas between the metal electrode and the plasma column.

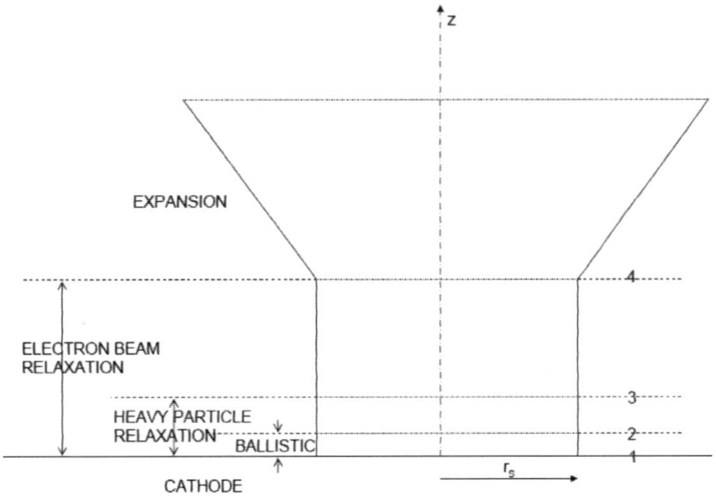

Figure 6. Model of the vacuum arc cathode spot [42].

In the model of the massive cathode spot can be considered stationary during his lifetime. For each zone: cathode, ballistic, heavy particle relaxation, electron beam relaxation and expansion, the author gives the various equations needed to describe the physical phenomena existing. It is possible to give the most features, for a complete set see [42].

IOP Conf. Series: Materials Science and Engineering **29** (2012) 012006 doi:10.1088/1757-899X/29/1/012006

The energy flow S arriving on the surface of the cathode is assumed distributed according to a Gaussian radial distribution [44]:

$$S = S_0 \exp\left(-\frac{r^2}{r_s^2}\right) \tag{21}$$

S_0 represents the average energy flux entering the electrode surface. The calculation of thermal diffusion is performed with the assumption of a semi-infinite medium. The Joule effect in the electrode is also taken into account.

On the surface of the cathode current density resulting from the emission of electrons is given by the TF emission. The balance of energy flow is given very fully, including all transfers: ions falling on the surface, electrons emitted, backscattered electrons, neutral emitted by evaporation, thermal radiation emitted and received. Ballistic area is described by calculating the electric field existing by integration of Poisson's equation, taking into account all the charged particles present in the area. In the heavy particle relaxation zone, atoms emitted from the surface undergo numerous collisions, their distribution function which initially may be treated as a half-Maxwellian will evolve according to the rating z. The conservation of momentum between the limits 1 and 3 is done taking into account the force exerted by the electric field between these two limits.

In the electron beam relaxation zone the electrons emitted by the cathode, accelerated by the cathode voltage drop will undergo collisions with the species causing the appearance of new electrons, and creating ions. The end of this zone (boundary 4) is defined by a uniformity between the local electrons and those from the initial beam. Three assumptions are required to write the equations for this area:

- the density of heavy particles is uniform,
- temperatures of electrons (T_e) and heavy particles (T_i) are different but nearly constant,
- the electric field is relatively weak and does not give any energy to particles.

The used equations are diffusion equations, species conservation and the equation of state gives the pressure:

$$p = k\left[(n_n + n_i)T_i + n_e T_e\right] \tag{22}$$

n_n, n_i, n_e are respectively the densities of neutral ions and electrons.

The plasma expansion zone was supposed conical to simplify the calculations, inside there is Joule heating, and quick speed flows. To be more accurately described it is necessary, in addition to the classical equations of motion using the Saha equation.

This model derives a complete system of 17 equations with 17 unknowns related, however, several approaches have been used to solve such a system. The hydrodynamic model [38, 44, 45] only takes into account that 9 equations with 9 unknowns related, but empirical values must be taken to the cathode fall and erosion. The heavy particle model [46, 48] uses 15 equations including the equation of Saha. However, the full model solved without empirical values is called kinetic model , the results are purely theoretical [49, 50]. For massive electrodes materials with mid-range thermo-physical properties (Cu, Ni, Ag ...) equations lead to a solution, it is not the case for tungsten and mercury that have properties more extreme. For spots crossed by high current ($200A$), two modes appear as high and low values of current density J, they are modes 1 and 2 and they correspond to Ecker's modes 0 and 1. Table 2 lists some of these results.

Table 2. Spots Parameters for the Massive Cathode

Material/ Current (A)	Mode	s	T_s (K)	E (V/m)	J(A/m^2)
Cu/200	1	0.450	3720	5.4×10^8	5.3×10^8
Cu/200	2	0.997	5110	5.75×10^9	1.6×10^{12}
Ag/200	1	0.487	3530	6.97×10^8	7.8×10^8
Ag/200	2	0.99	4990	7.2×10^9	5.88×10^{12}
Ni/200	1	0.356	3470	1.1×10^9	1.8×10^7
Ni/15	1	0.4	3890	4.9×10^8	3.8×10^8
Ni/15	2	0.775	4650	2.7×10^8	2.76×10^{10}

The hydrodynamic model also applies to protrusions, eg for a cylindrical protrusion radius and height of $2\mu m$ I.I. Beilis shows that for a given erosion, calculation is possible versus time, for example with copper for a spot mode 1 the current density will vary from $J = 10^{10} Am^{-2}$ at $t = 1\mu s$ to $J = 8\times10^9 Am^{-2}$ at $t = 2\mu s$.

For refractory materials, the zone of space charge due to electrons (see Figure 7) creates an electric field that directs the electrons towards the metal electrode in the area $0 < z < z_m$ and accelerates them in the opposite direction in the area $z > z_m$. It appears as a virtual cathode from which electrons are emitted, the current density obtained is given by: $J_e = J_{eo}\exp(-\Delta V / kT)$, and in the equation resulting from the integration of the Poisson law it is necessary to replace V_C by $V_m = V_c + \Delta V$. The resolution of the hydrodynamic model is then possible and results are available for metals such as W [51] or Gd [52].

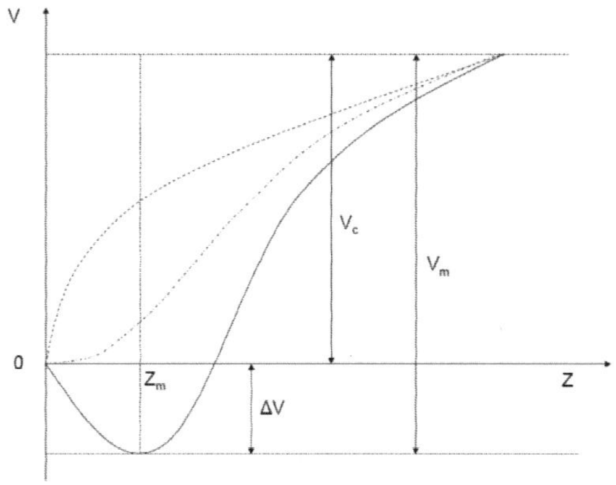

Figure 7. Electrical potential with respect to the cathode in the space charge sheath as a function of distance from the cathode. Upper curve is typical for materials with moderate thermophysical properties, while the middle curve, characteristic of a space charge limited diode, represents Gd, and the lower curve, showing a virtual cathode represents W [42].

IOP Conf. Series: Materials Science and Engineering **29** (2012) 012006 doi:10.1088/1757-899X/29/1/012006

For volatile materials such as mercury, the equations of the model does not allow the emission of electrons, to overcome this shortcoming, the author adds an additional area called "plasma cathode" between the metal electrode and the "ballistic zone".

Two types of spots appear: spots of type I and spots of type II , their characteristics are briefly summarized in Table 3 [53]. Type II characterizes spots with very low speed, with special conditions, they are usually in thermal equilibrium [54].

Table 3. Spot Types [53].

Characteristic	Mobile Type I		Slow Type II
	Type I-1	Type I-2	
Conditions for Occurence	contaminated surfaces	clean (degassed surfaces)	small electrode, ambient gas, hot cathodes, anchoring at inclusions
Velocity (m/s)	10-1000	1-150	<1
Spot Current (A)	<10	5-50	5-50
Lifetime (μs)	<0.01	<0.1	>10
Appearance of Craters	discontinuous	overlapping	overlapping
Crater Radius (μm)	≤1	3-10	>10
Erosion Rate (μg/C)	<10	10-100	>100
Available Theories	none	Non-stationary	Quasi-stationary

In spots of type I, only the spots of type I-2 are more present and more reproducible in the breaking vacuum arcs [55].

- *Model for high pressure arcs [56]*

The model developed is primarily related to the arcs produced in a high pressure gas (eg mercury) with electrodes primarily refractory (W). All equations for the conservation of species and the flows, are treated in a space between the electrode to the arc column as shown in Figure 8 [57].

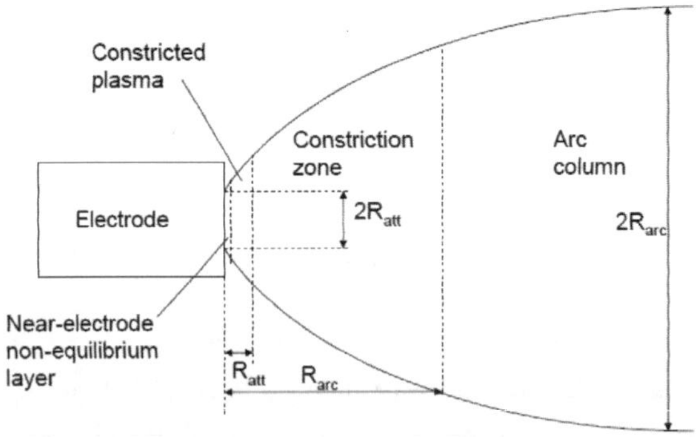

Figure 8. Structure of the near-electrode perturbation region

The constricted plasma zone is considered to be close to LTE which is a crucial assumption to calculate the characteristic parameters, the Saha equation can then be applied. Simplifying assumptions are added, in particular the author shows that the Joule heating and energy loss by radiation are the dominant phenomena, this give the relation $JE = w_{rad}$. As in the models studied in the previous paragraphs a law electron emission and a power balance are set to the cathode surface. Figure 9 shows the areas that appear when we study the behaviour of the species given eg ionization phenomena leading to differences in temperatures (respectively T_h for the heavy particles, T_e for electrons) and densities of particles (n_i for ions, n_e for electrons).

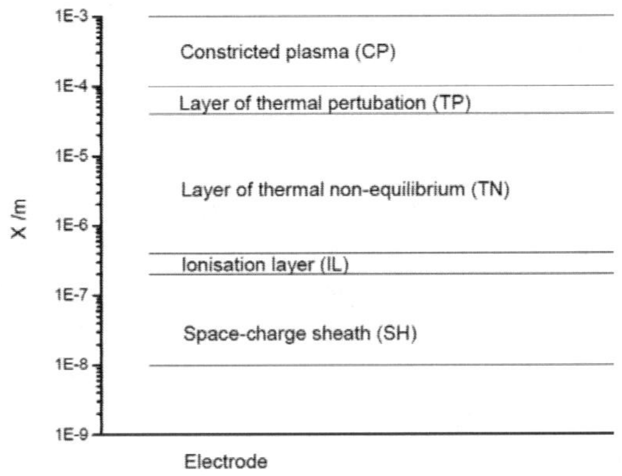

Figure 9. Structure of the near-cathode non-equilibrium-layer, Hg plasma, W cathode, p=30 bar, $J = 10^7 Am^{-2}$ [57].

The distribution of different parameters is the result given by the unified modelling [58], among characteristic values obtained, it should be noted for example, the electric field maximum value of $2.5\times10^8 V/m$, a slow evolution of the temperature of heavy particles ($< 10^4 K$), the passing by a maximum value of the electron temperature ($\cong 7\times10^4 K$) shows their acceleration by the electric field, then their thermalization to the temperature of the plasma column. Table 4 shows the behaviour of different species and the validity of certain assumptions.

Table 4. Fine structure of the near-electrode non-equilibrium layer.

	$n_e = n_i$	$n_e = n_{Saha}$	$T_e = T_h$	$JE = w_{rad}$
Constricted plasma	Yes	Yes	Yes	Yes
Layer of thermal pertubation	Yes	Yes	Yes	No
Layer of thermal non-equilibrium	Yes	Yes	No	No
Ionisation layer	Yes	No	No	No
Space-charge sheath	No	No	No	No

This new approach to description of the near-cathode layer, developed in the work [56], presents a difference with the traditional approach (for example, [42] and many others). This new approach is based on the modeling of the whole of a near-cathode region in the framework of a single set of

IOP Conf. Series: Materials Science and Engineering **29** (2012) 012006 doi:10.1088/1757-899X/29/1/012006

equations without simplifying assumptions, such as a thermal equilibrium, ionization equilibrium and quasi-neutrality, and without dividing the near-cathode region *a priory* into different sub-layers.

Under experimental conditions the mode of attachment on the cathode electrode can vary and be in the form diffused mode or constricted fashion. In [59] the authors show that the mode changes are varied in nature and will occur with different time scales, for the near-cathode plasma velocities of moving particles give them the time change well below the microsecond, while the change of state of the metal electrode is comparatively much longer [60, 61]. The model in numeric form gives the solution for different types of plasma and electrodes with different natures and different shapes, for the spot model or for the diffuse model it is possible to achieve results directly at the address given in [62].

4. Conclusion

The thread that has been followed here shows the evolution of theoretical work on the cathode zone of an electric arc. Problems related to the electron emission at the electrode, now appear well resolved, if the values of physical quantities that exist on the surface of the material are well defined. However, the problem of taking into account the change of the potential barrier for ions very near the surface should be reviewed. In the organization of the area then between electrode surface and the plasma column the space-charge sheath is fairly well defined, and most authors regard it as an area where the particles (electrons, ions and neutral) moved without interaction between them, and where the electrons are accelerated by the electric field. The following area takes charge of many tasks: thermalization of electrons and heavy particles, ionization of neutral, and plasma expansion. We have seen changes in the number of subdivisions of this area in order to better describe and put it into equations, the results are difficult to control experimentally for the reasons given in the introduction. It would be preferable to adopt a mesh for the area and solve the equations at each node needed, which amounts to much finally multiply the number of zones. The time factor appears timidly and most models are for steady. This item is to pose the problem of initiation of the arc, whether by creating a glow discharge by breakdown or rupture of a melted bridge, the origin of energy to ensure the early passage of electrons in the non-metal also remains a problem to solve.

References

[1] Richardson O W 1921 *The emission of Electricity from Hot Bodies* (Longmans Green and Company London)
[2] Raizer Y P *1991 Gas Discharge Physics (*Berlin: Springer)
[3] Murphy E L and Good R H 1956 *Phys. Rev* **102** 1464
[4] Testé Ph and Chabrerie J-P 1996 *J. Phys. D: Appl. Phys.* **29** 697
[5] Coulombe S and Meunier J-L 1997 *J. Phys. D: Appl. Phys.* **30** 2905
[6] Gayet R, Harel C and Jouin H 1996 *J. Phys. D: Appl. Phys.* **29** 3063
[7] Lee T H and Greenwood A 1961 *J. Appl. Phys.* **32** 916
[8] Reece M P 1958 *Nature* **181** 475
[9] Mackeown S S 1929 *J. Appl. Phys* **102** 1464
[10] Benilov M S and Marotta A 1995 *J. Phys. D: Appl. Phys.* **28** 1869
[11] Jeanvoine N and Muecklich F 2009 *J. Phys. D: Appl. Phys.* **42** 1
[12] Dushmann S 1949 *Vacuum Technique* (John Wiley & Sons, Inc., New York 1949)
[13] Lee T H 1960 *J. Appl. Phys* **31** 924
[14] Nottingham W B 1936 *Phys. Rev.* **49** 78
[15] Charbonnier F M, Strayer R W, Swanson L W and Martin E E 1964 *Phys. Rev. Lett.* **13** 397
[16] Swanson L W, Crouser L C and Charbonnier F M 1966 *Phys. Rev.* **151** 327
[17] Paulini J, Klein T and Simon G 1993 *J. Phys. D: Appl. Phys.* **26** 1310
[18] Ecker G 1971 *Z. Naturforsch.* **25a** 935
[19] Ecker G 1970 *G.E.-R & D-Report* N°70 – C 213
[20] Thiell G and Braun J C 1976, *Rev. Phys. Appl.* **11** 609

[21] Ecker G 1973 *Z. Naturforsch.* **28a** 417
[22] Fursey G N and Vorontosov-Vel'yaminov P N 1968 *Sov. Phys.-Tech. Phys.* **12** 1370
[23] Fursey G N and Zhukov V M 1976 *Sov. Phys.-Tech. Phys.(USA)* **21** 310
[24] Mitterauer J 1972 *Proc. 2th INT. Conf. Gas Discharges (London England) p* 215
[25] Mitterauer J 1973 *Acta Phys. Austr.,* **37** 175
[26] Beilis I I, Kanzel V V and Rakhovsky V I 1974 *Proc. 6th All-Union Conf. Electrical Contacts* (Moscow USSR)
[27] Kanzel V V and Rakhovsky 1974 *Proc. VIth Int. Symp. Discharges and Electrical Insulation in Vacuum* (Waterloo Canada)
[28] Kesaev I G 1963 *Sov. Phys.-Tech. Phys.* **8** 447
[29] Mesyats G A, Litvinov E A and Proskurovsky D I 1970 *Proc. IVth Int. Symp. Discharges and Electrical Insulation in Vacuum* (Oxford England)
[30] Mesyats G A 1971 *Proc. 10th Int. Conf. Phen. Ionised Gases* **1** 333
[31] Rakhovsky V I 1976 *IEEE Trans. Plasma Sci.* **PS-4** 81
[32] Mitterauer J and Till P 1987 *IEEE Trans. Plasma Sci.* **PS-15** N°5 488
[33] Hantzschke E 1982 *Beitr. Plasmaphys.* **22** 435
[34] Zhou X and Heberlein J 1994 *Plasma Sources Sci. Tecchnol.* **3** 564
[35] Finkelnburg W and Maecker H 1956 *Handbuch der Physik* **22** 254
[36] Richley E and Tuma D T 1982 *J. Appl. Phys.* **53** 537
[37] Griem H R 1964 *Plasma Spectroscopy* (New York: McGraw-Hill)
[38] Beilis I , Djakov BE, Jüttner B and Pursch H 1997 *J. Phys. D: Appl. Phys.* **30** 119
[39] Beilis I 1974 *Sov. Phys.-tech. Phys.* **19** 251
[40] Beilis I 1985 *IEEE Trans. Plasma Sci.* **13** 288
[41] Beilis I 1988 *Izv. Sibir. Otdel. AKad. Nauk* **7** 58
[42] Beilis I 1995 *Handbook of Vacuum ArcScience and Technology* ed. R.L. Boxman and al (New Jersey: Noyes) **3** 208
[43] Hantzsche E 1995 *Handbook of Vacuum ArcScience and Technology* ed. R.L. Boxman and al (New Jersey: Noyes) **3** 192
[44] Litvinov E A, Parfyonov A G and Shlemev D L 1992 *Proc.XVth Int. Symp. Disch. Electr. Insul. Vacuum (Darmstadt)* p 326
[45] Beilis I I, Lyubimov G A and Rakhivski V I 1972 *Sov. Phys. Doklady* **17** 225
[46] Beilis I I 1977 *IIIrd Int. Symp. Switch. Arc Phen.,(Lodz)* 194
[47] Beilis I I 1985 *IEEE Trans. Plasma Sci.* **13** 288
[48] Beilis I I 1986 *High Temp.* **24** 319
[49] Beilis I I 1982 *Sov. Phys. Doklady* **27** 150
[50] Beilis I I 1988 *Izvest. Siberian Akad. Nauk. Ser. Tekh. Nauk.* **7(2)** 64
[51] Puchkarev V F, and Murzakayev A M 1990 *J. Phys. D* **23** 26
[52] Paranin S N, Polishchuk V P, Sychev P E, Shabaschov V I and Yartsev I M 1986 *High Temp.* **24** 307
[53] Puchkarev V F 1995 *Handbook of Vacuum ArcScience and Technology* ed. R.L. Boxman and al (New Jersey: Noyes) **3** 257
[54] Rakhovsky V I 1976 *IEEE Trans. Plasma Sci.* **PS-4** 81
[55] Hantzsche E, Jüttner B and Pursch H J 1983 *J. Phys. D: Appl. Phys.* **16** 173
[56] Almeida N A, Benilov M S and Naidis G V 2008 *J. Phys. D: Appl. Phys.* **41** 245201
[57] Benilov M S 2008 *J. Phys. D: Appl. Phys.* **41** 144001
[58] Almeida N A, Benilov M S and Naidis G V 2007 *Proc. 28th ICPIG* (Prague, July 2007*)* p 1797
[59] Benilov M S, Cunha M D and Faria M J 2009 *J. Phys. D: Appl. Phys.* **42** 145205
[60] Hartmann T, Günther K, Lichtenberg S, Nandelstädt D, Dabringhausen L, Redwitz M and Mentel J 2002 *J. Phys. D: Appl. Phys.* **35** 1657
[61] Mitrofanov N K and Shkol'nik S 2007 *Tech. Phys.* **52** 711
[62] http://www.arc_cathode.uma.pt

IOP Conf. Series: Materials Science and Engineering **29** (2012) 012007 doi:10.1088/1757-899X/29/1/012007

Vacuum cathode arc root for a current of 10A on a silver contact

A Lefort [1,2] and M Abbaoui[1,2]

(1) Clermont Université, Université d'Auvergne, LAEPT, BP 10448, F-63000 CLERMONT-FERRAND
(2) Clermont Université, Université Blaise Pascal, LAEPT, BP 10448, F-63000 CLERMONT-FERRAND

E-mail: Andre.Lefort@univ-bpclermont.fr, Mhammed.Abbaoui@univ-bpclermont.fr

Abstract. The cathode spot is theoretically analyzed here for a 10A current and a silver contact in vacuum. The main equations leading to the values of physical characteristics are detailed and resolved according to the radius r in the domain 0.5-1.5μm. The choice of this field is imposed by the limits of the current density and time of appearance of the metal vapour. The findings show that when they are created the spots have cathode radius that fall to low values of field. Under the action of saturated vapour pressures higher radius of the spot will grow but its growth is limited by its lifetime.

1. Introduction

The interruption of electric current in vacuum has been under consideration since the 1920s, however, for technical reasons the first devices have emerged in the 1960s. In typical applications, at the contact separation, is formed an arc which is extinguished at the zero crossing of current, dielectric strength of vacuum insulation will induce the contact insulation and the actual interruption of the current. The arc corresponding to such devices is characterized by the absence of anode and a column consisting essentially of atoms and metal ions and electrons. The cathode consists of several spots separated from each other and very mobile on the surface of contact. The first serious theoretical study of these cathode spots is attributed to Lee and Greenwood [1], then we find the works of Mitterauer and Till [2] and an analysis of both theoretical and experimental works done by Boxman, Martin and Sanders [3]. Two types of analysis occur, those on a stationary phenomenon first appeared [4], and those showing the dynamics of phenomena over time (eg [5]). However, both the experimental design that theoretically, the radius of the cathode spot remains undefined. The measurements of traces on the electrodes can include cathode spot movements and then give too low values, they can also depend of the resolution of the experimental tools used for the traces analysis. The measurements of light emitted at this level give, because of time-excitation species transmitting, values too high. The works of B. Jüttner [6] show a more complex structure of cathode spots, they are composed of fragments each carrying currents whose intensity can go up to and whose lifetime varies from a few nanoseconds to a few tens of nanoseconds.

The analysis presented here relates to a cathode spot carries a current of 10A, operating on a silver electrode located in a vacuum. We assume that the steady state is reached, this will be the subject of discussion from the results obtained. Initially we apply the theory of cathode spot based on the energy exchange between the base of the arc column and the surface of contact material, the first calculation leads to the value of the flow of metallic vapor leaving the metal electrode surface. In a second step we calculate by applying the law of Clapeyron, the flow of vapor given this time by metal evaporation in a vacuum, that is to say in the presence of its saturated vapor. With a value of cathode spot radius given in the two calculations and with the temperature as single parameter, it is determined when the two flows become identical, all other parameters of cathode spot are then determined.

2. Theoretical analyse

In this section we give the main equations related to the operation of the cathode spot, and the method of solving the system of equations obtained.

2.1. The main equations

The metal surface of the cathode ensures the passage of electric current between the arc column and the metal electrode. We assume that the current density is uniform throughout the emission site (spot) which is plan, circular with radius . Hence, by definition:

$$J = \frac{I}{\pi r^2} \tag{1}$$

The works of Langmuir [7] indicate the presence of an electric field E very intense, appearing on the electrode surface, orthogonal to it and in front of the space charge located in an area called the relaxation zone ions, beginning at approximately 0.001µm of the surface of the electrode and ending at approximately at 0.01µm of the same surface [8]. Mackeown equation [9] gives the value of the electric field:

$$E^2 = \frac{4}{\varepsilon_0} J\left[(1-s)\left(\frac{M}{2e}\right)^{1/2} - s\left(\frac{m}{2e}\right)^{1/2}\right] V_C^{1/2} \tag{2}$$

with ε_0 permittivity of vacuum, s electronic fraction, e electron charge in absolute, M mass of a metal ion, m electron mass and V_C cathode fall. At the metal surface, the current flow is provided by two opposing flows: a stream of electrons emitted by the surface and a flux of ions accelerated by the electric field, the absolute values of these current densities corresponding to these two streams are respectively J_e and J_i. The electronic fraction is defined by the relationship:

$$s = \frac{J_e}{J} = \frac{J_e}{J_e + J_i} \tag{3}$$

Under the effect of electric field E and the surface temperature T, the metal surface emits electrons which flow is usually given by the relationship of Richardson-Schottky or Fowler-Nordheim [10] into areas of specific values of temperature and electric field. The relationship of E.L. Murphy and R.H. Good [11] gives the value of the emitted current density for all values of E and T, and gives by approximation relation values previously cited. Its overall formulation is as follows:

$$J_e = J_{MG} = e\int_{-W_a}^{\infty} D(W,E)\,N(W,T)\,dW \tag{4}$$

with:

W_a: potential energy of the electron in the metal bottom of the conduction band;

$D(W,\ E)$: probability of crossing the potential barrier for an electron with energy;

$N(W,T)$: number of electrons entering the barrier per unit area and per unit time.

After determining these magnitudes relation (4) becomes:

$$J_{MG} = \frac{4\pi mkTe}{h^3}\left\{\int_{-W_a}^{W_l} \frac{\ln\left[1+\exp\left(-\frac{W+U}{kT}\right)\right]}{1+\exp\left(\frac{8\pi\sqrt{2m}}{3}\frac{e^{5/4}}{(4\pi\varepsilon_0)^{3/4}\times E^{1/4}\times h}y^{-3/2}v(y)\right)}dW + \int_{W_l}^{\infty}\ln\left[1+\exp\left(-\frac{W+U}{kT}\right)\right]dW\right\} \quad (5)$$

with:
k : Boltzmann constant;
h : Planck constant;
U : work function.

When energy W becomes equal to or greater than the magnitude W_l, transmittance $D(W,E)$ of the potential barrier becomes equal to unity. Its value depends on the electric field E:

$$W_l = -\sqrt{\frac{e^3 E}{8\pi\varepsilon_0}} \quad (6)$$

In equation (5) the parameter y is defined by:

$$y = \frac{1}{|W|}\sqrt{\frac{e^3 E}{4\pi\varepsilon_0}} \quad (7)$$

and the function $v(y)$ is calculated using elliptic integrals $K(y)$ and $G(y)$:

$$v(y) = -\sqrt{\frac{y}{2}}\left\{-2G\left(\sqrt{\frac{y-1}{2y}}\right)+(y+1)K\left(\sqrt{\frac{y-1}{2y}}\right)\right\} \quad \text{if } y>1 \quad (8)$$

$$v(y) = -\sqrt{y+1}\left\{G\left(\sqrt{\frac{1-y}{y+1}}\right)-yK\left(\sqrt{\frac{1-y}{y+1}}\right)\right\} \quad \text{if } y<1 \quad (9)$$

with:

$$G(x) = \int_0^{\pi/2}(1-x^2\sin^2\theta)^{1/2}d\theta \quad (10)$$

$$K(x) = \int_0^{\pi/2}(1-x^2\sin^2\theta)^{-1/2}d\theta \quad (11)$$

The metal ions accumulated in an area defined above, fall on the metal surface and give their energy to the surface, the synthesis of the work of several authors ([12,13] eg), leads to the flow of energy corresponding:

$$P_i = (1-s)\left(\frac{5kT_i}{2e}+V_C+V_i-\frac{U}{e}+\frac{W_{ev}}{e}\right)J \quad (12)$$

with:
T_i : ion temperature in relaxation zone,

V_i : ionization potential of the metal contact,

W_{ev} : energy needed to vaporize a metal atom at temperature.

To determine the values of W_{ev} we used the enthalpy change shown in figure 1 [14] for silver at atmospheric pressure, the dotted curve is relative to the enthalpy of the material in its liquid form at temperatures above $2433K$ that corresponds to pressures above atmospheric pressure. The energy W_{ev}, for saturated vapor pressure more than atmospheric pressure, is determined, for the temperature T given, by the difference between the data for liquefaction curve (dashed) and the curve for the enthalpy of saturated vapor.

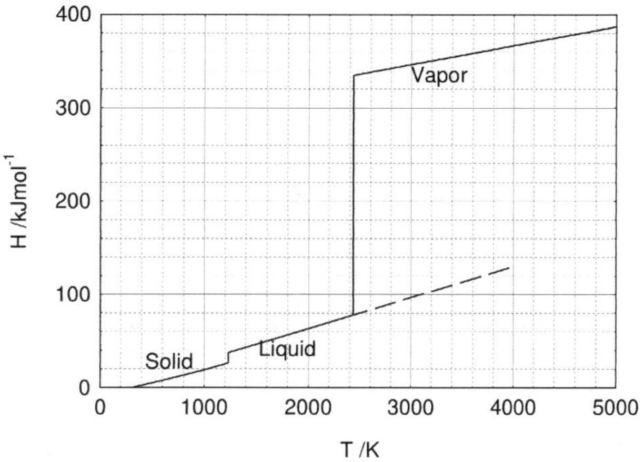

Figure 1. Enthalpy variations versus temperature for silver [14]

The energy flux P_i given by the relation 12 is transmitted to the electrode surface, it will dissipate in three terms [15, 16]:

$$P_i = P_{th} + P_{ec} + P_{rs} \tag{13}$$

P_{th} is the flow of energy transferred thermally,

P_{ec} is the flow of energy for electrons to leave the metal surface,

P_{rs} is the flux of energy radiated by the surface area occupied by the spot.

As a first approximation we consider that the flow P_{th} is dominant, and the equation used for the transfer of energy flow to the metal is reduced to:

$$P_{th} = P_i \tag{14}$$

Solving the equation 14 [17,18] gives the temperature evolution of the metal electrode in each of its points and the phase changes that are involved and particularly the value of vapor flow Φ_{th}.

Clapeyron's law relates the pressure of saturated vapour at the temperature of the surface of the liquid that is vaporized:

$$p_1 = p_0 \times \exp\left[-\frac{L_v}{R}\left(\frac{1}{T_1} - \frac{1}{T_0}\right)\right] \tag{15}$$

with:

L_v : value of W_{ev} at atmospheric pressure,
R : perfect gas constant,
p_0 : pressure of saturated vapor at temperature T_0,
p_1 : pressure of saturated vapour at temperature T_1.
The resulting vapour stream is obtained by the law of statistical physics:

$$\Phi_{vap} = \frac{p_1 \pi r^2}{\sqrt{2\pi M k T_1}} \tag{16}$$

2.2. Method of calculation
The organization chart shown in figure 2 summarizes the approach used. The current value I is initially set, the results are based on the cathode radius spot. For a given value of r, the assumed uniform current density J is calculated. Initially the value of the surface temperature T is fixed. For this value of temperature the current density J given by the equation of Murphy and Good (equation 5) is calculated, as well as the electronic fraction s (equation 2), for different values of electric field E, until the same values for J_{MG} and J. The energy flux P_i is calculated, in first hypothesis the temperature of the ions T_i and of the surface T are considered identical, this flux is applied to electrode surface, when spraying occurs, the vapour flux $\Phi_{th}(P_i)$ emitted is determined and compared to that given by the relationship after the Clapeyron formula (Equation 16). The temperature value T is incremented until the equality of the two vapor flux:

$$\Phi_{th}(P_i) = \Phi_{vap}(T) \tag{17}$$

Thus all physical quantities characteristic of cathode spot defined above, for a given value of radius r, are determined.

2.3. The equations applied to silver
Table 1 gives the values of the potentials associated to silver.

Table 1. Relative potentials to silver.

Potential	Values /V
Cathode fall V_C	12.3 [19]
Ionization V_i	7.56 [20]
Work function U	4.63 [20]

Given the values of physical constants necessary, equation 2 leads to the value of s when the electric field E and the density of total current J are known:

$$s = 1 - \frac{E^2}{1.1858 \times 10^9 \times J} \tag{18}$$

To obtain this equation we made the approximation: $\sqrt{m} << \sqrt{M}$.
Given the values given in figure 1, the magnitude of W_{ev} in equation 12 is obtained (in J / atom) by the numerical relation:

$$W_{ev} = \left[\left(285.606 + 2.24776\times10^{-2} \times T\right) - \left(\begin{array}{l} -1.40612\times10^{-12}\times T^3 + 8.20117\times10^{-9}\times T^2 \\ +0.0334562\times T - 3.58406 \end{array} \right) \right] 1.6606\times10^{-21} \quad (19)$$

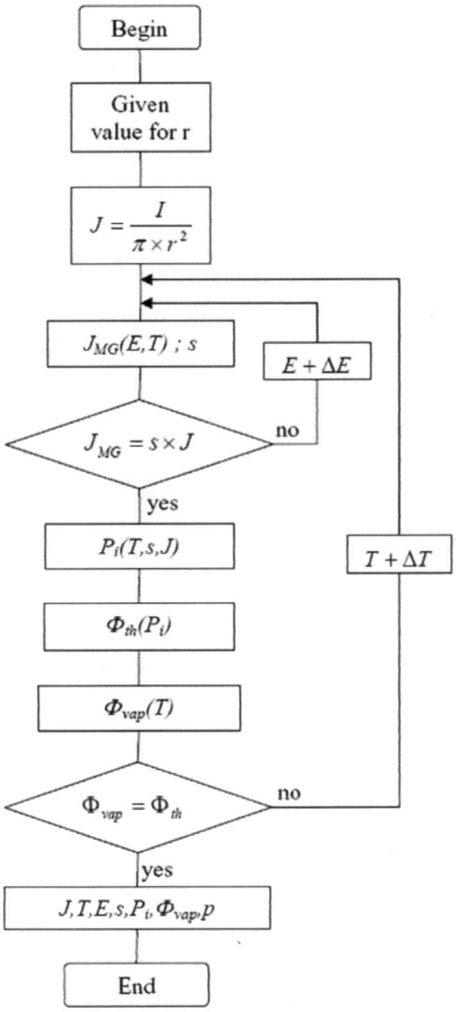

Figure 2. Organization chart used, the current value is 10A.

The functions in parentheses correspond to the curves giving the enthalpy of vaporization and enthalpy of fusion for silver.
When the flux of energy P_i is determined, it is applied to the material surface, a first phase change shows the molten metal, then a second phase change gives rise to the issue of a vapour flow. An example of this flow of vapour is given in figure 3 for two values of the radius of the spot cathode $0.5\mu m$ and $1.5\mu m$. The moments of appearance of vapour are significantly different, they are respectively $0.5ns$ and $16.2ns$, after a short transition phase two curves show a linear part whose slope

leads to the value of the mass flux emitted $\Phi_{th}(P_i)$. This value is compared to the mass flux emitted by a circular area of radius r extent, in the presence of its saturated vapour at temperature T :

$$\Phi_{vap}(T) = 1.4466\times10^4 \times \frac{r^2}{\sqrt{T}} \times \exp\left[-\left(\frac{1}{T}-\frac{1}{2433}\right)\times3.1611\times10^4\right] \quad (20)$$

When equality 17 is satisfied the calculation is finished for the given value of the radius r.

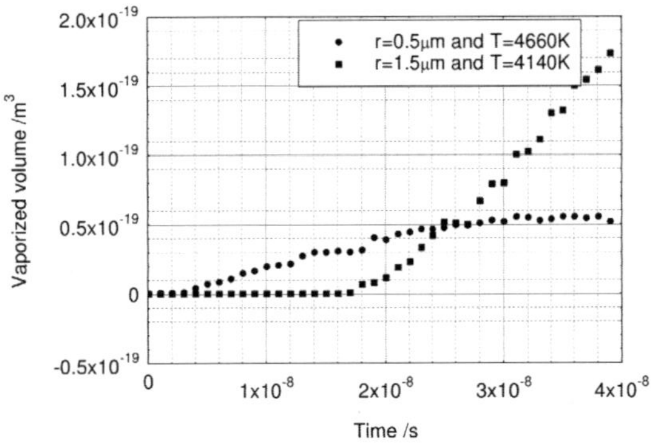

Figure 3. Vaporized volumes with energy flux respectively equal to $3.25\times10^{12}Wm^{-2}$ and $1.00\times10^{12}Wm^{-2}$ for radius equal to $0.5\mu m$ and $1.5\mu m$.

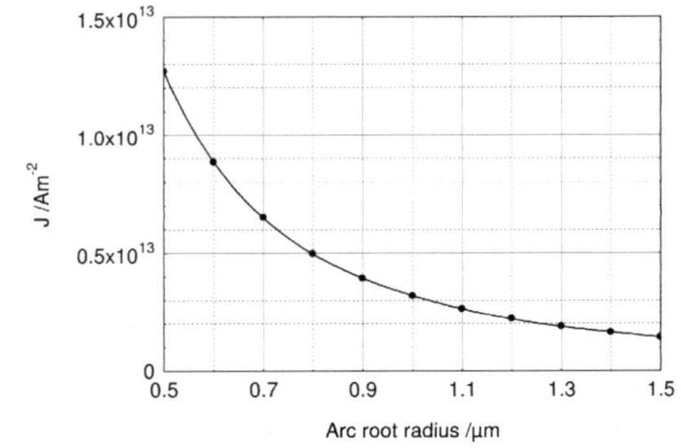

Figure 4. Variations of total current density.

3. Results

Given the literature review performed, it seems reasonable to limit the variation of r to the interval: $0.5\mu m - 1.5\mu m$, the analysis results will reinforce this choice. The current density, whose values are shown in figure 4 satisfies the criterion of maximum value equal to $1\times10^{14}Am^{-2}$ [2].

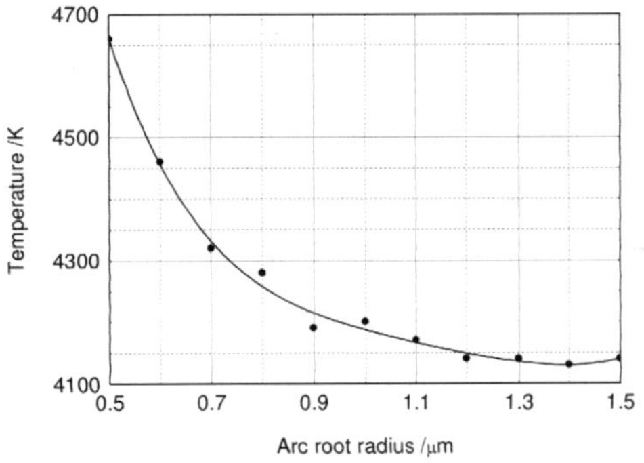

Figure 5. Surface temperature variations.

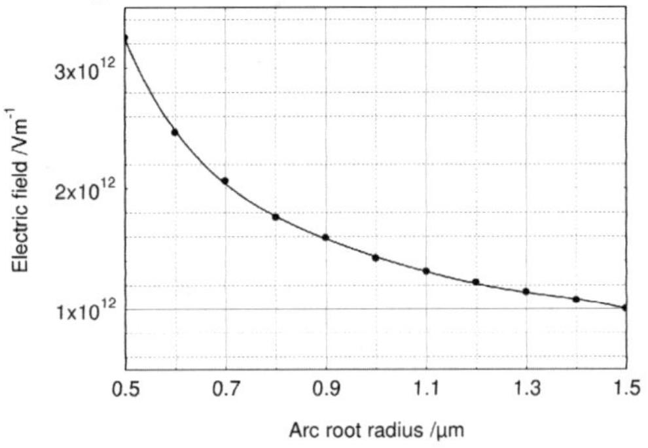

Figure 6. Surface electric field versus arc root radius.

Figure 5 gives the surface temperature of the liquid bath, it decreases with increasing radius of the cathode spot, this result seems logical it is to the same meaning as in figure 6 which gives the values of the electric field. As the total current density decreases it is logical that the emitted current density decreases giving a decline in values with temperature T and electric field E. However Figure 7 showing the variation of the electronic fraction could go in the opposite direction of this result, because s varies in the same direction as the ion current density J_i, i.e. in the same direction as the flow of ions arriving on the surface of the electrode, but the variations of are limited (approximately

2%). This is explained by the fact that the electric field decreases. The energy flux density arriving on the metal surface, represented in figure 8 decreases as the radius increases, which is quite logical given

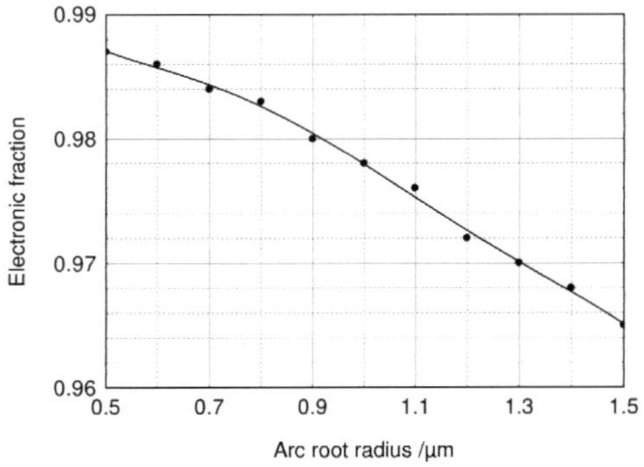

Figure 7. Electronic fraction variations versus arc root radius.

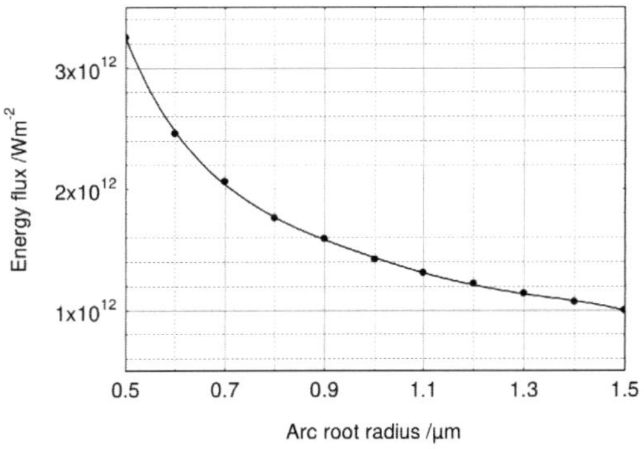

Figure 8. Energy flux variations with arc root radius.

the variations of temperature and electric field. The power P_t (in W) supplied to the electrode surface is easily calculated by the equation:

$$P_t = P_i \pi r^2 \tag{21}$$

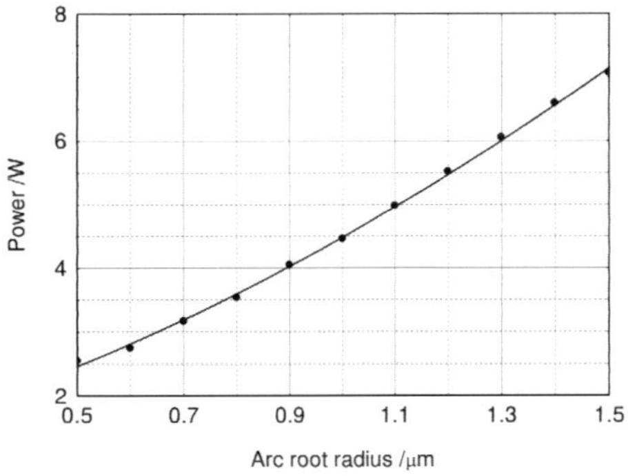

Figure 9. Total power applied through the cathode spot at the electrode surface.

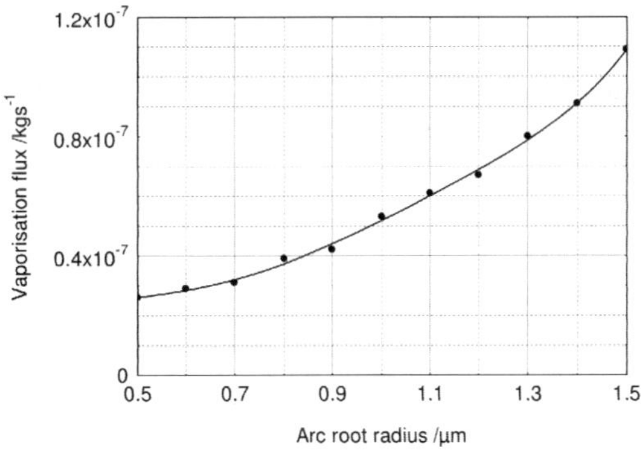

Figure 10. Variations of the vaporisation flux versus arc root radius.

Figure 9 shows the variations of this power with r, evolution is increasing although Figure 8 shows a falloff in the energy flow, this is the result in a low decrease in energy flux and change with r^2 of power P_t (Equation 21). The flow of vapour versus the radius r, shown in figure 10, shows a significant variation (43%) of the flow of vapour emitted from the low values and high values of the radius. The result (figure 11) giving the values of saturation pressure of the metal which evaporates shows high values (several hundred times the atmospheric pressure) which decrease with increasing

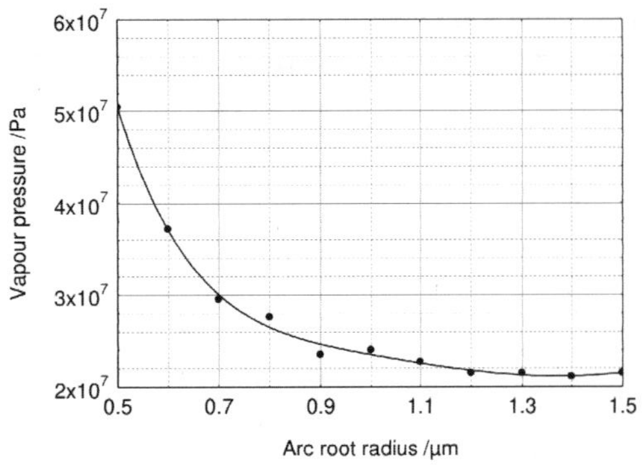

Figure 11. Vapour pressure variations versus arc root radius.

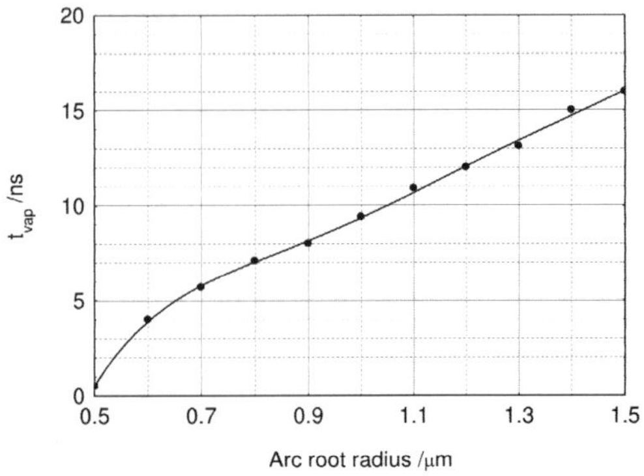

Figure 12. Time vapour emergence versus arc root radius.

radius r. For the range of values of the radius chosen, the value of time necessary to get the first phase change (solid-liquid) on the surface of the electrode varies between $0.1ns$ and $1ns$, this phenomenon is not critical to the operation of cathode spot. The second phase change (liquid-vapour) which corresponds to the emission of metal vapour is essential for the functioning of the cathode spot because it is necessary that the metal vapours are issued from electrode, in order that positive metal ions can fall on the electrode. Figure 12 gives the time t_{vap} for the second phase change can occur when the flow of energy P_i was applied uniformly on the electrode.

4. Discussion

The results obtained allow us to place several intervals for the operation of a cathode spot moving on a silver electrode in the presence of vacuum.

Firstly, the minimum value chosen for the radius and equal to $0.5\mu m$ seems difficult to cross to the lowest values for a steady state, but this is probably questionable for explosive regime in which the values for the various physical quantities calculated can reach values higher than those obtained from the results of section 3.

Returning to the end of the preceding paragraph, it is likely that the steady state as we have used for the cathode spot does not exist, but it is possible to consider a range of cathode spot for the first nanoseconds of its existence would have a value within the region of $r = 0.5\mu m$, see smaller, which then evolve towards higher values. As results are changing with the radius monotonically it does not limit this development.

The values of the vapour stream leaving the electrode, given by figure 10, are between $2.6\times10^{-8} kg/s$ and $10.9\times10^{-8} kg/s$, which for a current of $10A$ gives an erosion between $2.6\mu g/C$ and $10.9\mu g/C$. These values are lower than that given by Daalder [21] gives $40\mu g/C$ for copper. The values of this author are experimental, they include other forms of erosion such as the expulsion of droplets of molten metal, these measures are also post-mortem, they can be considered as limit values. This comparison leads to an assumption stronger for higher values of radius, or even higher than those selected for this study, the analysis of figure 12 invalidates this hypothesis. For a cathode spot radius $r = 1.5\mu m$ the time required for the appearance of the metal vapour is $16ns$, this period is comparable to the lifetime of a cathode spot. For a spot radius $r = 0.5\mu m$ time of the occurrence of metal vapour is equal to $0.5ns$ which is entirely consistent with the assumptions of a cathode spot. The results of figure 9 show that the energy involved is minimal for the lowest values of r, which reinforces the previous conclusion.

In conclusion of this analysis, we must remember that after the very short stage of creating the cathode spot (of duration less than $1ns$) the spot will be established with a radius around the value $r = 0.5\mu m$. Under the action of high pressures of the saturated vapour of the metal (see Figure 11) the radius of the spot will grow until the end of its life, the lives of spot below $16ns$ the corresponding radius is less than $r = 1.5\mu m$. The temperature of the surface of the electrode will change with a value between $4700K$ and $4100K$.

5. Conclusion

In this analysis of the cathode spot situated on a silver electrode with a current of $10A$ in a vacuum all physical quantities involved are defined, except the radius r. The point resolution of the problem that was raised is given by the comparison of flow sprayed when the liquid metal is subjected to bombardment of positive metal ions returning to the surface, and the natural flow given by the same liquid metal in the presence of its saturation vapour. The discussion showed that the values of radius r will evolve over time since their creation by explosion or rupture of molten bridge from the lowest values taken up till higher values, and this versus life time of cathode spot.

References

[1] Lee T H and Greenwood A 1961 Theory for the cathode mechanism in metal vapor arcs *J. Appl. Phys.* **32** 916
[2] Mitterauer J and Still P 1987 Computer simulation of the dynamics of plasma-surface interactions in vacuum arc cathode spots *IEEE Trans. Plasma Sci.* **PS-15** 488
[3] Jütter B, Puchkarev V F, Hantzsche E and Beilis I 1995 Cathode spots in Handbook of vacuum arc science and technology ed R L Boxman P J Martin and D M Sanders (New Jersey: Noyes) 73
[4] Ecker G 1973 The existence diagram-A useful theoretical tool in applied physics *Z. Naturforsch.* **26a** 935
[5] Mesyats G A 2000 Cathode Phenomena in a Vacuum Discharge: The Breakdown, the Spark and the Arc Moscow: Nauka
[6] Jüttner B 2001 Cathode spots of electrics arcs *J. Phys. D: Appl. Phys.* **34** R103
[7] Langmuir I 1923 The effect of space charge and initial velocities on the potential distribution and thermionic current between parallel and plane electrodes *Phys. Rev.* **21** p 419
[8] Beilis I 1995 Theoretical modelling of cathode spot phenomena in Handbook of vacuum arc science and technology ed R L Boxman PJ Martin et al (New Jersey: Noyes) p 208
[9] Mackeown S S 1929 The cathode drop in an electric arc *Phys. Rev.* **34** no.4 p 611
[10] Fowler R H and Nordheim L 1928 Electron emission in intense electric fields *Proc. Roy. Soc. (London)* **A119** p 173
[11] Murphy E L and Good R H 1956 Thermionic emission, field emission and the transition region *Phys. Rev.* **102** p 1464
[12] M.S. Benilov and A. Marotta 1995 A model of the cathode region of atmospheric pressure arcs *J. Phys. D: Appl. Phys.* **28** 1869
[13] Jeanvoine N and Muecklich F 2009 FEM simulation of the temperature distribution and power density at platinum cathode craters caused by high voltage ignition discharges *J. Phys. D: Appl. Phys.* **42** p 1
[14] Barin I 1993 Thermochemichal data on pure substance VCH Verlagsgesellscchaft mbH Germany
[15] Lee T H 1960 Energy distribution and cooling effect of electrons emitted from an arc cathode *J. Appl. Phys.* **31** p 924
[16] Lee T H and Greenwood A 1961 Theory of the cathode mechanism in metal vapour arc *J. Appl. Phys.* **33** p 916

[17] Rossignol J, Abbaoui M and Clain S 2000 Numerical modelling of thermal ablation phenomena due to cathode spot *J. Phys. D: Appl. Phys.* **33** p 2079
[18] Abbaoui M and Lefort A 2009 Arc root interaction with the electrode: a comparative study of 1D-2D axi symmetric simulations *Eur. Phys. J. Appl. Phys.* **48** p 11001
[19] Barinov V N, Goncharov V K and Turomsha E P 1988 Burning voltage of high-current vacuum arcs on clean metals *Sov. Phys. Tech. Phys. (USA)* **33** N°8 p 937
[20] Lide D R 2001 Handbook of Chemistry and Physics (CRC Press LLC USA) p 175
[21] Daalder J E 1976 Component of cathode erosion in vacuum arcs *J. Phys. D: Appl. Phys.* **9** N° 16 p 2379

IOP Conf. Series: Materials Science and Engineering **29** (2012) 012008 doi:10.1088/1757-899X/29/1/012008

A simple generator of a stationary flow of non-equilibrium atmospheric-pressure plasma.

P Andre[1], Yu Barinov[2], G Faure[1] and S Shkol'nik[2]

[1] Clermont Université, Université Blaise Pascal, LAEPT, BP 10448, F-63000 CLERMONT-FERRAND, FRANCE
[2] A F Ioffe Phys. Techn. Inst. Rus. Acad. Sci., Politechnicheskaya 26, St Petersburg 194021, RUSSIA

E-mail: pascal.andre@univ-bpclermont.fr

Abstract. A simple design of a plasma discharge set up, providing a stationary flow of non-equilibrium atmospheric-pressure plasma is proposed. The non-equilibrium state of plasma is maintained by using a cathode with a weakly conducting liquid like tap water. The anode is formed with a cooled metal ring-shaped inserted in a dielectric barrel. This assembly is fixed at ~ 1 cm above the water surface. The velocity of the air flow through the anode hole may be changed from 0 m/s up to ~ 10 m/s. The studied discharge in the open air gap between the tap water surface and the anode ring is powered by direct current I ~ 0.1 A. The voltage on the air gap is ≥ 1 kV. Measurements of current-voltage characteristics at different flow velocities were carried out. Spectra of radiation of the discharge column and the torch above the anode were measured. The measurement results were compared with the results of radiation spectra modeling for non-equilibrium plasma of humid air. The comparison made possible to determine the characteristics of plasma. The characteristics of the plasma depending on the speed of the airflow and the current are studied.

1. Introduction.

In recent years a great interest arose to the different methods of generation of non-thermal (cold) atmospheric pressure plasmas [1]. Various constructions of the cold plasma sources were elaborated and investigated [2]. The activity is high because the sources of non-thermal atmospheric pressure plasma have many various applications. The main ones are gas cleaning, material processing and medicine.

The pulsed and AC discharges are widely used as sources of non-thermal high-pressure plasma. The discharges supplied by DC sources seem more favorable for some applications because the DC power sources are simpler and cheaper. But, it is not an easy task to realize a DC discharge with classical method burning at atmospheric pressure and generating cold plasma. It is necessary to realize the discharge of glow type with a relatively low current density. The usage of the non-metallic (low conductive) liquid as the electrode may be helpful and is a promising method to solve this task. Under certain conditions the discharge with liquid non-metallic electrodes (one or both), the DLNME, can burn steadily in the diffuse form with relatively low current density at DC supply in open air. The first publications on investigations of such discharges appeared in the early 1980s (see [3] and the references in it). Later it was proven that this discharge really generated plasma out of thermal equilibrium [4, 5].

In the paper the simple source of non-thermal atmospheric pressure plasma on the basis of the DLNME with tap-water cathode and ring metallic anode is presented. The discharge is supplied by DC power source. The working gas is atmospheric air which is pumped through the anode. The electrical and the spectral characteristics of the discharge are measured. The parameters of plasma are investigated by means of the comparison of measured molecular spectra with modeling spectra. This method was used in the past [4, 6] and appeared to be a powerful tool for investigation of low-temperature plasma of different kinds of electric discharges and flames. We enhance the modeling molecular spectra by a new calculation code that allows us to simulate the spectra with the contribution of various chemical species at the same time. Consequently great possibilities are opened by this method in the investigations of non-equilibrium plasma.

2. Experimental set-up.

Source assembly was posed vertically. It consists of the tap-water cathode and the dielectric flange with water-cooling ring metal anode. The flange is fixed above the water-cathode. The quartz cylinder is connected to the upper surface of the flange. The air of surrounding atmosphere is evacuated from the top of the quartz cylinder. The gas consumption was measured by the flowmeter.

The cathode is a metal cylinder. Inside, the tap-water is supplied from the bottom and runs out to the top freely. The high voltage is applied between this metal cylinder and the metal anode. The power source can supply the DC current $I \leq 2A$ at voltage $U \leq 10kV$. The ballast resistor is connected in series with the discharge assembly to limit the current in the case of the water electrode breakdown. The discharge ignition is performed by interrupting the circuit after the short-time closing of the interelectrode gap with a special metallic conductor.

The discharge was photographed by Olympus E-520 camera. The side view of the working source is presented on figure 1.

Figure 1. Side view of the experimental device. 1- cathode; 2- the dielectric flange with the water-cooled ring anode inside. The plasma jet can be clearly seen above the flange

The voltage of discharge assembly (between the metal cylinder and the metal anode) and the discharge current were measured during experiment by oscilloscope Tektronix TDS 3012. The voltage of the discharge itself (the voltage between the surface of water in the cathode attachment and the metal anode) can be estimated by taking into account the voltage drop ~ 1kV across the water-cathode. The later was measured when a metal disk (with diameter equal to diameter of discharge attachment to the water) was brought into contact with the water surface near the assembly axis and DC current equal to the mean value of discharge current was supplied.

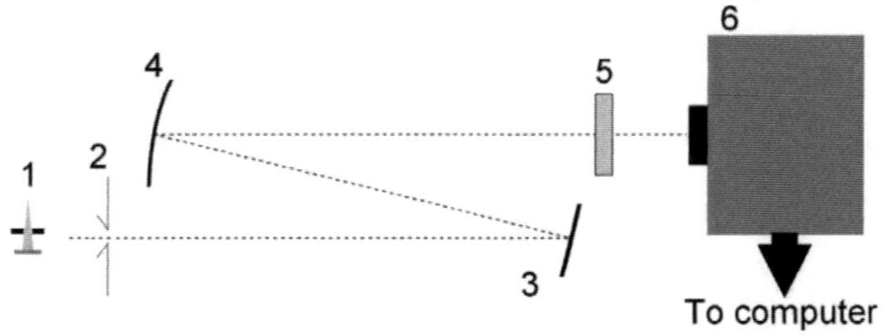

Figure 2. Block diagram of spectroscopic measurements: 1 – the discharge column; 2 – diaphragm; 3 – plain mirror; 4 – spherical mirror; 5 – plain-parallel quartz plate; 6 – monochromator with photodetector.

The spectroscopic measurements were done side on. A block diagram of spectroscopic set-up is shown on Figure 2. To avoid the chromatic aberration in the ultraviolet spectral range, a spherical mirror was used. The spherical mirror produced a reduced image of the discharge on the monochromator entrance plane. The monochromator is equiped with the grating of 1200 grooves/mm and provided the inverse dispersion 1.3nm/mm. The two crossing slits were used. The vertical slit was set at 0.1mm. The horizontal slit was set at 0.2mm. The scanning of the discharge image across the slits was carried out using plane-parallel quartz plate. This plate can be inclined in two perpendicular directions. A multichannel optical analyzer with a linear detector was used to record the spectra. The detector could record in a single measurement a part of spectrum of 22nm in length. The spectral resolution with a vertical entrance slit 0.1mm is about 0.2nm. The spatial resolution along the discharge axis Z was ~ 0.5mm (Z=0mm corresponding to the cathode surface).

Spectral sensitivity of our set-up was measured with the help of calibrated band lamp with a sapphire window. Equipment used provided the possibility to calibrate our experimental set-up in the spectral range ($240nm \leq \lambda \leq 850nm$).

3. Measurement results.

At the fixed voltage of the source which supplied the discharge, the voltage and the current of discharge and the discharge appearance depend on the gas consumption (figure 3). At zero consumption, the discharge is attached to the bottom of the anode. The voltage and the current are somewhat noisy (Figure 3a). At weak consumption, the anode attachment goes inside the orifice. The voltage and the current exhibit a sawtooth-like behavior (figure 3b). When increasing the consumption, the discharge attachment moves to the top of the anode channel and even begins to appear at its exit. The amplitude of sawtooth increases. The oscillograms obtained in the regime in which the spectroscopic measurements were done are shown in Figure 3d. This regime is characterized by the gas consumption G ~ 3 m^3/h. The amplitude of high-frequency noise seems to be independent of the gas consumption.

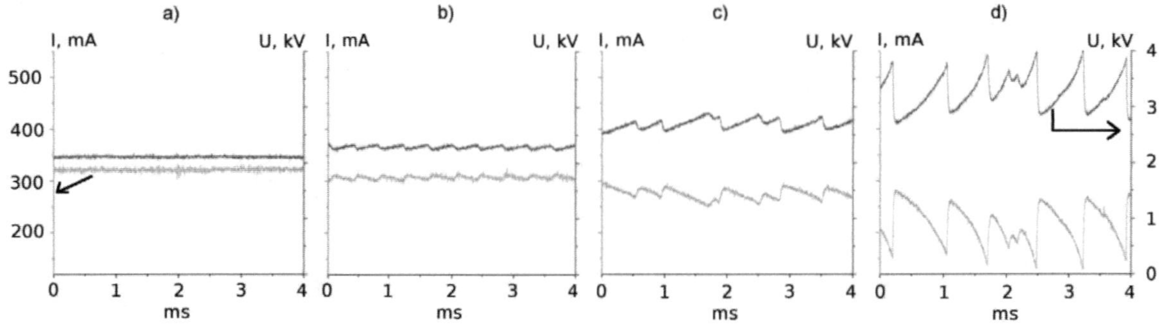

Figure 3. The current and the voltage oscillograms of the discharge with the different gas consumption G. a) to d) are from G = 0 to G ≈ 3 m^3/h.

The spectroscopic measurements show that the discharge emits mainly in violet and ultra violet spectral ranges. The analysis of the spectra showed that the main contribution to the radiation was made by OH radical (A-X transition), NO radical (A,B,C,D-X transitions), NH radical (A-X transition) and molecular nitrogen N_2 (the second positive (2+) system, C-B transition). The emission of OH is most intensive in the spectrum. The structure of the aforementioned bands is in a varying degree distorted by radiation of molecular oxygen O_2 (the Schumann-Runge, SR, system). The emission of O_2 is much weaker than the emission of NO and OH and in the violet part of spectrum it is comparable with emission of N_2. The examples of the most essential parts of spectra are shown in the Figure 4. In the range 240-260 nm, the emission is mainly due to molecule NO and to a weak emission of O_2. The figure 4b shows the strong emission of OH, the spectra of the other species (O_2 and N_2) emitted in this interval are not visible due to the strong emission of OH. NH emission can be observed in the figure 4c with N_2 emission. The spectrum of the figure 4d is mainly formed by the N_2 2+. OH, O_2 and NH also emit in this range 360-382 nm but more weakly than N_2.

The measurements of radial distributions of the intensity of radiation of different plasma species showed that these distributions are sufficiently well approximated by axisymmetric curves. Having performed the Abel inversion procedure, one can see that the emission coefficients attain their maximum on the axis of the discharge. In the cathode attachment, immediately near the surface of water, the half-widths D (the width at half height) of radial distributions of emission coefficients of different species attain maximum value and decrease quickly in the direction of anode. At the distance Z>=2mm from the cathode the discharge channel turns into a cylinder. The spatial resolution along the discharge axis Z provided in this work is not sufficient to investigate the cathode region of the discharge where strong spatial gradients in Z-direction exist. Also, we should remember that work [7] put into evidence the filament structure of the cathode region of discharge of similar type. In our work, some measurements took an acquisition time of a few seconds. Therefore, emphasis was put on the measurements of the discharge channel, with a postponing of the investigation of the cathode region.

Axial dependencies of the emission intensity of the most intensive molecular bands are presented in Figure 5. It can be seen that the emission of the discharge attains its maximum near the entrance into the anode. Also, the emission is intensive enough at the exit of the anode. The plasma jet, originated in the anode orifice, runs more than 10 mm away from the anode.

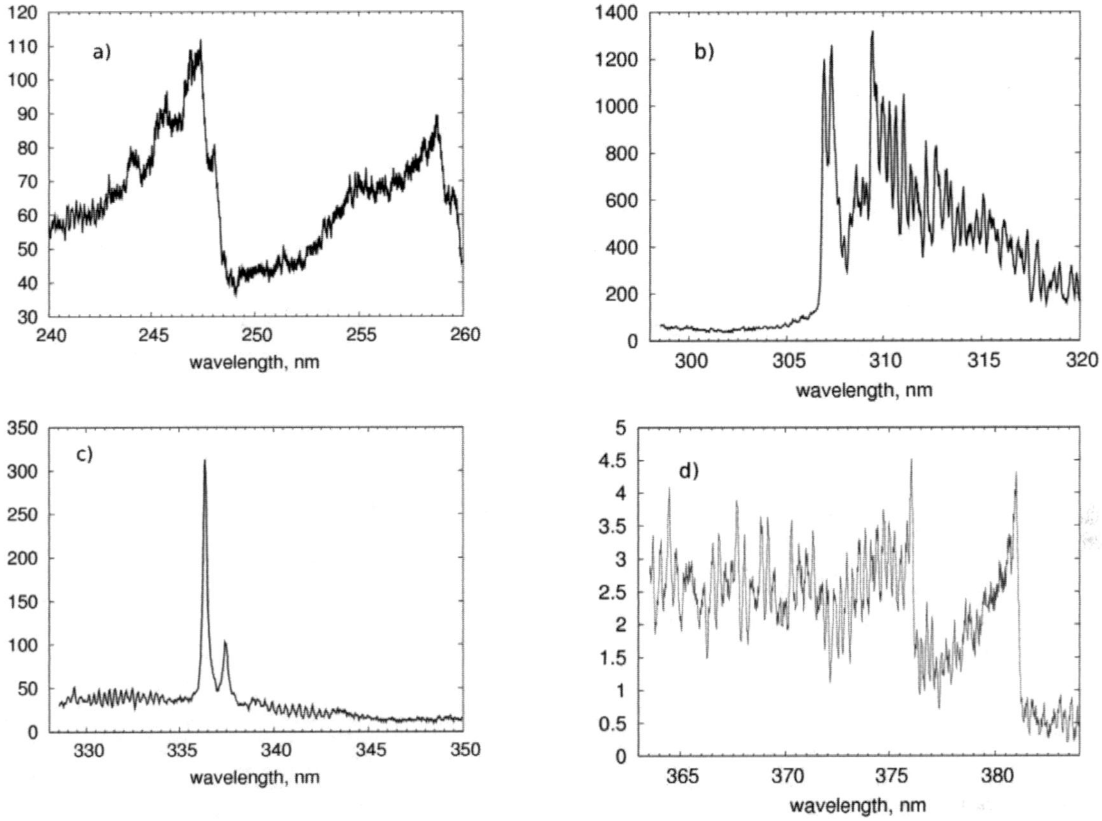

Figure 4. The examples of the results of measurements of most essential parts of spectra. The parts of spectra in which NO (a), OH (b), NH (c) and N_2 (d) emission gives the main impact into the intensity (Z = 3 mm).

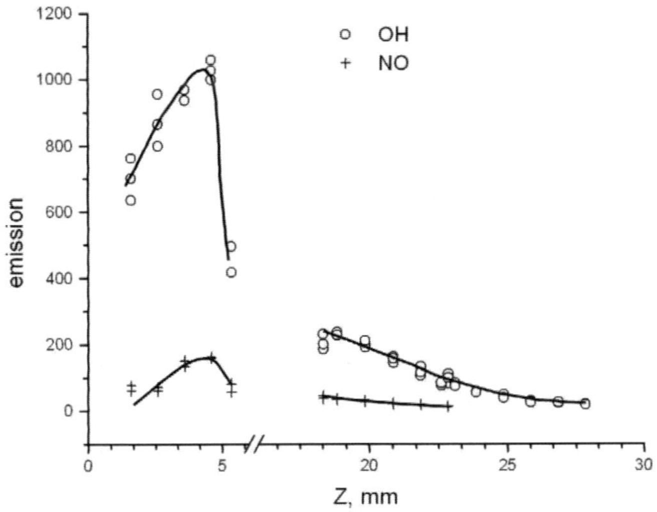

Figure 5. Axial distributions of molecular bands integrated intensity versus the height: OH is integrated within 306.7-307.6 nm) and NO is integrated within 256.5-258.3 nm)

4. Spectra modelisation and comparison with the experiment.

In this second part of the paper, modeling of the radiation from the systems of five molecules (N_2, O_2, OH, NO and NH), which account for the main contribution to the emission spectrum of the discharge in humid air, in the conditions of thermal nonequilibrium:

- second positive system of N_2 (2+),
- Schumann-Runge system of O_2 (SR),
- ultraviolet system of OH (UV)
- systems of NO (γ, β, δ, ε),
- system of NH A-X

is described.

We assume a Boltzmann population distribution for all the internal quantum levels but with different characteristic temperatures for the internal models (rotation, vibration, and excitation). Moreover, the assumption that all kinds of temperatures are the same for all species is debatable. These temperatures cannot be assumed to be the same from one species to another. Therefore, for each kind of molecules we will consider individual temperatures T_{exc}^{diat} (the temperature of distribution in the electronic states), T_{vib} (the temperature of distribution in vibration states) and T_{rot} (the temperature of distribution in rotation states). This assumption is widely used (see, for example, [8]) and allows researchers to easily use our results. One can find the description of general calculation method as well as data which are necessary to perform the calculation in [6].

The modeling results are used for investigation of the state of the molecular component of the plasma in the column of the discharge and the jet which is formed in the anode orifice by means of comparison of the modeled spectra with those obtained in the experiment. As it was shown in our previous work [6], the emission of NO and OH cannot be used for the determination of the gas and the electrons temperatures of a cold plasma with high humidity. So, relatively weak emission of N_2 2+ system is of great importance because it can be used for gas temperature determination. The opinion that the rotation temperature obtained from N_2 2+ system emission is a good estimation of the gas temperature is confirmed by the results of number investigations (see, for example, [9] and the references therein). Also, this part of the spectrum can be used for the estimation of the degree of the non-equilibrium, because the temperature of excitation of N_2(C) is close to the electrons temperature [5].

But only one glance at part of the spectra near 370 – 380nm is enough to understand that the bands of N_2 2+ system (sequence $\Delta v = -2$) are strongly overlapped by the tails of other species bands (Figure 4d). Also, the spectra is influenced by emission from the O_2 SR system. The modeling of emission of NO, OH, NH, O_2 and N_2 and the comparison of simulations and measurements show that the part of spectra which is of main interest is formed by emission of four molecules: OH, NH, O_2 and N_2. The impact of NH emission is weak; the impact of OH and O_2 emission is of significance, but is weaker than N_2 (Figure 6). So the parameters of energy distributions of excited N_2 molecules can be determined.

The analysis of the part of spectra near 370 – 380nm measured in the gap between tap-water cathode and metal anode at different distances Z from the water surface shows that beginning from Z > 2mm the gas temperature and the electron temperature change but a little. The gas temperature is ~ 2000K. The degree of non-equilibrium θ, the ratio of the gas temperature to the electron temperature, is not less than two ($\theta \approx 2.1$).

In the jet originated from the anode orifice, from the 370–380nm spectra the gas temperature appears to be slightly higher (~ 2200K), but the electron temperature a little lower, so the thermal nonequilibrium is somewhat less ($\theta \approx 1.9$).

One more significant conclusion of the comparison between the modeling results and the measured spectra concerns the high concentration of chemically active radicals in the jet. For example, the concentration of OH is estimated to be ~ $10^{22}m^{-3}$ at a distance ~ 1cm away from the anode.

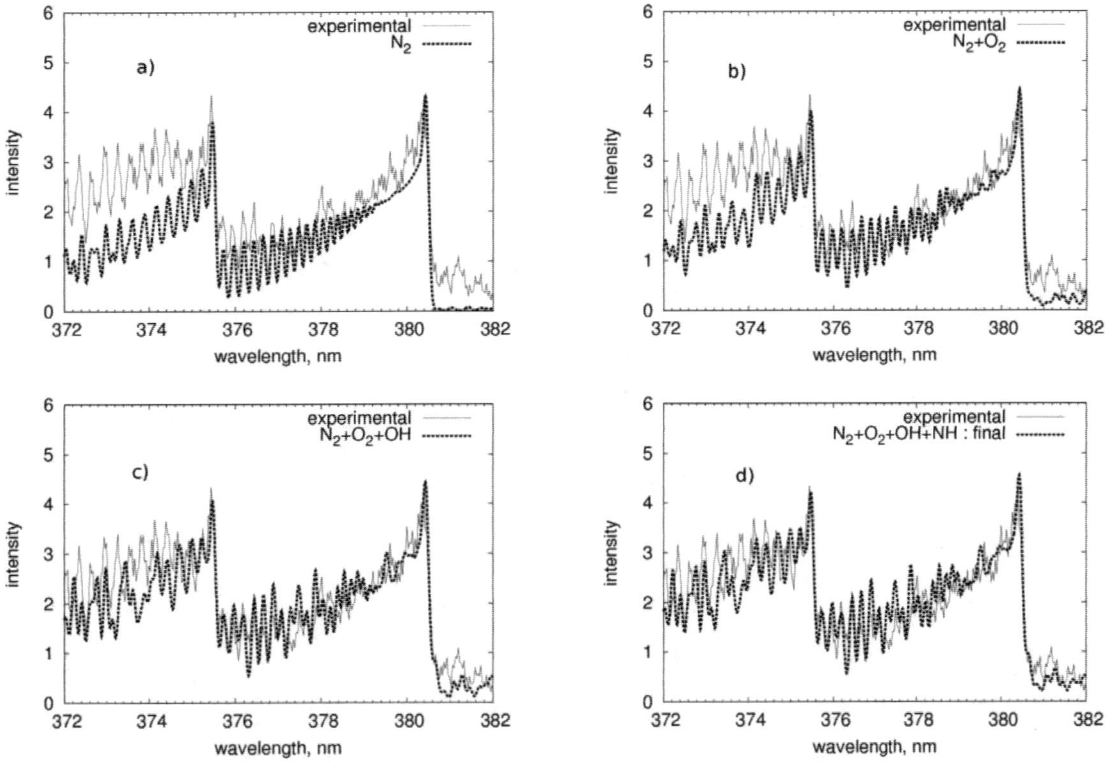

Figure 6. Comparison of the measured and the modeled spectra. Experimental spectrum (Z = 3 mm) – thin line; modeled spectrum – bold dash line. a) N_2 ; b) N_2 and O_2 ; c) N_2, O_2 and OH ; d) N_2, O_2, OH and NH emission is modeled.

5. Discussion

It can be seen from Figure 3 that at zero gas consumption $G = 0$ m^3/h, the discharge voltage is minimum and the voltage drop across the discharge itself (between the water surface and the metallic anode) is ~ 1kV at current ~ 0.3A. An increase of gas consumption leads to the appearance of "sawtooth" structure and also an increase of the voltage mean value. This is a consequence of the discharge channel lengthening under action of the gas flow. The current-carrying channel is ejected from the anode orifice and is curved to connect to the anode. A strongly lengthened channel becomes unstable and is bridged (shunted) by the breakdown of the near-anode layer inside the orifice. The channel becomes shorter and the process of lengthening begins again. This process is well-known in arc-plasmatrons. But in arc discharges, the electric field is much lower, so the amplitude of the "sawtooth" structure is much lower.

The photograph of the discharge (Figure 1) and the spectroscopic measurements (Figure 5) show that at not so high gas consumptions (~ $3m^3/h$) the device under investigation generates a plasma jet which is ejected from the anode orifice for the distance of about centimeter. The comparison of simulated and measured spectra shows that the plasma in the jet is in thermal nonequilibrium. The ratio of the gas temperature to the electron temperature is of about two ($\theta \approx 2$). The concentration of chemically active radicals is high (for example, the concentration of OH is estimated to be ~ $10^{22}m^{-3}$).

6. Conclusion

In this paper a first step to characterize a discharge with a tap-water cathode in flowing air is carried out, with a first comparison in the range of [365-382] nm between simulated and measured spectra. The first temperature measurements have been carried out, evidencing thermal nonequilibrium. We find an electronic and rotational temperature ratio above 2.1 between the cathode and the anode, and below 1.9 in the plasma jet. The rotational temperature was found to be respectively around 2200 K and 2000 K.

These promising results show that the discharge with tap-water cathode can be used to generate the jet of plasma in thermal nonequilibrium at atmospheric pressure. Application of the non-metallic liquid as the cathode secures the main advantage of this technique – the plasma source can be supplied by DC power generator

This investigation is a preliminary part of the overall work to be carried out. Depending on the application of the device, the optimization of the plasma jet parameters should be done by means of selection of proper values of the discharge parameters and the device construction.

References

[1] Laroussi M and Akan T 2007 *Plasma Process. Polym.* **4** 777–88
[2] Tendero C, Tixier C, Tristant P, Desmaison J and Leprince P 2006 *Spectrochim. Acta* B **61** 2–30
[3] Gaysin F M, Son E E 1989 Electrophysical Processes in the Discharges with Solid and Liquid Electrodes (Sverdlovsk: Ural University) (in Russian)
[4] Andre P, Barinov Yu, Faure G, Kaplan V, Lefort A, Shkol'nik S and Vacher D 2001 *J. Phys. D: Appl. Phys.* **34** 3456-3465
[5] Andre P, Aubreton J, Barinov Yu, Elchinger M F, Fauchais P, Faure G, Kaplan V, Lefort A, Rat V and Shkol'nik S 2002 *J. Phys. D: Appl. Phys.* **35** 1846-1854
[6] Andre P, Barinov Yu, Faure G and Shkol'nik S M 2011 J. Phys. D: Appl. Phys. **44** 375202
[7] Bruggeman P, Ribezl E, Maslani A, Degroot J, Malesevic A, Rego R, Vierendeels J and Leys C 2008 *Plasma Sources Sci. Techn.* **17** 025012
[8] Staack D, Farouk B, Gutsol A F and Fridman A 2006 *Plasma Sources Sci. Techn.* **15** 818-827
[9] Verreycken T, van Gessel A F H, Pageau A and Bruggeman P 2011 *Plasma Sources Sci. Techn.* **20** 024002

IOP Conf. Series: Materials Science and Engineering **29** (2012) 012009 doi:10.1088/1757-899X/29/1/012009

Investigations of a microwave plasma source operating with air, N_2, CO_2 and argon gases

D Le Quang[1], Y Babou[1] and P Andre[2]

[1] Aeronautics and Aerospace Department, von Karman Institute for Fluid Dynamics, Chaussée de Waterloo 72, 1640 Rhode-Saint-Genèse, Belgium
[2] Electric arc and thermal plasmas (LAEPT), Université Blaise Pascal, 24 avenue des Landais Aubière – France

Email: lequang@vki.ac.be

Abstract. The present contribution will report on the development of a Microwave Plasma Torch (MPT) at the von Karman institute and on the characterization of subsonic plasma flow of various plasmagene gases (air, N_2, CO_2 and Ar) at atmospheric pressure. The aim of this work is to explore plasma flow regimes for a large envelope of operating conditions. Using a conventional arrangement for MPT, including cylindrical open-ended dielectric quartz tube, plasma generated with molecular gases have been found to be homogeneous while with argon filamentary structures have been observed. Electrical characterization of the source has been performed for some operating conditions using forwarded and reflected microwave power monitoring. Comprehensive characterization of the flow by means of high speed imaging has been performed to estimate the effect of the oscillation of the delivered power on plasma jet unsteadiness. Analysis in time and frequency domains of the light intensity recorded at 1 kHz rate is presented. Low resolution Optical Emission Spectroscopy (OES) diagnostic was performed in order to identify typical radiative signature of air, N_2 and CO_2 plasma jet. High resolution OES was performed on pure N_2 plasma suitable for thermodynamic characterization using N_2^+ First Negative and N_2 2nd Positive systems. Making the common assumption that the internal energy levels of molecular species are populated according to a Boltzmann distribution and making no assumption concerning the total species concentration similarly to [1], comparisons between measured and calculated radiative signature of molecular systems evidence for the achievement of thermal equilibrium and slight departure to Saha equilibrium.

1. Introduction

High temperature plasma flow jet processes are extensively employed in a broad range of industrial applications requiring high energy density medium and elevated thermal fluxes. In the aerospace domain, high temperature subsonic flow jet are commonly used to investigate gas surface interactions [2], [3], [4], [5] to rely Thermal Protection Systems (TPS) design procedure with higher confidence. High temperature expanded jets are generated to provide test flow fields for high-speed compressible flows appropriate to investigate on ground kinetic processes involved during atmospheric entries. The technology adopted to warm up the initial cold mixture can be a stringent source of undesirable phenomena. Indeed, arc plasma jet processes can be affected by unsteady phenomena arising from arc root motion and by the contamination with cathode erosion products. Among electrode-less plasma sources, Inductively Coupled Plasma ICP torches offer high purity processing necessary for metrological applications. Nevertheless, efficiency of the coupling energy into the plasma does not exceed usually 50 % and can drop significantly for power above 100 kW [6]. Microwave plasma sources using electromagnetic surface waves to sustain a discharge are known for several decades to be efficient devices to produce high temperature plasmas [7]. The simplest surface wave discharge (SWD) device is the so called Microwave Plasma Torch (MPT) consisting of a wave launcher coupled to an open-ended cylindrical dielectric tube. The fact that a surface wave can propagate along the interface between a plasma column and its confining dielectric tube has been identified by Trivelpiece and Gould [8]. Although discharge extent is constrained by the wavelength of the applied microwave electromagnetic field, MPT are employed since a long time in many industrial and laboratory applications [9], [10], [11], [12] when small scale high temperature plasma sources are required. When operated in ambient air at atmospheric pressure with an open-ended dielectric tube, MPT can typically produce high temperature homogeneous plasma flow close to thermal equilibrium (up to 7000 K) over a wide range of gas flow rate (0-100 LPM) and input power (100 W - 10 kW), starting from various atomic and molecular initial cold gas mixtures (Ar, He, H_2, N_2, Air, CO_2). The MPT is still under considered in the aerospace domain, although it is a valuable alternative to small scale ICP torches and arc jets to support research and development tasks. The motivation of the work undertaken and outlined below is driven by the development of a new MPT at von Karman institute to provide a reservoir of high temperature gas at equilibrium. The present work addresses MPT setup and preliminary characterizations of the plasma flow generated with various operating conditions and gases. The effect of operating conditions (power, flow rate, mixture) on the produced plasma flow dynamical features and thermodynamic state are investigated by means of optical, electrical and spectroscopic diagnostics. Plasma flow behaviour is presented in Section 3 for various operating conditions. The dynamical features are examined in Section 4. The thermodynamic state of the plasma is discussed in Section 5. Conclusion and perspectives are given in Section 6.

2. Microwave sustained plasma source

Surface waves are employed in microwave induced plasma devices to sustain plasma discharges within a large range of operating conditions and arrangements. A compact and flexible source figure 1 was built in VKI on the basis of a magnetron head (MH3.OW-S1 Richardson Electronics) operating at 2.45 GHz with SM1150T power supply to produce microwave power in the kilowatt range transmitted through a rectangular waveguide (WR 340 adapted for λ=12.24 cm wavelength) to the gas flowing in a quartz tube of 30 mm inner diameter similarly to reference [10].

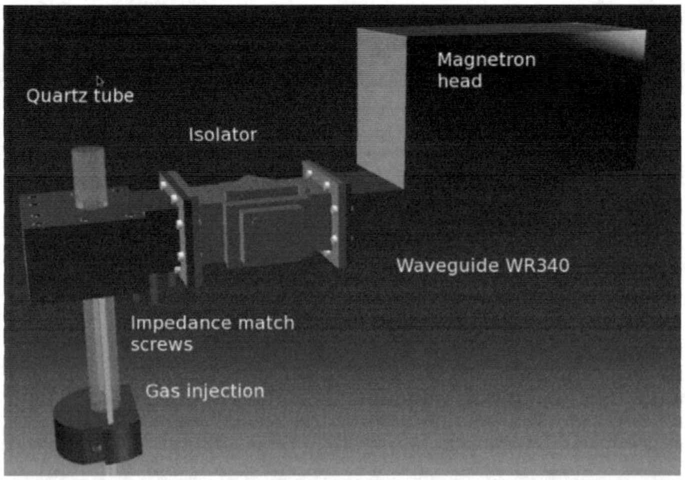

Figure 1. Representation of the Microwave Plasma Torch.

Microwave power forwarded through the waveguide is converted into surface wave power when crossing the quartz tube. The plasma is sustained by the electromagnetic wave propagating along the discharge, guided through the cylindrical dielectric quartz tube. Plasma discharge is created, starting from an initial breakdown which evolves towards steady state discharge inside the tube. The initial breakdown taking place in the discharge tube close to the waveguide is triggered manually using iron rode Figure 2 intensifying locally the electric field and allowing the first electronic collisions. The characteristic impedance presented by the plasma column to the wave might be adapted to improve the energy transfer process from the launcher to the travelling wave. For this purpose, impedance matching screws are located on the waveguide which allows adapting the impedance circuit by changing the geometry of the waveguide section. When the steady state is achieved the plasma column length reached depends on the amount of power delivered to the launcher. The efficiency of power transfer from the microwave generator to the plasma in surface wave discharge depends mainly on the coupling efficiency at the interface between the waveguide and the quartz tube. The launching efficiency is the fraction of the total power leaving the launcher that is converted into surface wave power flux. The reflected power lost is damped by a circulator preventing from any damages on the magnetron head. The amount of radiated power into space is negligible compared with that transferred to the discharge. A preliminary investigation on microwave leakages ensured safe environment for the operator.

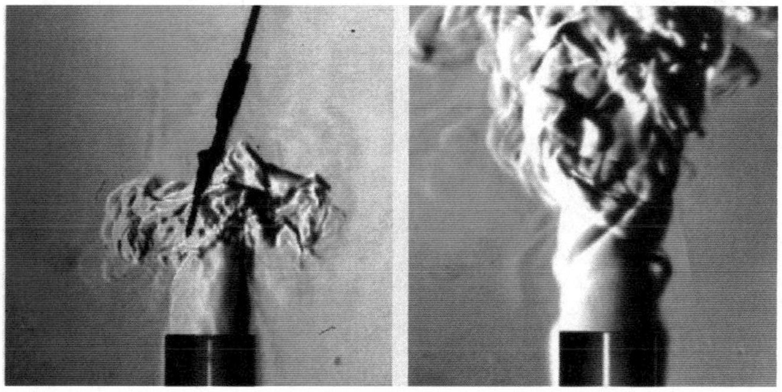

Figure 2. Schlieren pictures above the quartz tube during air plasma ignition (left) and established flow (right).

The efficiency can be measured directly into the waveguide power feeding line, using standard power sensor. The waveguide is equipped with two SMA connectors allowing forwarded and reflected power measurements. The waveguide is terminated with a water cooled part with typical flow rate of 4 LPM, allowing large running time of the facility. Various configuration of the torch have been investigated for a wide range of operating flow rates (0-100 LPM), injected power (1 kW - 3.5 kW) and geometrical configurations (injection strategies, quartz tube length). In this contribution, we investigate subsonic plasma jet at atmospheric pressure, obtained using swirl gas injection. The experimental diagnostics employed to perform electrical characterization of the source, flow regime and thermodynamic state characterization of the plasma jet are presented in Figure 3.

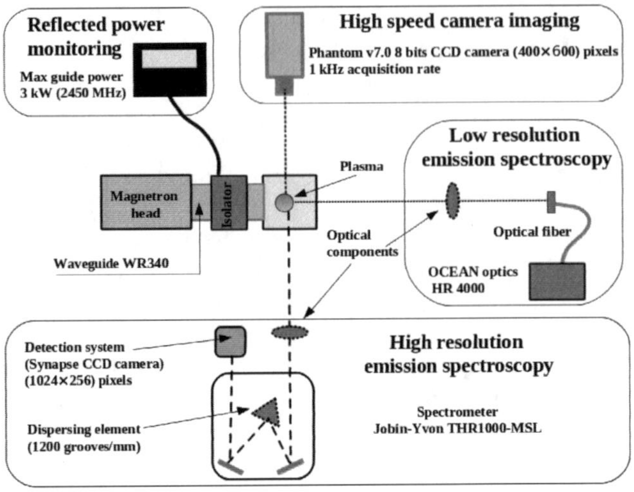

Figure 3. Scheme of diagnostics applied to MPT.

3. Plasma flow features

Photographs of air plasma jet obtained at 5, 15 and 90 LPM, given Figure 4(a), show considerable change in the plasma flow topology going from classical axisymmetric jet to less conventional helicoidal shape ended with an ovoid region where internal recirculation zone is resulting from vortex breakdown. Indeed, as described in [13], at high flow rate, adverse pressure gradient and viscous diffusion yield to the establishment of divergent vortex core at the exit of the tube. Similar trends are observed in N_2 and CO_2 plasmas as depicted in Figure 4(b) and 5(a). In contrary to operation with molecular gases, sustained Argon discharge is characterized by filamentary structures as observed by Hong & al. [11]. As shown in Figure 5(b), incrementing the power leads to an increase of filaments length and branching.

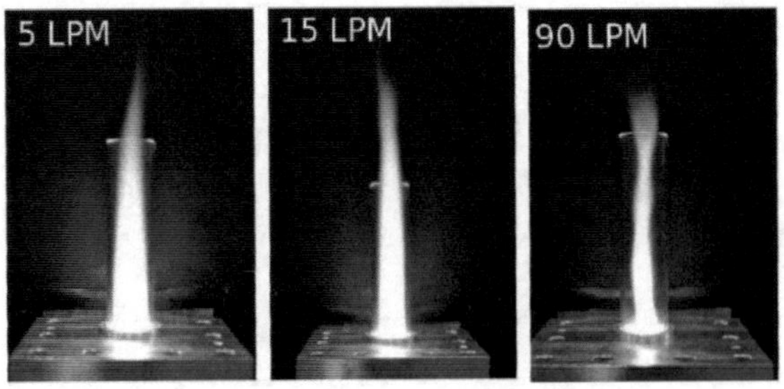

(a) Air plasma at 2 kW

(b) N_2 plasma at 2 kW

Figure 4. Pictures of air and N2 plasma flow.

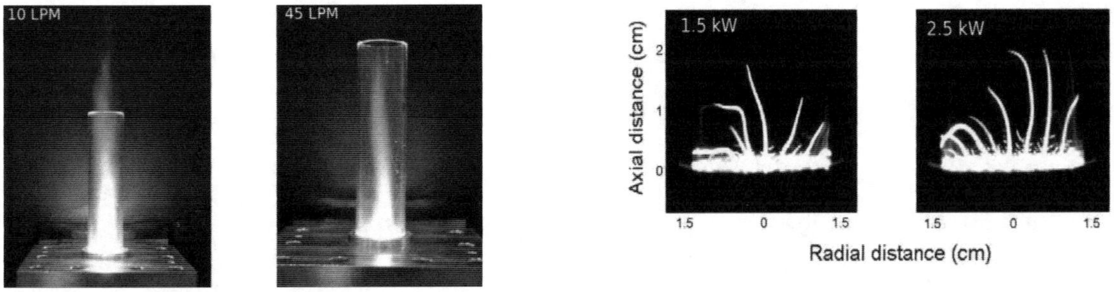

(a) CO_2 plasma at 2 kW. (b) Argon gas streamer.

Figure 5. Pictures of CO_2 and Ar plasma flow.

Forwarded power and gas flow rate injection play an important role on the structure of the jet. High speed camera frames of air plasma jet in Figure 6 illustrate typical behaviour of the plasma within various applied operating conditions range as also described in [9].

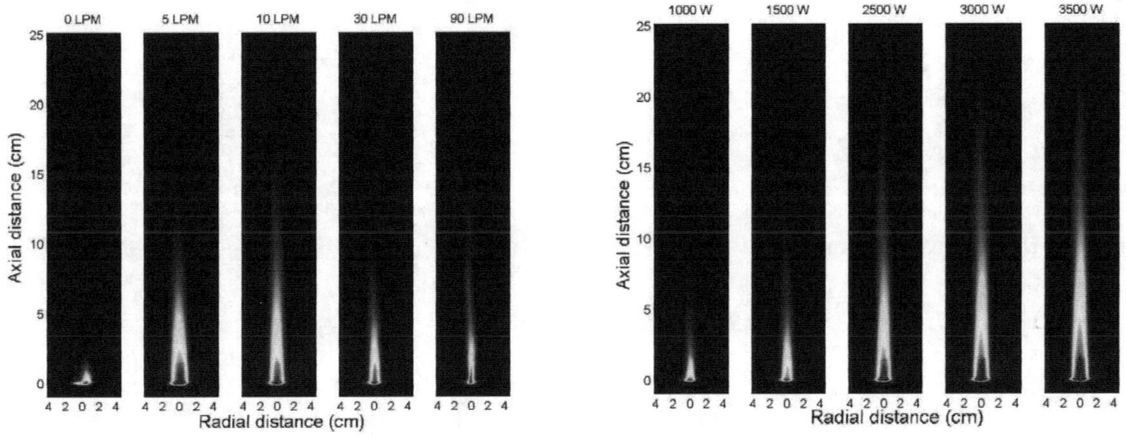

(a) Camera frame of an air plasma jet at 2 kW. (b) Camera frame of an air plasma jet at 15 LPM.

Figure 6. Camera frame of an air plasma jet at 2 kW.

An additional microwave power contributes to expansion of the plasma in both radial and axial directions associated with an increase of the light emission intensity. It is consistent with the observations in [14]. In [14], the author also observed that the expanding plasma volume is the major phenomenon induced by an increase of the power, as the plasma heating remains weakly affected. This assertion will be assessed in Section 5. The plasma column is constrained to a smaller diameter while increasing the injected swirling flow rate. This behaviour results from the increase of the swirl flow velocity leading to higher convection cooling of the flow on the edge of the quartz tube. Additional observations that are not presented here, have shown that axial gas injection leads to highly unstable plasma jet in contrary to swirl gas injection which is used in the following reported investigation.

4. Plasma unsteadiness investigation

4.1. *High speed camera imaging*

The fluctuating part of the intensity has been investigated by means of high speed imaging measurements using Phantom v7.0 8 bits CCD camera at the acquisition rate of 1 kHz. The examination of the unsteady feature of the flow emission provides useful information about the effect of microwave power oscillation on thermal inhomogeneity taking place in the plasma flow. Simple processing have been applied to high speed frame records to evidence for the presence of regions inside the jet that are hotter or colder than the steady contribution. The evolution and transport of thermal pockets have been investigated using the following relation:

$$I_p(k) = I(k) - \frac{\sum_{k=1}^{N} I(k)}{N}$$

where N is the number of frame accounted to achieve the statistical convergence criterion on the mean intensity. Ip(k) represents the fluctuating part of the k_{frame} I(k).

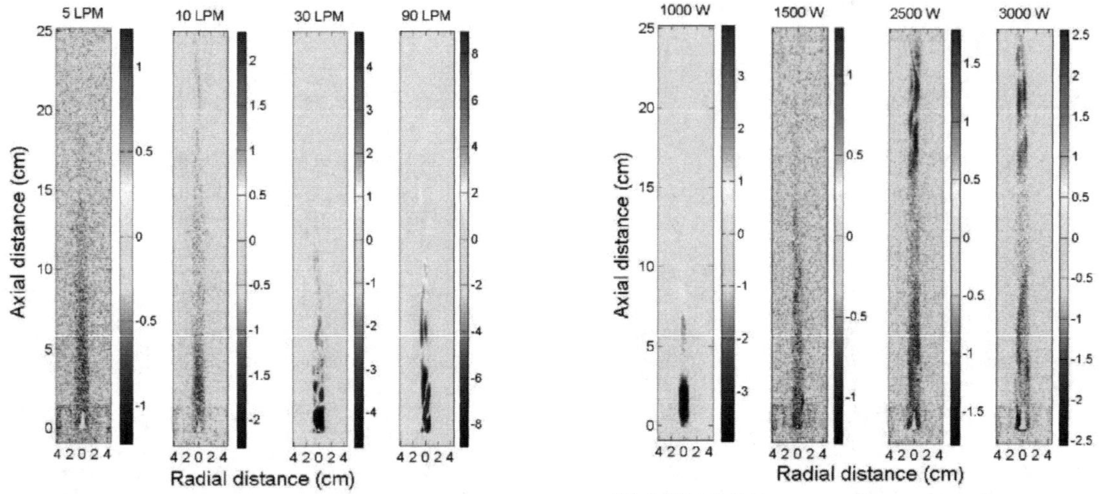

(a) 2 kW input power for various flow rates.

(b) 15 LPM flow rate for various input Powers.

Figure 7. Emission fluctuation of an air plasma jet for various operating conditions.

$I_p(k)$ frames are given for air plasma jet at different flow rates in Figure 7(a) and various input power in Figure 7(b). Amplitude of thermal inhomogeneities increases as the flow rate is increased. Their maximum is found just above the waveguide and tends to disappear along the axis of jet. Also, alternative thermal pockets tend to stretch in the axial direction as the flow rate is increased. Similarly, increasing the power leads to the generation of longer thermal pockets as shown in Figure 7(b). At high power, the axial extent of the jet allows to distinguish between confined plasma situation and the plasma exiting the tube. In Figure 7(b), the end of the quartz tube is located 15 cm above the waveguide. The amplitude of the oscillations is accentuated at the exit of the tube. In the case of quasi homogeneous plasma column emission (15 LPM - 1500 W), the plasma radiation oscillation does not exhibit any particular change exiting the tube. The FFT of optical emission fluctuation was executed to evidence the heating modulation at well defined frequencies. Figure 8(a,b) shows the FFT operated on a spatial pixel located at 1 cm from the waveguide at the plasma center line for an air mixture at different power and flow rates. At 5 LPM, the spectrum is mainly composed of well defined frequencies above 50 Hz (50, 100 and 300 Hz) related to the coherent dynamic of the jet. Increasing the flow rate leads to the damping of these peaks vanishing into noisy background rising towards low frequency range. Similar behaviour is depicted for N_2 and CO_2 plasmas. The spectral signature of the dynamics at 5 LPM flow rate is composed of peaks above 50 Hz (50, 100, 150 and 300 Hz for N_2 and 50, 100, 150, 200, 250, 300, 350, 400 and 450 Hz for CO_2).

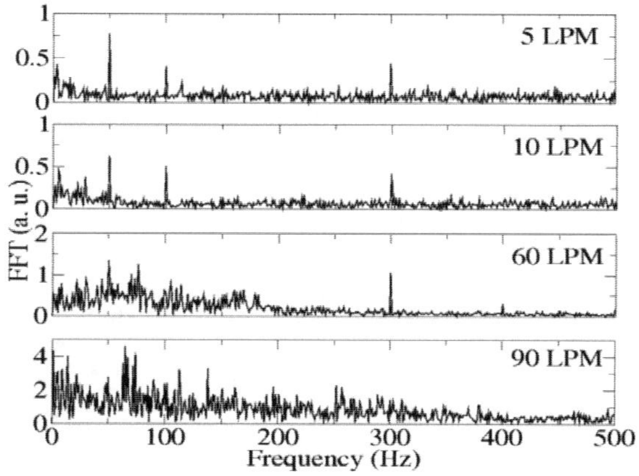

(a) 2kW air plasma jet.

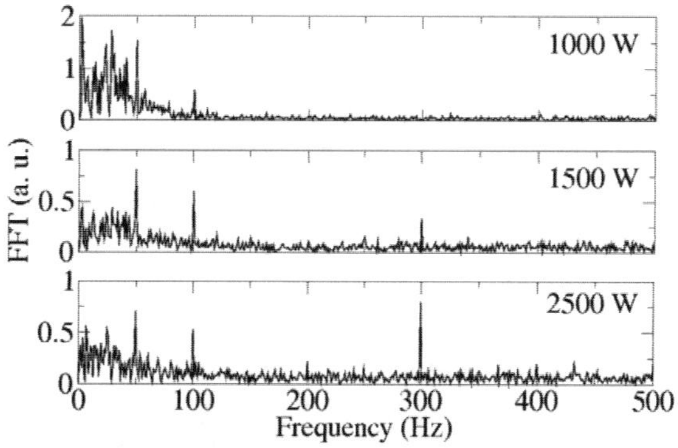

(b) 15 LPM air plasma jet.

Figure 8. FFT of an air plasma jet emission fluctuation on the axis at 1 cm from the waveguide.

These coherent peaks result from the modulated Joule heating of the feeding circuit. It has been observed that operating with a forwarded power slightly above the minimum threshold value required to sustain the plasma leads to a highly unstable plasma flow. The instability of the flow resulting from thermal inhomogeneities generated by the power modulation is damped by increasing the power. Figure 8(b) evidences the effect of the input power on the plasma jet dynamics. As the power is increased, the spectral dynamics of the oscillations is mostly driven by peaks above 100 Hz as moderate frequencies contribution tends to fade. The coherent oscillation of plasma emission is damped along the axis of the jet, leading to the only contribution of low frequencies above 6 cm axial distance as shown Figure 9.

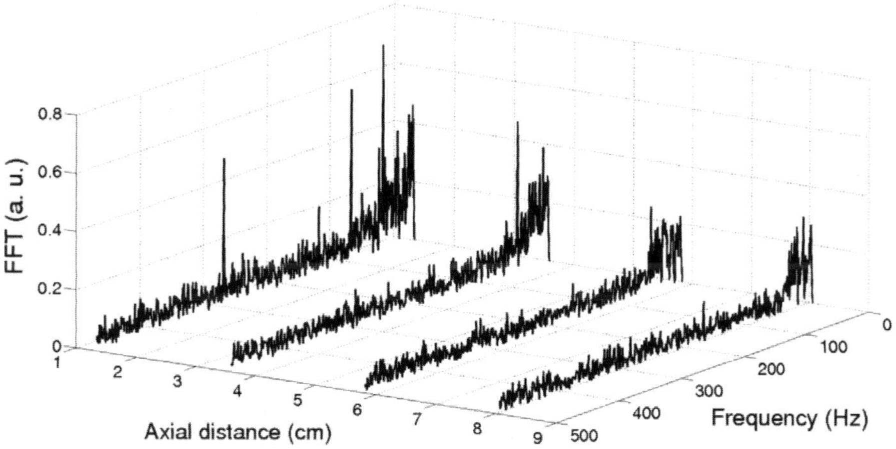

Figure 9. FFT of an air plasma jet emission fluctuation as a function of the distance from the waveguide.

4.2. *Forwarded an reflected power monitoring*

Forwarded and reflected microwave powers are monitored at different flow rates and power for an air plasma. Alter RD8400 power sensor is used to measure the microwave signal at 2.45 GHz previously attenuated at 60 dB by the circulator device. It allows measuring the maximum waveguide power with a full scale 0-10 Vdc output voltage. Alternative measurements have been performed on the forwarded and reflected signals. Temporal and frequential forwarded and reflected power signal are plotted in Figure 10(a, b) and 11(a, b).

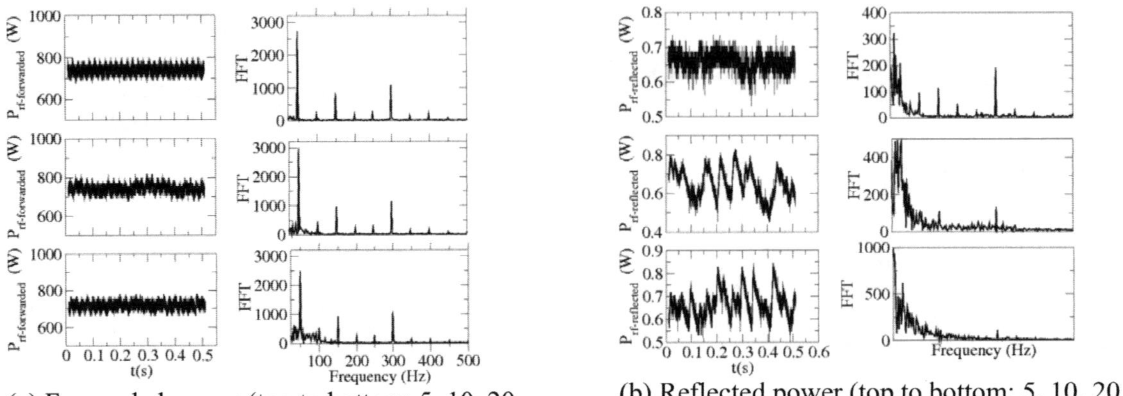

(a) Forwarded power (top to bottom: 5, 10, 20 LPM).

(b) Reflected power (top to bottom: 5, 10, 20 LPM).

Figure 10. Microwave power and associated FFT signal for air plasma at different flow rate.

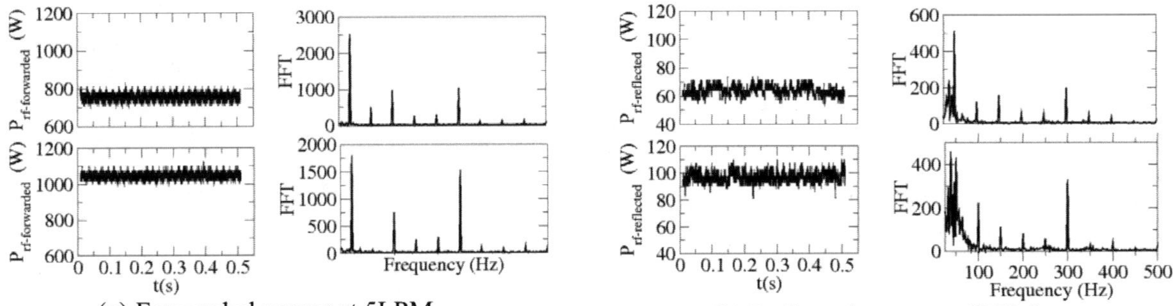

(a) Forwarded power at 5LPM.

(b) Reflected power at 5LPM.

Figure 11. Microwave power and associated FFT signal for air plasma at different power.

The effect of the flow rate variation on the reflected power is considerable. As the flow rate increases, the amplitude of the oscillations evolves from a coherent situation to random variation associated to noisy dynamic increasing towards low frequencies below 100 Hz. It is in agreement with plasma optical emission light dynamic oscillation. Coupling microwave power to the plasma requires the impedances between the discharge and the microwave power source to be matched. A common means to achieve this match is to insert a metallic element into the waveguide, which is executed here using screws. Since impedance mismatch has phase and amplitude components, both position and depth of insertion of the matching element must be adjusted. Three screws, manually adjusted, are positioned inside the waveguide and the effect of the depth of the insertion has been studied. For different configurations, the efficiency of the plasma source is illustrated on Figure 12, where the reflected RF power is plotted versus the forward RF power. The dielectric properties of the plasma vary with changes in temperature and chemical composition. Nevertheless a single configuration has been chosen for the various experiments conducted here, corresponding to the black curve in Figure 12.

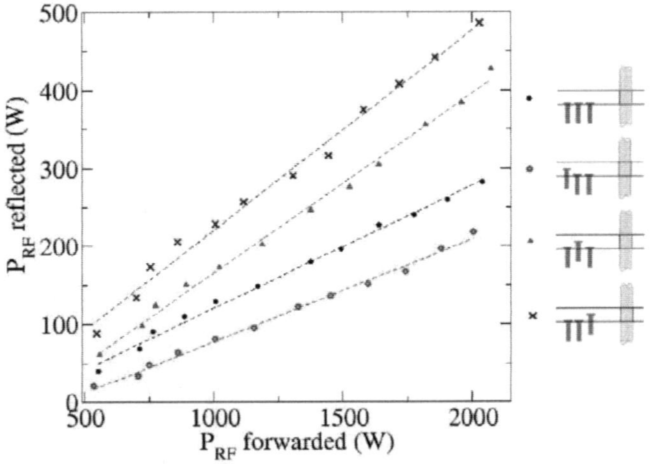

Figure 12 . Reflected power versus forwarded power for different configurations of the impedance matching screws.

5. Thermodynamic equilibrium assessment

Preliminary work has consisted in the examination of radiative transitions of air, N_2 and CO_2 plasmas recorded by means of a low resolution OES bench. Acquired raw data are displayed in Figure 13(a) for various operating power. The displayed spectra are characterized by significantly different contribution depending on the plasmagene operating gas. Air radiative signature is mainly constituted of molecular emission of O_2 Schumann-Runge in the spectral domain below 600 nm and of NO systems (β, γ, δ, ε) peaking at 250 nm. N_2^+ First Negative and N_2 Second Positive systems are the main contributors in pure N_2 plasma between 290 nm and 450 nm. CO_2 plasma emission is constituted of both molecular transitions from C_2 Swan system and atomic lines of oxygen and carbon. Air plasma emission spectrum is found to be less appropriate for temperature determination by means of spectra fitting approach in comparisons with N_2 and CO_2 radiative signature. Indeed, air plasma emission in the visible range exhibits a poor dynamic detrimental for accurate fitting and the NO systems lying in the UV are self absorbed preventing from a direct access to local emission. A comprehensive analysis of CO_2 and CO_2-N_2 plasma produced by a similar MPT has already been conducted previously [10], therefore investigations of the thermodynamic state by means of OES diagnostic reported here are focused on N_2 plasma.

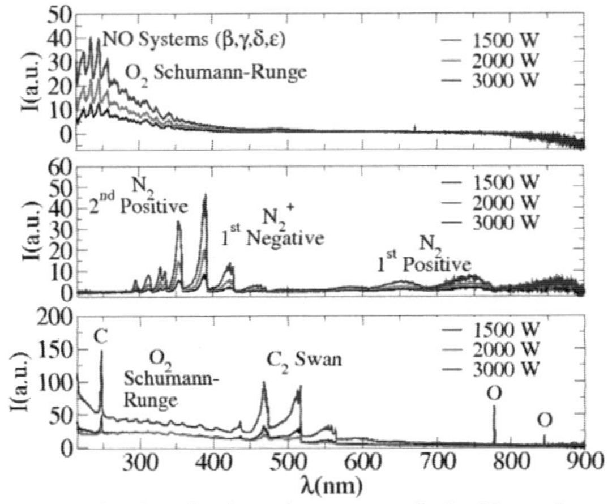

(a) Typical radiative signature of air, N_2 and CO_2 plasmas using different input power at 10 LPM.

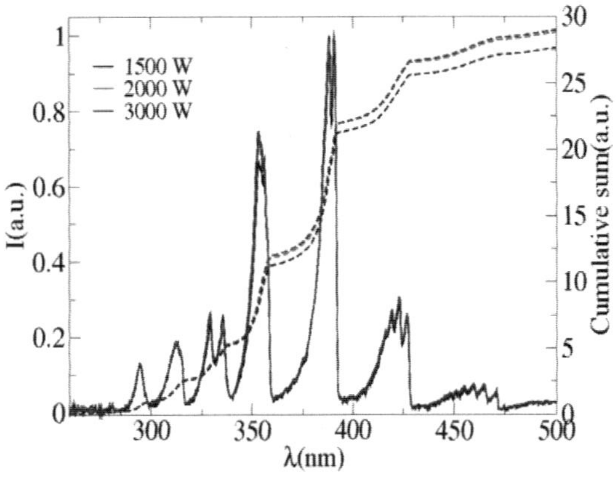

(b) Normalized N_2^+ First Negative system of pure N_2 plasma using different input power.

Figure 13. Low resolution spectra.

In Figure 13(b), normalized radiative signatures of N_2 plasma are presented at different input power. The poor dynamic variation in the shape of the spectra allows us to assess for a weak change in temperature.

The line of sight plasma jet emission is collected through an aperture and then focused with a fused silica lens onto the entry of a spectrometer Jobin-Yvon THR1000 equipped with a 1200 grooves per millimeter grating. The spectrometer is combined with a CCD camera Synapse with a frame of 1024 × 256 pixels of 26 μm side. Emission issued from the jet axis was spatially resolved in the axial direction only. The whole axial profile of the jet was imaged on the CCD array of the camera with a spatial resolution of approximately 800 microns. The spectral resolution was measured to be 0.066 nm, setting the entrance slit of the monochromator to 10 microns. Plasma emission was recorded in the spectral range 350-395 nm, constituted mainly of N_2^+ First Negative $\Delta v=+1$ and 0 vibrational band sequences. Axial emission profile exhibits a monotonic variation peaking just above the waveguide and decreasing along the axis. In this case, Abel transformation can not be applied. For simplicity we assumed in the following that optical line-of-sight emission results from homogeneous isothermal medium in the radial direction. Relative calibration has been performed in the considered spectral range combining the response of a halogen and a deuterium source (DH-2000 Mikropack). Spectra measured at the base of the N_2 plasma jet produced at 1500 W and 5 LPM are evinced in Figure 14(a). The characterization of the N_2 plasma in terms of temperature and relative concentration was made using the spectra measured in the range 350-360 nm by means of common fitting approach consisting in the minimization of the Root Mean Square Error RMSE between the experimental spectrum and the calculated spectrum. Simulated spectrum is calculated assuming a Boltzmann distribution of the energy levels and with no assumption on the concentration of N_2 and N_2+. The fitting is performed accounting for the contribution of N_2 Second positive system by leaving free the relative concentration of N_2 and N_2^+. The equilibrium temperature T_{LTE} and the concentration ratio $\alpha_{exp} = N_2/N_2^+$ are determined pragmatically using spectral intensities normalized to unity and through the minimization of the RMSE defined as:

$$RMSE(T_{LTE},\alpha_{\exp}) = \left(\frac{1}{N} \sum_{i=1}^{N} \left(\overline{S}_{\lambda_i}^{\exp} - \overline{S}_{\lambda_i}^{theo} \left(T_{LTE},\alpha_{\exp} \right) \right)^2 \right)^{1/2}$$

where S^{exp} (resp. S^{theo}) designates the normalized experimental spectrum (resp. calculated spectrum) and N is the number of experimental points in the considered spectral range. The theoretical spectra have been computed on the basis of the advanced tool developed and available at the EM2C laboratory, partially presented in [10] and so-called High Temperature Gas Radiation (HTGR) code.

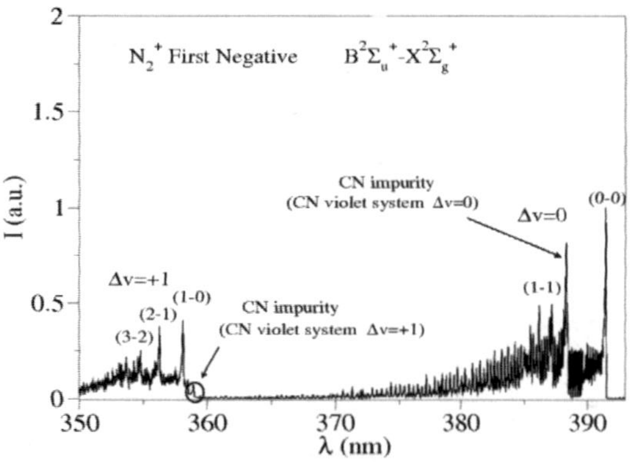

(a) Experimental spectrum of of pure N_2 plasma in the spectral range 350-393 nm.

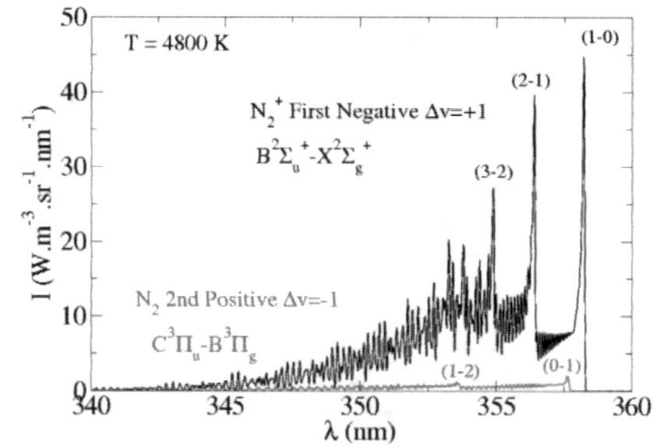

(b) Spectra calculated at equilibrium T_{LTE}=4800 K.

Figure 14. Measured N_2^+ First Negative systems Δv =0,+1 (a) and calculated N_2^+ First Negative Δv =+1 and N_2 Second positive Δv =-1 systems (b).

Following the recommendation of [10] appropriate corrections are applied to the theoretical emission spectra generated at high resolution, in order to elaborate simulated spectra compliant with the experimental data features (apparatus function, air refractive index, experimental background). Possible downgrading of the spectral resolution was also considered to reduce errors rising from uncertainties on spectral positions. The high resolution N_2^+ First Negative $\Delta v=0$ vibrational bands emission measurements displayed in Figure 14(a) are contaminated with overlapping CN Violet system emission, indicating the presence of carbonaceous components in the quartz tube. The unexpected CN emission, evidenced at 359 and 388 nm, deteriorate considerably the $\Delta v=0$ vibrational bands measurements but however remains weak for the band sequence $\Delta v=+1$. Also in the considered temperature range, the N_2^+ $\Delta v=+1$ vibrational bands are overlapped with small but not negligible contribution of the N_2 Second Positive $\Delta v=-1$ vibrational bands. This situation is evidenced in Figure 14(b) where the spectra, calculated at 4800 K using the HTGR code with concentration values reported in Figure 17(a), are plotted in the range 340-360 nm at experimental spectral resolution of 0.066 nm. Omission of the N_2 emission contribution in the range covered by N_2^+ $\Delta v=+1$ bands is too detrimental to the fitting procedure reliability as it is evidenced in Figure 15 where best fits are compared considering only N_2^+ First Negative system and including the N_2 Second Positive system.

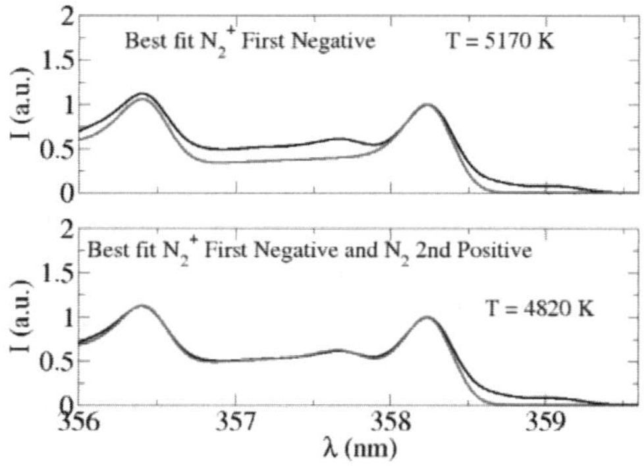

Figure 15. Best fit with experimental spectrum (black) using only N_2^+ calculation (top) and including N_2 2nd Positive calculation (bottom) at 0.3 nm spectral resolution.

The fitting was applied to spectra measured at the plasma jet base downgrading the spectral resolution to 0.3 nm to disclose clearly the effect of the N_2 Second Positive system exclusion. The effect of the spectral resolution applied on experimental and elaborated spectra during the fitting procedure is shown Figure 16(a). A very good agreement is found between temperature profiles derived using 0.066 nm and 0.10 nm spectral resolutions with a temperature of about 4800 K at the plasma jet base. Resulting axial temperature profiles are plotted in Figure 16(b) for different microwave power.

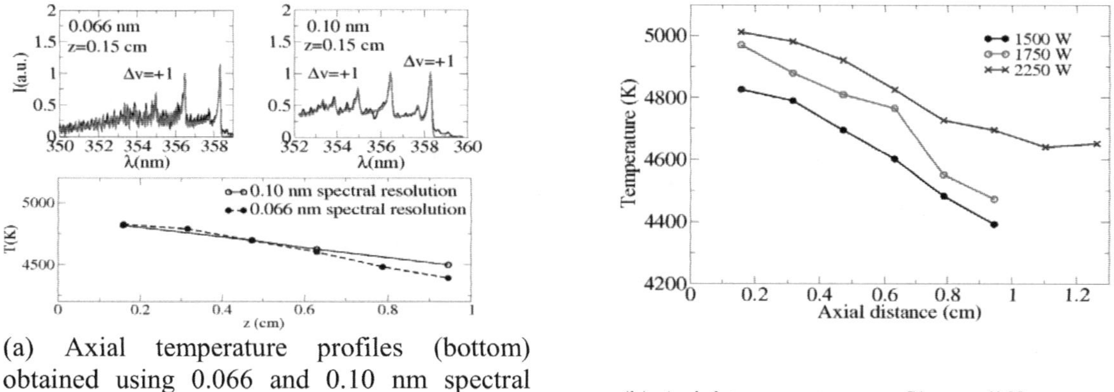

(a) Axial temperature profiles (bottom) obtained using 0.066 and 0.10 nm spectral resolution associated with experimental (black) and calculated (red) best fits (top).

(b) Axial temperature profiles at different power.

Figure 15. T_{LTE} axial temperature profiles.

As previously stressed for air mixture [9] [10], the N_2 plasma temperature is only slightly affected by the power variation. Indeed a power elevation of 50% results in less than 5% temperature increase. Additional microwave power contributes to expanding the plasma volume rather than heating the plasma to a higher temperature, as discussed in [14]. The resulting concentration ratio α_{exp} is plotted in Figure 17(b) for 1500, 1750 and 2250 Watts, with the associated α_{eq} ratio determined using equilibrium concentration calculated at measured temperatures.

Equilibrium concentrations of N_2 plasma, computed at atmospheric pressure using CEA code [15], are reported in Figure 17(a). α_{exp} and α_{eq} increase both of one order of magnitude from the base up to 1 cm downstream, mainly due to the significant N_2^+ concentration drop in conjunction with N_2 concentration weak variation in the considered temperature range as shown Figure 17(a). Saha equilibrium assumption gives the ratio of densities of the same species at different stage of ionization whatever the excited state involved. The ratio α_{eq}/α_{exp} remains nearly constant along the axis (~ 0.6) indicating a slight departure to Saha equilibrium. Also the ratio is weakly affected by the power setting similarly to gas temperature.

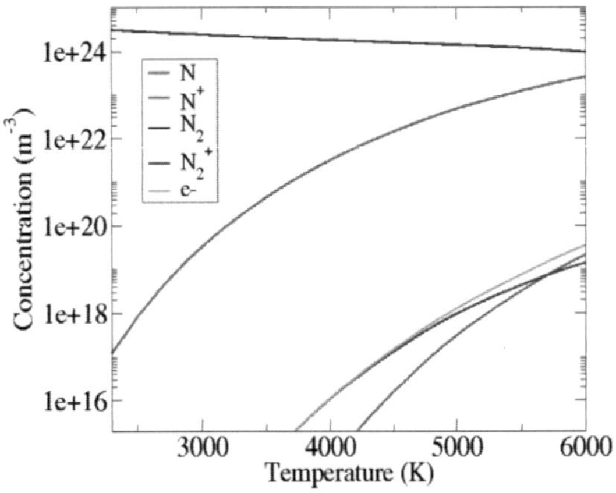

(a) Species concentration of a pure N_2 plasma at 1 bar.

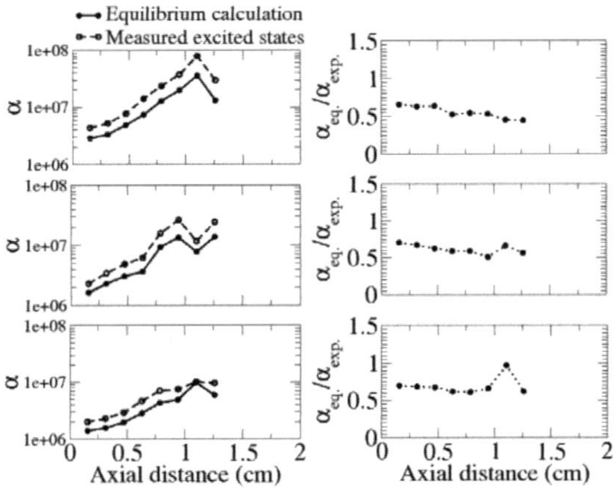

(b) Experimental and calculated α ratios (left). $\alpha_{eq} / \alpha_{exp}$ (right).

Figure 15. Concentration calculated at LTE (a) and chemical equilibrium departure assessment (b).

6. Conclusion and perspectives

Description of microwave sustained plasma jet structures has been established within a large range of operating conditions (0-100 LPM, 1-3.5 kW) for air, N_2, CO_2 and Ar gases. Flow rate and input power contribute highly to the jet dynamics and shape. The fluctuating emission of the plasma jet has been monitored by means of high speed camera imaging. Plasma jet unsteadiness was clearly consistent with the powering circuit oscillations. Thermodynamic characterization of N_2 plasma has been carried out by means of high resolution Optical Emission Spectroscopy using N_2 Second Positive Δv=-1 and N_2^+ First Negative Δv=+1 vibrational bands in the spectral range 350 -360 nm. Thermal equilibrium assumption was fairly assessed and a slight departure to Saha equilibrium has been evidenced using N_2/N_2^+ measured ratio compared with equilibrium concentrations. Deeper investigations including the measurement of N_2 First Positive system would be worthwhile to evaluate the accuracy of excited state kinetic models. The MPT might be considered as a valuable source to provide appropriate reservoir of high temperature gas close to equilibrium. Additional campaign has ensured the operability of the source in supersonic regime and will be presented in future contribution.

References

[1] Andre P, Barinov Yu A, Faure G and M. Shkol'nik S 2011 Modelling radiation spectrum of a discharge with two liquid non-metallic (tap-water) electrodes in air at atmospheric pressure *J. Phys. D: Appl. Phys.* **44** 375203-16

[2] Playez M, Fletcher D G, Marschall J, Fahrenholtz W G, Hilmas G E and Zhu S 2009 Optical Emission Spectroscopy During Plasmatron Testing of ZrB2-SiC Ultrahigh-Temperature Ceramic Composites *J. of thermophysics and heat transfer* **23** 279-85

[3] Panerai F, Chazot O, Tagliaferri E and Rossi G 2009 Testing of the EXPERT thermal protection system junction in a plasma wind tunnel *AIAA paper* 2009/7243

[4] A. Kolesnikov, A. N. Gordeev and S. A. Vasilevskii *2009* Capabilities of RF-Plasmatron IPG-4 for re-entry simulation *J. of Technical Physics* **50** 181-98

[5] Balat-Pichelin M, Badie J M, Berjoan R, Boubert P 2003 Recombination coefficient of atomic oxygen on ceramic materials under earth re-entry conditions by optical emission spectroscopy *Chemical Physics* **291** 181-94

[6] Fauchais P, Vardellea A and Denoirjeana A 1997 Reactive thermal plasmas: ultrafine particle synthesis and coating deposition *Surface and Coatings Technology* **97** Issues 1-3

[7] M. Moisan and Z. Zakrzewski. Plasma 1991 Sources based on the propagation of electromagnetic surface waves *J. Phys. D: Appl. Phys.* **24** 1025

[8] Trivelpiece A W, Gould R W 1959 *Space charge waves in cylindrical plasma columns. J. of Applied Physic* **30** 1783

[9] Green K, Borrs M C, Woskov P P, Flores III G J, Hadidi K and Thomas P 2001 Electronic Excitation Temperature Profiles in an Air Microwave Torch *IEEE Transactions on Plasma Science* **29** 309

[10] Babou Y 2007 Transferts radiatifs dans les plasmas de mélange CO2-N2: base de données spectroscopiques, étude expérimentale et application aux rentrées atmosphèriques *Thèse Laboratoire d'Enérgètique Moléculaire et Macroscopique, Combustion (EM2C) du CNRS et de l'ECP* Paris France

[11] Hong Y C, Uhm H S and Cho S C 2008 Argon Microwave Discharges Sustained at Atmospheric Pressure: Suppression of Plasma Filaments with Molecular Gases *J. of the Korean Physical Society* **53** 3220-23

[12] Wijtvliet R, Felizardo E, Tatarova E, Dias F M, Ferreira C M, Nijdam S, Veldhuizen E V, Kroesen G 2009 Spectroscopic investigation of wave driven microwave plasmas *J. of Applied Physics* **106** 103301

[13] Brown G L and Lopez J M 1990 *Axisymmetric vortex breakdown part II, Physical mechanisms Journal of Fluid Mechanics* **221** 553-76

[14] Timofeev A V 1997 Theory of microwave discharges at atmospheric pressures *Plasma Physics Reports* **23** 158-64

[15] McBride B J, Gordon S 1996 Computer program for calculation of complex chemical equilibrium compositions and application II: User manual and program description *NASA Lewis Research Center Technical Report E-8017-1* Cleveland Ohio USA

Plasma propulsion for geostationary satellites for telecommunication and interplanetary missions

M Dudeck[1], F Doveil[2], N Arcis[3] and S Zurbach[4]

[1]Institut Jean Le Rond d'Alembert, Université Pierre et Marie Curie
4 place Jussieu, 75252 Paris, France
[2]Laboratoire de Physique des Interactions Ions Molécules
Avenue Escadrille Normandie-Niemen, 13397 Marseille, France
[3]French Space Agency, CNES
18 avenue Edouard Belin 31401 Toulouse, France
[4]Snecma - Group Safran
Forêt de Vernon, 27280 Vernon, France

E-mail : michel.dudeck@upmc.fr, fabrice.doveil@univ-provence.fr

Abstract. The advantages of electric propulsion for the orbit maintenance of geostationary satellites for telecommunications are described. Different types of plasma sources for space propulsion are presented. Due to its large performances, one of them, named Hall effect thruster is described in detail and two recent missions in space (Stentor and Smart1) using French Hall thrusters are briefly presented.

1. Introduction

Geostationary telecommunication satellites and geostationary imagers developed by ESA and CNES since 1977 require propulsion systems throughout the mission from its separation from the third stage of the launcher to de-orbiting. At this separation with the third stage, the satellite is injected on a transfer orbit with an apogee at around 36000km from Earth surface. At the apogee of this orbit an applied thrust of 400-500N gives an increment of velocity of ~1,9m.s^{-1} and moves the satellite to a quasi-circular and quasi-equatorial orbit. One or two months are after necessary to make circular and equatorial the orbit, to open solar panels, to verify the instruments, to modify the satellite attitude and to set the satellite at its working longitude.

The satellite has to be maintained at this working position during all the time of the mission (15 to 20 years). However the satellite moves in accordance with Lunar and Sun trajectories, non homogeneity of the Earth and radiative pressure of the Sun. At the altitude of the geostationary orbit (GEO) the drag effect is negligible. The predominant effect is the Lunar-Solar interactions which are greater than the Earth non-homogeneity and the Sun radiation effects. The perturbation effects are briefly commented in the section 2. Consequently, thrusters are required to bring back the satellite to its initial working place and to perform daily North-South and East-West corrections. The thrusters run during one hour each day and deliver a thrust in the range 80-100mN. The variation of velocity is 50m.s^{-1} (North-South) and 5m.s^{-1} (East-West) per year. Sometime it is be necessary to move the satellite during the mission to avoid a collision with an object moving in space. Because of human activities, more than 13000 objects larger than 10cm are in space, more than 200 000 from 1 to 10cm [1]. In the case of a risk a thruster is switch on to move the satellite and to avoid a destructive collision.

Finally the geostationary satellite is de-orbited at the end of its mission to let a place to another satellite (approximately 400 satellites can stay on the GEO orbit). The de-orbitation with an increment of velocity ΔV=3m/s is realized to send the satellite on a higher orbit (of 300km). The de-orbitation is not realized by a return in the low Earth atmosphere in order to have not more fragments close to the Earth.

The choose of the best propulsive device is function of a set of various criteria for the thrusters (cost, mass, size, qualification tests), for the propellant (risks during propellant filling, launch and journey in space, possibilities of supply, physical properties, and cost). Different parameters are used in order to define the behaviour of a thruster. They are recalled in the section 3.

All the operations in space are currently performed with chemical propulsion engines (mainly orbit change from the orbit transfer) or electric propulsion engines (for the other missions). Electric propulsion using ion ejection by an electromagnetic field was suggested in 1906 by R.H.Goddart as a future means of moving in space [2, 3, 4]. The field of electric thrusters is very large and the suggested plasma sources are numerous. However, only a few have been tested, validated and used in space (Pulsed Plasma Thruster - PPT, Field Electron Emission Plasma - FEEP, Hall Effect Thruster - HET and Gridded Ion Thruster - GIE). Other propulsion systems such as Magneto Plasma Dynamics - MPD or Variable Specific Impulse Magnetoplasma Rocket - VASIMR are futuristic concepts. The main electric propulsion devices are shortly presented in the section 4 and theirs performances (thrust, specific impulse, global efficiency) are indicated.

To maintain the position of satellites in geostationary orbit and to perform North-South corrections, the propulsion by Hall effect thruster (cross-electromagnetic field or closed electron drift thrusters) is now recognized as one of the most interesting concept because of its performances mainly in the economy in propellant mass consumption (about 400kg for a satellite 4t-class for a mission of 15 years). The Hall effect thrusters concept is explained in the section 5.

Hall effect plasma propulsion system from Snecma - Group Safran (France) was on board two European spacecrafts: CNES/DGA/France Telecom Stentor probe and ESA probe Smart1 that was successfully sent to a Lunar orbit (Sept. 2003-Nov. 2004) in order to show the interest of Hall Effect Thruster for interplanetary journeys. These two missions are briefly described in the section 6.

2. Perturbation effects

It is easily shown that there is a geostationary circular orbit in the equatorial plane of Earth with a radius R_0 from its center of 44 244.5km. However, this simple result requires a two bodies model (Earth-satellite), a homogeneous and spherical Earth, no gravitational effects from other planets and no pressure effects due to solar wind and solar radiation. The perturbation effect on a geostationary satellite is not negligible and to maintain it at its working location for communications it is necessary to run a thruster around one hour per day during all the time of the mission. However, all the effects have not the same importance.

The Sun emits a flux of electrons and ionized atoms of hydrogen (solar wind) with a mean velocity of $500 km.s^{-1}$ (velocity up to $900 km.s^{-1}$) with a density of *10* particles per cm^3. During 1s, 5.10^{12} particles strike a $1m^2$ surface set perpendicularly to the flux and give a pressure around $10^{-8} Pa$. The effect of the solar wind is negligible on the solar panels of a geostationary satellite but is of interest on very large areas: it is the concept of solar sails (advantages: no fuel, large autonomy). One can indicate the launch of the micro-satellite FASTSAR on Nov.20, 2010 equipped of the solar sail nanoSail-D2 that was deployed on Jan. 20, 2011.

The pressure to a surface exposed to the radiation from the Sun for is expressed from the impulse p of the photons as $p_{rad} = \frac{dp/dt}{A} = \frac{\Phi(W.m^{-2})}{c} = \frac{\sigma T_s^4 D_s^2}{4cD_{S-E}^2}$ where A the surface, σ is the Stephan-Boltzmann constant, T_s the temperature of the photosphere of the Sun which is considered as a black-body with a mean temperature of *5 780 K*, D_s the diameter of the Sun (*1 391 000km*), c the light velocity, and D_{S-E} the Sun-Earth mean distance *(149 597 870km)*. Then, the radiation pressure is $4.610^{-6} Pa$ with pure absorption $C_R = 1$ and normal incidence and as example, its produces a force around $115\mu N$ on the $25m^2$ solar panels of the geostationary satellite Intelsat 10-2. The pressure from the radiation of the Sun is more than *100* times greater than the pressure from the solar wind. One can note the SORCE (Solar Radiation and Climate Experiment) NASA satellite launch in 2004 for an analysis of the Sun radiation in a large spectrum from UV to near-IR and for a measurement of the total Sun irradiance.

The Earth composition being not homogeneous, the gravitational potential is not isotropic in space. A more important effect is the flattening at the poles. The equatorial diameter is around *43km* larger than the pole diameter. In a first approximation, the Earth can be seen as an ellipsoid with symmetry around the North-South poles axis and with an equatorial radius r_e of *6 378,137km* and a polar radius r_p of *6 356,752km*. The gravitational potential is expressed as $U(r,\phi) = -G\frac{mM_T}{r}(1 + (\frac{r_e}{r})^2 J_2.P_2(\sin\phi) + ...)$ where G is the gravitational constant, m and M_T the masses respectively of satellite and Earth, r the distance to the center of Earth, J_2 a non dimensional constant $J_2 \approx \frac{2}{5}\frac{r_e - r_p}{r_e}$, ϕ the latitude angle and P_2 the polynome of Legendre of order two, $P_2(\sin\phi) = \frac{1}{2} - \frac{3}{2}\sin^2\phi$. Using the expression of $U(r,\phi)$, one can deduce the radius of the geostationary trajectory as $R = R_0[1 + \frac{1}{2}(\frac{r_e}{R_0})^2 J_2]$. The change of radius is lower than *1 km*. Morever, the Earth flattening has consequences of the moves of the ascending and descending nodes for a non equatorial trajectory.

The more important effects are the Sun and Lunar gravitational influences. The ratio of the gravitational forces from Sun and Earth for a geostationary satellite is $2.68.10^{-2}$ while the same ratio to the Moon and Earth is $1.4.10^{-4}$ about 100 time less.

3. Thrust parameters

A set of parameters has to be considered to evaluate the properties of a propulsive system in order to choose the best for a given space mission or only for a part of a space mission. More economic parameters (cost, availability of ergol, independency versus furnishers) and ecologic criteria (risk in space and during the launch, toxicity), they are some usual parameters linked to the physical properties of the propulsive system, they are thrust, specific impulse, efficiency, power per mN, divergence of the plume, and total impulse.

- The axial thrust T is defined as the produce of the propulsive mass flow rate $\dot{m}$ by the exhaust velocity V_e of the propellant along the thruster axis.

$$\text{Axial thrust } T(N) = \dot{m}V_e \qquad (1)$$

A high thrust allows a decreasing of the time of an orbit change but requires a large propellant consumption.

- The specific impulse *Isp* is defined as the ratio of the thrust to the mass flow and to the intensity of gravity at the surface of Earth.

$$\text{Specific impulse } Isp(s) = \frac{T}{\dot{m} g_0} \tag{2}$$

where g_0 is the strength of the Earth gravitational field at the Earth surface. The specific impulse is measured in seconds. The well known relation (rocket relation) from C.Tsiolkovski (1857-1935) $\Delta M = M_0(1 - e^{-\frac{\Delta V}{V_e}}) = M_0(1 - e^{-\frac{\Delta V}{g_0 Isp}})$ where ΔM is the consumption of propellant, M_0 the initial mass, ΔV the required velocity variation of the spacecraft for a defined mission, shows that a high specific impulse allows to decrease the mass of propellant and consequently to decrease the mass of the satellite or for the same mass to have more on-board scientific instruments. In the rocket equation the term of difference of pressure is neglected. Using (1) the specific impulse is proportional to the exhaust velocity $Isp = \frac{V_e}{g_0}$: a high exhaust velocity induces a high specific impulse and then a low propellant consumption. The specific impulse can considered as the time for a virtual thruster to deliver a thrust of *1N* with a mass flow of *0.1kg/s*.

- The global efficiency of the thrusters

$$\text{Efficiency } \eta = \frac{T^2}{2\dot{m}P} \tag{3}$$

where *P* is the input power required by the thruster.

- Power *p* per unit of thrust

$$\text{Power per } mN \quad p(W/mN) = \frac{P}{T} \tag{4}$$

This parameter *p* which indicates the input power consumed per unit of thrust has to be minimized. It is measured in *W/N* or in *W/mN*.

- Plume half angle

The plume half angle θ containing 90% (sometimes 95%) of the propulsive flux:

$$\int_0^{\theta} j(\theta')\sin\theta' d\theta' / \int_0^{\pi/2} j(\theta')\sin\theta' d\theta' = 0.9 \tag{5}$$

where $j(\theta')$ is the density of flux. This parameter has to be low because a best focusing increases the efficiency of the thrust and minimizes possible interactions between plume and solar arrays depending of the position of the thruster.

- Total impulse *I*,

The previous parameters are time depending and they can change during the mission in space. The total impulse is defined as,

$$\text{Total impulse } I(mN.s) = \int_0^t T(t)dt \tag{6}$$

where *t* is the time operation of the thruster in space. In case of a constant thrust the total impulse is: $I(mN.s) = T\Delta t = \dot{m}V_e\Delta t$.

4. Solar electric propulsion

Solar power is used as power source in near-Earth orbits up to the planet Mars. Solar panels are coupled with batteries for geostationary satellites.

IOP Conf. Series: Materials Science and Engineering **29** (2012) 012010 doi:10.1088/1757-899X/29/1/012010

4.1. *Solar panel for Electric Propulsion - SEP*

A spacecraft needs power to run the on-board computer, sensors, scientific instruments, heating and cooling systems, and antenna. The spacecraft are equipped of solar panels which convert light coming from the Sun into electricity; solar panels have to provide enough power during all the time of the space mission of the spacecraft. The first spacecraft to use solar panels was the Vanguard 1 small satellite - 1.47kg - launched in USA in March, 1958.

For a power of 1366W.m^{-2} received near the Earth from the sun and with an efficiency of the solar panels of 18%, only 246Watts are available (for the satellite Intelsat-10.2 having 25m^2 of solar panels, the available power is around 6kW).

It is well-known that the cells made of amorph silicon allows an efficiency of 5-7% and the monocristalline silicone an efficiency of 14-16%. Gallium arsenide-based solar cells with a higher efficiency are developed for space applications. The most efficient solar cells currently in production are multi-junction cells. They use a set of several layers of gallium arsenide and silicon to absorb the largest spectrum of received light. Multi-junction cells are able to give an efficiency around 29%. For future missions using photovoltaic energy, it is desirable to reduce solar array mass, and to increase the power generated per unit area, new photovoltaic cell materials and solar concentrators that intensify the incident light are required. Intensive researches are currently carried out in laboratories to increase the efficiency of solar cells. A solar cell efficiency of 40.8% has been obtained by Boeing-Spectrolab (2008), 39.7 % in the frame of the Full Spectrum, FP-6 European project, and recently 41% in the FP-7 Hiper programme.

Morever, it is necessary to use batteries when the satellite is moving in the shadow part of its trajectory. Then, solar panels have to re-charge the batteries during the time of sun light. Solar power can only used in near-Earth orbits, no further than Mars. For example, they are used for:

- Venus Express, ESA, launched on Nov. 9th, 2005 (0.718-0.728 AU)
- SMART-1 ESA, launched on for a Lunar mission, Sept.27th, 2003
- Magellan, Mars Global Surveyor, and Mars Observer for Mars exploration used solar power (1.381- 1.666 AU)
- Stardust spacecraft (2 AU)
- Rosetta space probe, launched on March 2nd, 2004 used solar panels as far as the orbit of Jupiter (4.95-5.46 AU).

In order to prepare long journeys in space, nuclear reactors have been studied in Soviet Union in the Topaz programme. The Soviet Topaz-I unit has been on-board the satellite Kosmos 1818 launched on 1st Feb. 1987. The Topaz-II reactor has a capacity of about 10kW and new reactors with a power of 25kW have been designed. A 1MW nuclear reactor space power is currently studied in Russian from mid-2010 (NEP cargo ferry, [5]). Nuclear Electric Power generation has been studied in the frame of the Hiper (FP-7 European programme).

4.2. *Electric propulsion*

Electric propulsion is based on the ejection of charged particles by an electromagnetic field that allows a large ejection velocity and consequently a specific impulse (2000s) greater than the specific impulse from chemical thrusters (300s). The gain in mass is significative because of the cost of 1kg send in space (today from15 000 to 25 000€).

In the ablative *Pulsed Plasma Thrusters (PPT),* the ions are produced by successives sparks with a frequency of a few Hz between two high voltage electrodes set in front of a solid propergol (Teflon). PPT delivers a high specific impulse (800-1200s), a thrust around 1mN and operates with an electric power up lower than 100W [6, 7, 8]. The PPT are used for attitude control, micro-satellites and low thrust maneuvers. Zond2 was the first spacecraft launched in space with a PPT (Soviet Union - 1964) to Mars without success (6 thrusters for attitude control).

A *Field Emission Electric Plasma* thruster (FEEP) uses the flow of a liquid metal (caesium) through a slit limited by two metal surfaces. The extracted positive ions Cs^+ are accelerated by an electric field (potential ~10kV). The very low thrust (micro-N to milli-N) of FEEP engines is used to compensate the drag effect, for attitude control (ESA/NASA LISA Pathfinder [8] and Microscope missions) or for formation flying. A high specific impulse is obtained by this micro-propulsion engine: more than 10 000s. Alta SpA (Italy) [10] and Space propulsion (Austria) [11] are developing FEEP.

In a *Gridded Ion Engine* (GIE) also called Gridded Ion Thruster (GIT), the positive ions (Xenon) are obtained by electron impacts (Kaufman or radiofrequency ion thrusters), then extracted and accelerated by a set of 2 or 3 multi-aperture grids. This electrostatic ion engine delivers a thrust up to 670 mN and a Isp up to 9620 s for a power of 39.3kW, Gleen Research Center, NASA [12]. GIE is used for satellite station keeping and also for deep space trips. For the successful exploration of the Comete Borrelly (NASA Deep Space 1 mission; 1998-2001) the NEXT thrusters (90mN at a power 2,3kW) from the Jet Propulsion laboratory were used [13]. The Asteroid Explorer "Hayabusa" (JAXA, Japan) reached the asteroid Itokawa in 2005 and come back to Earth on June 13th, 2010 with collected dusts [14]. Its propulsion engine was a GIE 2,6kW. One can also note the use of a GIE thrusters (2 RIT and 2 UK10) to replace the satellite Artemis (2001) on the GEO orbit after a failure of the launcher [15]. The satellitePanAmSat-5 (PAS-5) launched in 1997 was equipped of a XIPS 601 HP (P=500W, Isp=2568s, T=18mN) (Hughes Space and Com. / Boeing Satellites Syst.). The satellite Astra 2A launched in 1998 (Hughes Space and Com. / Boeing Satellites Syst.) uses 4 ions thrusters XIPS 601 HP.

Arc-jet thrusters are electro-thermal engines. The gas (hydrogen, ammonia, hydrazine) enters in a chamber and cross the throat of a nozzle used as anode. An arc is sustained between this anode and a cathode set in the chamber. The gas is heated and expanded through the nozzle. Arc-jets generate a thrust in the range 0.01N - 0.5N with a specific impulse between 500 and 1000s, Busek Comp. Inc [16], Institut fur Raumfahrtsystem (IRS – Stuttgart university) [17,18,19]. A 750W ammonia arc-jet thruster has been manufactured at IRS (AMSAT P3-D satellite, 1994 Germany) [20].

Hall Effect Thrusters (HET) also named "closed-drift thrusters", "Stationary Plasma Thruster" (SPT) or "Propulsion par Plasma pour Satellites" (PPS) use a partially magnetized plasma discharge in a cross-electromagnetic field. To maintain the satellites in geostationary orbit with North-South and East-West corrections, Hall Effect Thruster is now recognized as one of the most interesting concept because of its performances mainly in the economy in propellant mass (saving about 400kg for a 4t-class satellite for a mission of 15 years). After the first HET on the Meteor meteorologic satellite (two SPT 60, USSR, 1972), a few hundreds of SPT was used on-board Russian satellites. A large family of HET with different input powers is currently developed:

- Fakel EDB, Russia: SPT 25, 35, 50, 60, 70, 100, 140, 200, 290

The stationkeeping of a few European satellites for telecommunication as Immarsat 4, Intelsat 10 (Astrium spacecrafts) is realized by SPT-100

- Snecma-Safran group, France: PPS-1350, PPS300, PPS-5000 and PPS20k for future satellites.

The European technological satellite Stentor [21] was equipped with two PPS1350 (T=70mN, Isp =1600s, η=55%, Snecma, France) and two SPT-100 (Fakel EDB, Russia). The ESA Smart1 Lunar mission [22] used a PPS1350G (Snecma, France). PPS1350 of Snecma will equip the Alphabus plateform (CNES-ESA) able of carrying a payload mass up to 1500kg

- Busek Co. Inc., USA: BHT-200 (P=200W, T=13mN, Isp=1375s), BHT-400, BHT-1000, BHT-1500, BHT-8000, BHT-20k (P=20kW, T=1090mN, Isp=2750s).

The first American HET in space (16 Dec. 2006) was the BHT-200 from Busek Co. United States (T=12.8mN, Isp=1390s) on the TacSat-2 technological microsatellite, Dec.16, 2006

- Pratt & Whitney Space Propulsion, NASA, USA:T-40, T-140, T-220
- Keldysh Research Institute, Russia: KM-32 and in collaboration with Astrium: ROS-99, ROS-2000

- Rafael Space Systems, Israel: IHET-300, two IHET-300 will operate on the Raphael Venus Satellite.

The *High Efficiency Multistage Plasma thrusters* (HEMP) is an engine developed by Thales (Germany) since 2000 [23, 24]. In this technology of thrusters, a permanent periodic magnet configuration focus the plasma along the thruster axis and the propellant (Xenon) is ionized by the electron flow. The magnetic field is radial and the axial electric field allows the acceleration of the ion to the channel exit. Two versions of HEMP are currently tested: HEMP-T-3050 (P=1.5kW, T=50-80mN, Isp=2000-3000s) and HEMP-T-30250 (P=7.5kW, T=250mN, Isp=2000-3000s. An HEMP thruster will be used on the ARTES 11 Small GEO ESA/Hispasat satellite (2012).

5. Basic concept of Hall Effect Thruster

Hall effect thrusters (schematic view shown in figure 1) are advanced electrostatic propulsion devices. The plasma discharge is sustained between two coaxial dielectric cylinders and between an external hollow cathode emitting electrons and an anode set in the bottom of the annular chamber. The discharge voltage is generally from 300V to 350V but high voltage up to 1000V have been tested in order to increase the specific impulse. The propellant is generally injected through the anode. Xenon is the more suitable propellant because of its low first ionisation level (12.23eV) requiring a low energy consumption, its high mass to maintain a satisfying level of thrust, its absence of toxicity and its thermodynamical properties. Due to the low pressure of the channel (~ 10^{-4} at the bottom 10^{-5}mbars at the exit), it is necessary to trap the electrons by a radial magnetic field (~ 0.02T at the channel exhaust for a PPS100). This magnetic field is created by external magnetization coils (inner and outer coils).

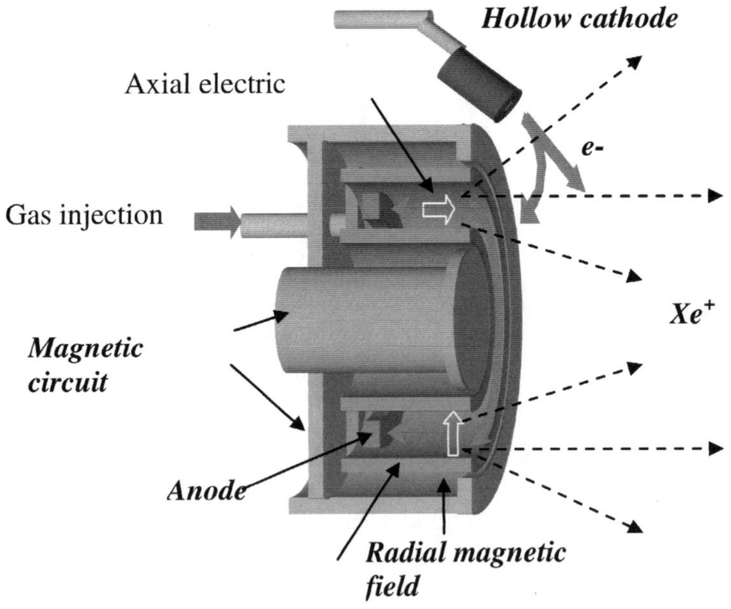

Figure 1. View of a Hall Effect Thruster [25].

The electron Larmor radius ($\sim 10^{-3} m$) is small compared to the chamber dimensions ($\sim 10^{-2} m$) and consequently electrons are confined and magnetized. Xe^+ ions are then created by inelastic electron-neutral collisions and only 10 to 15% of ion Xe^{++} is created in the discharge. In the channel the electrons have three origins: external hollow cathode, ionization process and wall. Typically, the plasma parameters are an electronic density of 10^{-11} cm^{-3} and an electron temperature around 20eV at the channel exhaust. Then the electron plasma frequency is around 3GHz and the cyclotronic electron frequency is around 300MHz (the hybrid electron frequency is closed to the electron plasma frequency).

The electrons have to move toward the anode and several mechanisms have been suggested to explain the electron transverse transport through the magnetic lines. For a low pressure discharge (10^{-4}-10^{-5} mbars) the electron-neutral xenon collisions frequency is insufficient to explain this transport in the channel because of the low neutral density at the channel exhaust. A.I.Morozov [26, 27] suggested a "wall" transport: electrons from the sheath scattering and secondary electrons allow an electron transport toward the anode by successive traps along the magnetic lines. The more recent and more satisfying explanation is done by the presence of high frequency fluctuations (1-10MHz) of the azimuthal electric field of the plasma discharge [27, 28]. A Particle-In-Cell model (azimuthal and axial *x* axis) developed by Adam *et al.* [29] have permitted to calculate the electron mobility and to show that the collisional mobility is not sufficient in the region of low neutral density (see figure 2). Micro-structures of the azimuthal electric field are observed (PIC model) and they are in interaction with electrons, they induce a sufficient electron mobility [30, 31].

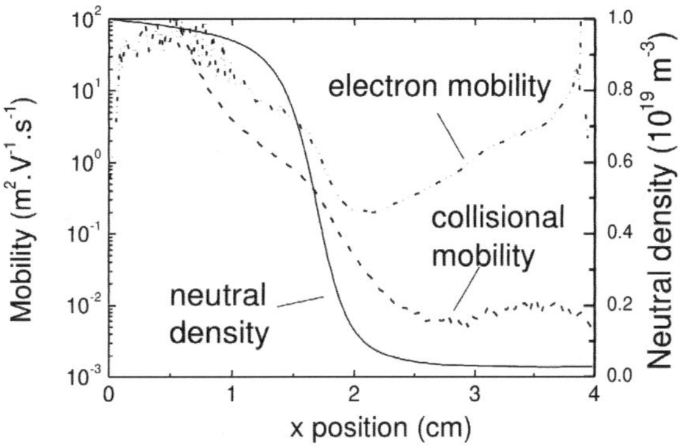

Figure 2. Electron and collisional mobility profiles
and neutral density profile from the PIC model [30,31]
x = 0cm: bottom of the channel, x = 2,5cm: channel exit, x =4cm: limit of the domain

The fall of the electron mobility at the channel exhaust in reason of the increase of radial magnetic field induces an axial electric field with a maximum value around 500V/cm (PPS1350). This axial electric field is generated without grids: it is one of the advantages of the HET compared to the GIE.

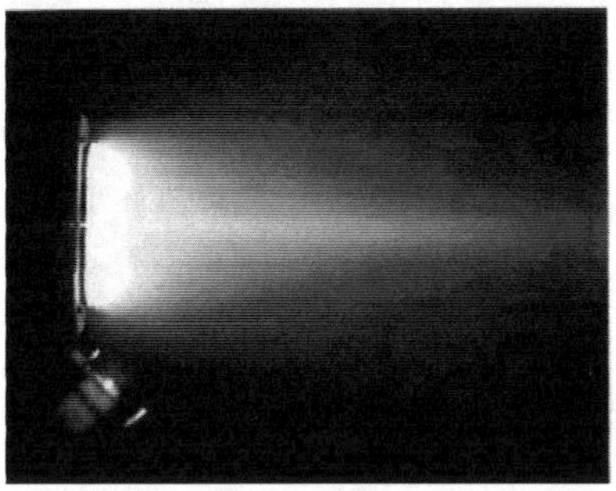

Figure 3. Xenon plasma plume from the PPS100-ML (laboratory model)
Pivoine facility at laboratoire d'Aérothermique, CNRS Orléans, France

IOP Conf. Series: Materials Science and Engineering **29** (2012) 012010 doi:10.1088/1757-899X/29/1/012010

The plasma discharge is obtained in a crossed ExB field and the electrons present an azimuthal drift velocity (~ 10^6m/s). The Xenon ions are axially accelerated out of the thrusters with a high velocity (15-20km/s) and this high velocity conducts to a high specific impulse (1650s for the PPS-100ML). Two ionic zones appear in the channel, the first one is the ionization zone inside the channel where ions are produced and the second one is the acceleration zone located inside and outside the channel. A best efficiency is obtained when these two zones are separated. There is an overlap of these two zones in a classic HET and their separation is the task of the two stage thruster MAG B [32] from MIREA (Moscow- Russia).

The Larmor radius of the ions is around 1m and then the ions are not magnetized inside the channel (no cyclotronic ion frequency). The extraction of the ions from the channel depends of the shape of the electric potential lense which is closed to the magnetic lense. Measurements performed by fluorescence induced by laser [33, 34] shown that a large part of the axial acceleration of the ions takes place out of the channel (see figure 4) and that the ions presents low frequency oscillations [36] (see figure 5) following the low frequency fluctuations of the discharge current in the range 10-30kHz. This mode of fluctuations is named "breathing mode" and due to the depletion of neutral in the channel [35] (see figure 5).

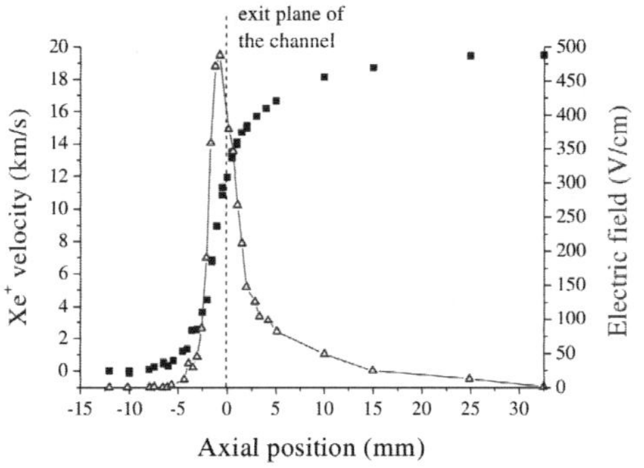

Figure 4. Xe^+ velocity measurement by Doppler shifted laser induced fluorescence and axial electric field [29].

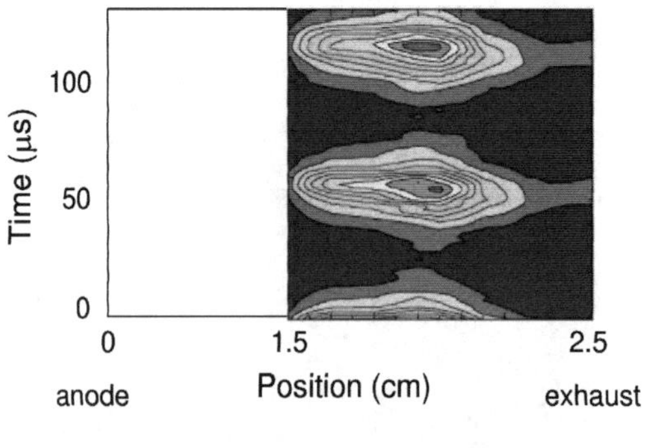

Figure 5. Low frequency ion fluctuations in the discharge channel from the bottom (anode) to the exhaust [36].

Another complex and fundamental point is related to the role of the plasma surface interactions on the behaviour of the plasma discharge of hall thrusters. A surface erosion appears in the region in contact with the acceleration zone where the ions have a divergent trajectory related to the thruster axis and impact the surface of the channel. This erosion is mainly due to ion sputtering. Morever, superimposed "anomalous" erosion process is observed with all Hall thruster (see figure 6). This erosion is characterized by regular visible ridges. Today, no satisfying theory is suggested to explain this "anomalous" erosion: effect of electrons? effect of electrons and ions?

Figure 6. Anomalous erosion on the walls of a Hall effect thrusters (document Snecma – Safran group)

The secondary electron emission rate (S.E.E.) is now recognized as having a significant contribution on the plasma discharge: values of the plasma discharge current and low frequency fluctuations (amplitude and frequency) [37, 38]. It seems from various tests with different materials, shown in figure 7, that a secondary electron emission (s.e.e.) rate lower than one is required to optimize the plasma properties. A s.e.e. rate greater than one induces a potential sheath saturation effect increasing the discharge current and the electron energy deposition on the walls. The role of the VUV radiation on the energy deposition on the walls is currently studied.

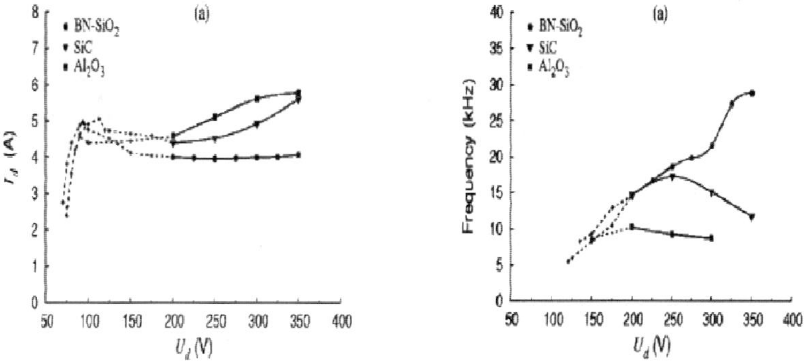

Figure 7. Discharge current and low frequency measurements for surfaces made of BN-SiO_2, SiC, Al_2O_3 [37,38]

The temperature of the walls changes the temperature of the neutral Xenon by thermal accommodation, but the Xe temperature has not a great influence on the plasma flow properties.

6. Two European missions

Snecma-Safran group (France) has developed a series of Hall effect plasma thrusters (PPS1350,PPS5000) and PPS1350 was equipped on-board the CNES/DGA/France Telecom Stentor probe and the ESA probe Smart1 that was successfully sent to a Lunar orbit (Sept 2003-Nov. 2004).

Stentor for "Satellite de Télécommunications pour Expérimenter les Nouvelles Technologies en Orbite" was a French experimental programme to validate advanced technologies for next generations of telecommunications spacecrafts (see figure 8). This programme was supported by CNES (French Space Agency), France Telecom (French Telecommunications Operator) and DGA (Ministry of Defence). Stentor was equipped with two PPS1350 from Snecma (France), two SPT100 from Fakel (Russia) and instruments as micro-quartz balances and electrostatic probes. Unfortunately, the Stentor satellite was lost in the launch failure on 11 Dec. 2002.

SMART1 for "Small Missions for Advanced Research in Technology" with a launch on 27th Sept. 2003 is the first mission with an use of Hall thrusters outside of Earth'orbit (see figure 9 for SMART 1 [39]). It is the first European mission to the Moon. With an initial mass of 366.5kg, 19kg of scientific instruments and only 82kg of Xenon, the probe reached the Earth-Lunar Lagrange point on 15th Nov. 2004 after a trip of 84Mkm and a Lunar orbit (3000-10000km) after a trip of 15 months from the launch. Finally, the probe impacted the Lunar surface on 3rd Sept. 2006 after a trio of 1072 days. This Lunar mission has been done with success thanks to a PPS1350G (T=68mN, Isp=1640s, Snecma, France). Only 80kg of Xenon has been consumed for a total thruster run of 4600h with a maximum of 240h of run without interruption.

Figure 8. Stentor - Dec.11th, 2002
(*copyright ESA*)

Figure 9. Smart 1 - Sep.27th, 2003
(*copyright ESA*)

7. Conclusion

Electric thrusters and especially Hall effect thrusters now present a great interest for space applications: for orbit maintenance of geostationary satellites for telecommunication as shown the large number of satellites using this propulsive device (~ 400) and in future for orbit transfers and interplanetary missions as already demonstrated by the success of the European Smart 1 mission. A large range of power for HET from a few 100W to 100kW or more, allows to use this propulsion device for a wide domain of missions in space. Recently, with the purpose of the preparation of future long journeys in space, a PPS20k-ML for a power up to 20kW has been studied, manufactured and tested [40] in the frame of the FP-7 Space European programme.

Since 1996, a large scientific effort has been developped in France in the domain of Hall effect thruster to deepen the knowledge of the complex and coupled phenomena appearing in the plasma discharge such as plasma fluctuation in a broad frequency spectrum, wall interactions, influence of the magnetic topography and electron conductivity. This effort is performed with three main objectives: greater lifetime, better performances, and innovation in Hall effect thrusters. This activity associates CNES, Snecma, French laboratories with scientific collaborations with foreign institutes as IPPLM (Warsaw, Poland), KhAI (Kharkov, Ukraine), ISFM (Gdansk, Poland), Univ. Charles (Prague, Rep. Tch) and IST (Lisbonne, Portugal). The research uses theoretical developments (Karney model), numerical simulations (PIC, MC, fluid models and wave equation) and experiments. These experiments are carried out in the national ground test facility Pivoine (ICARE laboratory, CNRS Orleans, France) [41] using a large set of plasma diagnositics. The research programme on Hall thrusters is developed in the frame of collaborative national scientific research programme (GDR "Propulsion par plasma dans l'espace" 1996-2011 and GIS "Propulsion par plasma dans l'espace" from 2012).

References

[1] Bonnal Ch 2010 Orbital debris: What to do with the Kessler syndrome ... *Space Propulsion Conf. 3-6 May* (San Sebatian Spain)

[2] Goddard R H 1970 *The papers of Robert H Goddard* ed E C Goddard and C E Pendray (New York: MacGraw-Hill)

[3] Wilbur P J, Jahn R G and Curran F 1991 *Space Electric Propulsion Plasmas IEEE Transactions on Plasma Science* **19** N°6

[4] Choueiri E Y 2004 A Critical History of Electric Propulsion: The First Fifty Years (1906-1956) *J. of Propulsion and Power* **20** N°2 pp 193-203

[5] Loeb H W, Feili D, Popov G A, Obukhov V A, Balashov V V, Mogulkin A I, Murashko V M and Nesterenko A N 2011 Design of High-Power High-Specific Impulse RF-Ion Thruster IEPC-2011-290 *32nd Int. Electric Propulsion Conf. September 11 – 15* (Wiesbaden Germany)

[6] Burton R L and Turchi P J 1998 Pulsed Plasma Thruster *J. of Spacecraft and Rockets* **14** N°5 716-735

[7] Popov G, Antropov N, Diakonov G, Orlov M, Tyutin V and Yakovlev V 2001 Experimental Study of Plasma Parameters in High-Efficiency Pulsed Plasma Thrusters *26th IEPC* paper 01-163

[8] Pillet N, Pouilloux B, Pelipenko P, Folliard J, Darnon F, Chesta E, Bousquet P, Alby F, Antropov N N, Diakonov G A, Jakovlev V N, Orlov M M, Trubnikov P M and Tyutin V K 2005 Pulsed Plasma Thruster Option for Myriade Deorbiting *European Conf. for Aerospace Sciences EUCASS 4-7 July* (Moscow Russia)

[9] Nicolini D 2007 LISA Pathfinder Field Emission Thruster System Development Progam *29th International Electric Propulsion Conference September 17-20* (Florence Italy)

[10] Marcuccio S, Saviozzi M, Priami L and Andrenucci M 2003 Experimental Performance of 1 mN-class FEEP Thrusters *28th Int. Electric Propulsion Conf. March 17-21* (Toulouse France)

[11] Genovese A, Budrini N, Tajmar M and Steiger W 2003 Experimental Performance of 1 mN-class FEEP Thrusters *28th Int. Electric Propulsion Conf. March 17-21* (Toulouse France)

[12] HIPEP 2004 *NASA report* NASA/TM-213194

[13] Nordholt J E et al 2003 Deep Space 1 encounter with Comet 19P/Borrelly: Ion composition measurements by the PEPE mass spectrometer *Geophysical Research letters* **30** 1465

[14] Kubota T et al 2009 Guidance and Navigation Scheme for Hayabusa Asteroid Exploration and Sample Return Mission *ESA workshop on GNC for Small Body Mission ESA/ESTEC January 14-15* (Noordwijk The Netherland)

[15] Killinger R et al 2003 Artemis Orbit Raising Inflight Experience with Ion Propulsion

IEPC paper 0096 *28th Int. Electric Propulsion Conf. March 17-21* (Toulouse France)

[16] Hohman K, Brogan T, Hruby V, Annen K and Brown R 2003 Two kilowatt bipropellant arcjet developments IEPC paper 0122 *28th Int. Electric Propulsion Conf. March 17-21* (Toulouse France)

[17] Heiermann J and Auweter-Kurtz M 2003 Numerical simulation of the Arthur-2 arcjet IEPC paper 0033 *28th Int. Electric Propulsion Conf. March 17-21* (Toulouse France)

[18] Bock D, Auweter-Kurtz M and Kurtz H 2005 1kW ammonia arcjet development for a science mission to the Moon *29th Int. Electric Propulsion Conf. IEPC October 31 – November 4* paper 2007-075 (Princeton University)

[19] Bock D, Roser H P, Herdrich G and Auweter-Kurtz M 2007 Subscale lifecycle test of thermal arcjet thruster Talos for the Lunar mission BW1 *30th Int. Electric Propulsion Conf. IEPC September 17-20* paper 2007-137 (Florence Italy)

[20] de Dalmau J and Gigou J 1997 Ariane-5: Learning from Flight 501 and Preparing for 502 *ESA Bulletin* Number 89 February

[21] Darnon F, Diris J-P, Hoarau J, Torres L and Grassin 2001 Th Plasma Propulsion on STENTOR Satellite: In Flight Acceptance Operations ans Experimental Program *27th Int. Electric Propulsion Conf. October 15-19* paper IEPC-01-167 (Pasadena California USA)

[22] Smart-1 2002 European Space Agency BR-191 June and Smart-1: A Solar-Powered Visit to the Moon 2003 *ESA bulletin* Number 113 February

[23] Kornfeld G, Koch N and Harmann H P 2007 Multi Stage Plasma Thruster Development for HEMP-T 3050 and HEMP-T 30250 IEPC-2007-110 *30th IEPC Conf. September 17-20* (Florence Italy)

[24] Koch N et al 2011 Development, Qualification and Delivery Status of the HEMPT based on\Ion Propulsion System for SmallGEO *30th IEPC Conf. September 11-18* (Weisbaden Germany)

[25] Gascon N 2000 Etude des propulseurs plasmiques à effet Hall pour systèmes spatiaux Performances propriétés des décharges et modélisation hydrodynamique *Thesis (in French)* December 15 University of Provence Marseille France

[26] Morozov A I and Savelyev V V Fundamentals of stationary plasma thruster theory *Reviews of Plasma Physics* 21 Ed By B B Kadomtsev and V D Shafranov Kluwer (Acad/Plenum Publishers New-York)

[27] Morozov A I and Savelev V V 2001 Theory of the near-wall conductivity *Plasma Physics Reports* **27** N°7 607

[28] Tsikata S 2009 Small-scale electron density fluctuations in the Hall thruster investigated by collective light scattering *Thesis* November 19 Ecole Polytechnique France

[29] Tsikata S, Lemoine N, Pisarev V and Grésillon D 2009 Dispersion relations of electron density fluctuations in a Hall thrusters plasma observed by collective light scattering *Physics of Plasmas* **16** 033506

[30] Adam J C, Heron A and Laval G 2004 Study of stationary plasma thrusters using 2D fully kinetic simulation *Physics of Plasmas* **11** 295-305

[31] Coche P and Garrigues L 2011 Study of stochastic effects in a Hall effect thruster using a test particles Monte-Carlo model *32nd Int. Electric Propulsion Conf. IEPC-2011-255 September 11-15* (Wiesbaden Germany)

[32] Perez-Luna J, M.Hagelaar G J, Garrigues L and Boeuf J. P 2007 Model analysis of a double stage Hall effect thruster with double-peaked magnetic field and intermediate electrode *30th International Electric Propulsion Conference paper IEPC-2007-124 September* (Florence Italy)

[33] Sadeghi N, Dorval N, Bonnet J, Pigache D, Kaldec-Philippe C and Bouchoule 1999 A Velocity measurements of Xe^+ in Stationary Plasma Thruster using LIF AIAA 99-2429 *35th AIAA/ASME/SAE/ASEE Joint Propulsion Conf. & Exhibit June 20-24* Los (Angeles CA USA)

[34] Mazouffre S, Gawron D, Kulaev V and Sadeghi N 2008 Xe+ Ion Transport in the Crossed-Field Discharge of a 5-kW-Class Hall Effect Thruster *IEEE Transactions on Plasma Sciences* **36** N°5 1967-1976

[35] Bœuf J P and Garrigues L 1998 Low Frequency Oscillations in a Stationary Plasma Thruster *Journal of Applied Physics* **84** 3541

[36] Pagnon D, Darnon F, Roche S, Béchu S, Magne L, Minea T, Bouchoule A, Touzeau M, Lasgorceix P 1999 Time Resolved Characterization of the Plasma and the Plume of a SPT Thruster – AIAA-99-2428 *35th AIAA/ASME/SAE/ASEE Joint Propulsion Conf. & Exhibit June 20-24* (Los Angeles CA USA)

[37] Gascon N, Dudeck M, Barral S 2003 Wall material effects in stationary plasma thruster I: Parametric studies of an SPT-100, *Physics of Plasma* **10** n°10 4123-4136

[38] Barral S, Makowski K, Peradzynski Z, Gascon N and Dudeck M 2003 Effect of wall material in stationary plasma thrusters II – Near-wall and in-all conductivity *Physics of Plasma* **10** n°10 4127-4152

[39] Koppel C, Marchandise F, Prioul M, Estublier D and Darnon F 2005 The SMART-1 Electric Propulsion Subsystem around the Moon: In Flight Experience *Proc. of 41st AIAA/ASME/SAE/ASEE Joint Propulsion Conf.* Reston VA Paper N° AIAA-2005-3671

[40] Zurbach S, Cornu N and Lasgorceix P 2011 Performance Evaluation of a 20 kW Hall Effect Thruster *32nd International Electric Propulsion Conference, IEPC-2011-020 September 11-15* (Wiesbaden Germany)

[41] Lasgorceix P, Perot C and Dudeck M 1997 PIVOINE ground test facility for ion thrusters testing *Proc. of the Second European Spacecraft Propulsion Conf. May 27-29*

IOP Conf. Series: Materials Science and Engineering **29** (2012) 012011 doi:10.1088/1757-899X/29/1/012011

Plasma Treatments and Biomass Gasification

J Luche[1], Q Falcoz[2], T Bastien[2], J P Leninger[2], K Arabi[1], O Aubry[1], A Khacef[1,†], J M Cormier[1] and J Lédé[2]

[1]GREMI, CNRS-Université d'Orléans, 14 Rue d'Issoudun, BP 6744, 45067 Orléans, France
[2]LRGP, Université de Nancy, 1 Rue Grandville, BP 20451, 54001 Nancy, France,

E-mail: ahmed.khacef@univ-orleans.fr

Abstract. Exploitation of forest resources for energy production includes various methods of biomass processing. Gasification is one of the ways to recover energy from biomass. Syngas produced from biomass can be used to power internal combustion engines or, after purification, to supply fuel cells. Recent studies have shown the potential to improve conventional biomass processing by coupling a plasma reactor to a pyrolysis cyclone reactor. The role of the plasma is twofold: it acts as a purification stage by reducing production of tars and aerosols, and simultaneously produces a rich hydrogen syngas. In a first part of the paper we present results obtained from plasma treatment of pyrolysis oils. The outlet gas composition is given for various types of oils obtained at different experimental conditions with a pyrolysis reactor. Given the complexity of the mixtures from processing of biomass, we present a study with methanol considered as a model molecule. This experimental method allows a first modeling approach based on a combustion kinetic model suitable to validate the coupling of plasma with conventional biomass process. The second part of the paper is summarizing results obtained through a plasma-pyrolysis reactor arrangement. The goal is to show the feasibility of this plasma-pyrolysis coupling and emphasize more fundamental studies to understand the role of the plasma in the biomass treatment processes.

1. Introduction

Biomass gasification processes produce synthetic gas (syngas) that can be used for further catalytic synthesis of Fischer-Tropsch fuels, SNG fuels, H_2 or for the direct production of heat and electricity. For these applications, high quality of syngas is required, including very low tar contents. Among the conventional gas purification technologies, plasmas offer an interesting alternative, as it is well known that plasmas can efficiently convert hydrocarbons into syngas.

Started at the end of 2006 and achieved in 2010, the Pyroplasm ANR French project [1] had combined a high temperature cyclone reactor and a plasma reactor to quantitatively study the abatement efficiency of condensable vapours produced by biomass pyro-gasification.

The cyclone reactor is well known for its gas-solid separation properties, but it can be also considered as a high temperature reactor for solid reactions. Fast heating followed by thermo-chemical reactions occur along the heated walls of the cyclone. The solid residues are then automatically separated at the bottom of the reactor and the fluid products leave the cyclone at its top [2].

† To whom any correspondence should be addressed

IOP Conf. Series: Materials Science and Engineering **29** (2012) 012011 doi:10.1088/1757-899X/29/1/012011

Previous studies have shown the potential of using a cyclone for biomass pyro-gasification. Depending on the operating conditions, the fractions of oils and gases produced can be very different [3-6]. As example, biomass produces 75% of oils at 880 K (reactor wall temperature) and 80% of gases at 1220 K. Such a multifunctional reactor is thus very interesting to carry out either gas production or bio-oil production.

Tar reduction or destruction during biomass gasification is a most of concern topic for current scientific research [6]. A lot of different technologies have been already studied: mechanism method such as cyclone, filters, electrostatic precipitator, scrubbers; catalytic or thermal cracking or plasma methods.

Plasmas offer a lot of possibilities for waste treatment and valorisation, bio fuel production, pollution control or gas treatment. In this work, the reactor used for pyro-gasification and the plasma reactor are separated and placed in series in order to study the plasma effect on tar removal.

In the first part of this paper, we present results obtained from plasma treatment of bio-oils. These oils are produced by using the cyclone-pyrolysis arrangement. The outlet gas composition is given for various types of oils obtained at various wall temperatures of the cyclone-reactor. Results obtained are quite difficult to discuss because the nature and the composition of oils are unknown. Complementary experiments were performed by using methanol considered as a model molecule for which a kinetic combustion mechanism is available. The discussion of the effects produced by the plasma can then be opened.

The second part summarizes results obtained with the hybrid plasma-pyrolysis system [8].

2. Bio-oils and methanol steam-reforming

One way of storing the biomass energy is to produce oils from the biomass. Oils have higher energy density than natural biomass. The implementation of a plasma process can be used for the bio-oils cracking and then produce hydrogen rich gas mixture.

This part of the paper presents results obtained from plasma treatment of bio-oils and methanol. These oils are produced by using a cyclone-pyrolysis arrangement [2-6].

2.1. Experimental techniques

The studies are performed using a simplified plasma reactor called Statarc [9] generating a more stable arc than the plasma torch used in Pyroplasm project. Laboratory scale experiments (figure 1) were carried out with liquid mixtures injected with a syringe pump in the Statarc reactor at atmospheric pressure and low temperature.

The low current plasma arc reactor includes two graphite electrodes in a quartz tube (400 mm length and 30 mm inner diameter). The electrodes gap is adjustable from 1 to 2 cm.

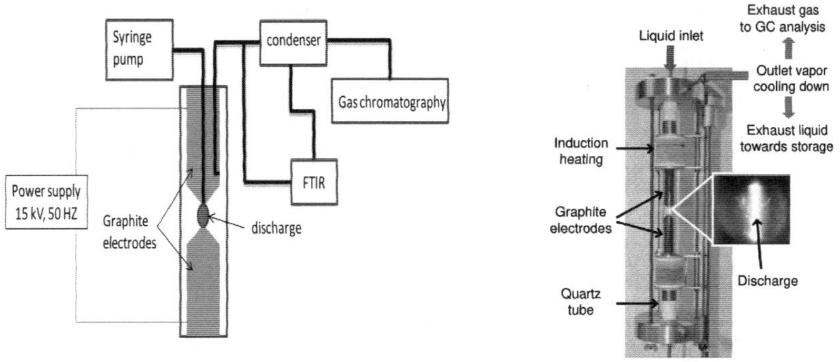

Figure 1. Simplified diagram of the experimental set-up and a photography of the Statarc reactor.

The reactor is powered by a 50 Hz high voltage transformer with leakage flux. The current and voltage are measured using specific probes. The signals are recorded on a digital oscilloscope as shown in figure 2. From the current and voltage data, the power can be calculated.

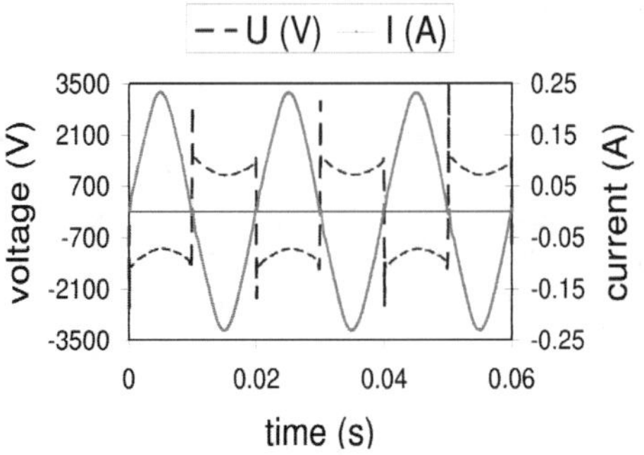

Figure 2. Voltage and current waveforms: liquid volume methanol to water ratio of 0.23.

Chemical analyses of dry exhaust gas are performed to measure species concentrations. The outlet gas is condensed using a cryogenic trap (-20°C) and analyzed using a gas phase chromatography analyzer (GC-Varian CP 3800) and Fourier transform infrared spectrometer (FTIR). FTIR analyses without cryogenic device showed that the CO and CO_2 are not trapped. So, the analyses carried out in dry gases are representative of plasma-treatment.

2.2. Results

2.2.1. Steam reforming of Bio-oils. Experiments are achieved with light or heavy bio-oils (HPL and HPH, respectively or Heavy tars and Light tars respectively) produced from fast pyrolysis process reactor (Pyroplasm facility) at different wall temperature of the cyclone reactor (900, 1000, and 1100 K). Characteristics of bio-oils used in this study are given in table 1.

Table 1. Bio-oils characteristics [5].

	T (K)	Heavy bio-oil (HPH)	Light bio-oil (HPL)
Viscosity (Pa.s)	295	≥ 0.15	≤ 0.01
Density (g/mL)		~ 1.20	~ 1.05
Molar elementary composition	900	$CH_{1.55}O_{0.67}$	$CH_{5.06}O_{3.03}$
	1000	$CH_{1.01}O_{0.23}$	$CH_{1.49}O_{0.20}$

Dry output gas is analysed and H_2, O_2, N_2, CO, CO_2, CH_4, C_2H_2, C_2H_4 and C_2H_6 species are quantified. The heaviest species are mainly CH_4 and C_2-hydrocarbons with mole concentrations ranged from 0.3% to 4%. Other species are negligible, so we only present the evolution of main produced species: H_2, CO, and CO_2.

In figure 3, exhaust gas mole fractions and H_2/CO ratio are displayed as a function of the wall pyrolysis reactor temperature used to obtain the light and heavy bio-oils. These results suggest that bio-oil reforming by non-thermal plasma is strongly influenced by four parameters:

- wall pyrolysis reactor temperature,
- bio-oil nature,

- bio-oil diluted or not with water,
- bio-oil flow rate.

Conversion rates of light and heavy bio-oils into syngas are 20-25% (wt) and 10-15% (wt), respectively. After the plasma treatment, significant H_2/CO concentration ratios are obtained in experiments using bio-oils produced at low temperature (900 K) and bio-oils diluted in water. In addition, higher H_2/CO ratios are obtained from light bio-oil plasma treatment compared to those obtained with heavy bio-oil. No significant influence of bio-oil flow rate variation is observed on syngas and CO_2 mole fractions.

From these results, two main comments can be done: The first one concern the comparison between the composition of a gas produced by pyro-gasification and a gas obtained from plasma bio-oil steam reforming. The second one concerns the comparison between the experimental and theoretical compositions.

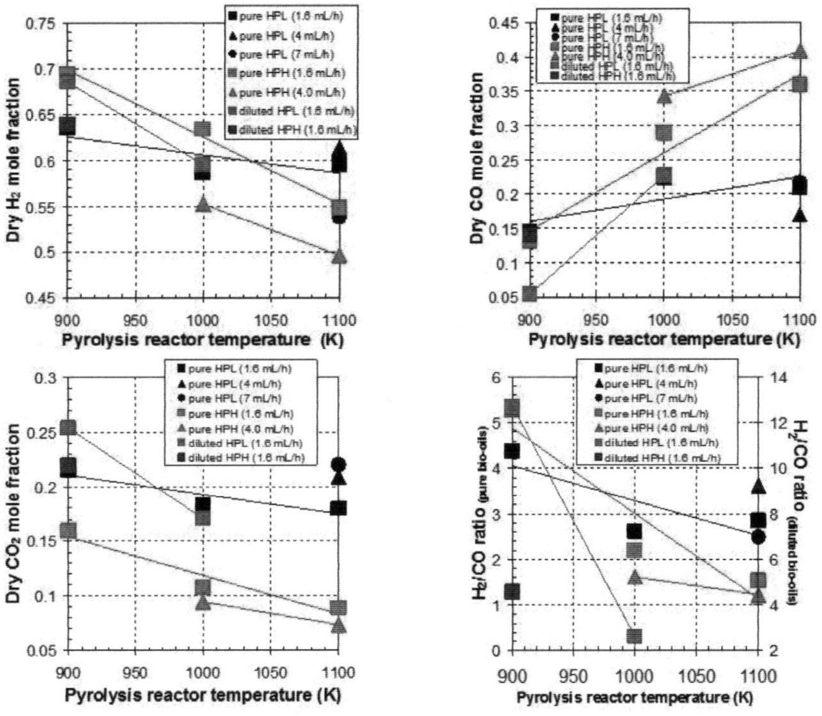

Figure 3. Plasma treatment results: H_2, CO, CO_2 mole fractions and H_2 to CO ratio as a function of the wall pyrolysis reactor temperature for different bio-oils at different flow rates.

2.2.2. Comparison between the composition of a gas from pyro-gasification and from plasma bio-oil steam reforming. The results obtained using the pyro-gasification facility [8] show that the compositions of the output gas mixtures are generally close to the following composition: 18% H_2, 63% CO, 12.5% CO_2, 2.5% CH_4, and 1% C_2-C_3 species. The synthesis gas obtained from a pyro-gasification includes many more carbon monoxide than hydrogen (H_2/CO # 0.3).

In experiments where the synthetic gas is obtained by plasma reforming of bio-oil, the H_2/CO ratio is higher, between 2 and 9. Thus, high hydrogen enrichment can be obtained using plasma steam reforming.

Although the gas mixtures at the reactor outlet include small amounts of methane and heavier hydrocarbons, CO and H_2 remain the most energizing components. Therefore the ratio of H_2/CO is a significant parameter of the process.

2.2.3. Comparison between experimental and theoretical compositions. Biomass composition can be described by the following overall formula: $C_6H_9O_4$. Consider a typical steam reforming reaction derived from reaction:

$$C_6H_9O_4\,(s) + 5H_2O\,(g) \rightarrow 3CO\,(g) + 3CO_2\,(g) + 9.5H_2(g) \quad (1)$$

From this overall reaction, CO and CO_2 concentrations in gas phase are equal and the theoretical corresponding value of H_2/CO ratio is about 3.16. Reaction (1) is not a general reaction; it is merely useful to give the order of magnitude of the main products. In the general case, the CO and CO_2 are produced at different values, so, general equation must take into account the water gas shift reaction (WGSR, reaction (2)):

$$CO\,(g) + H_2O\,(g) \rightarrow CO_2\,(g) + H_2\,(g) \quad (2)$$

As an example, in case of following reaction the H_2/CO value is 5.25

$$3CO\,(g) + 3CO_2\,(g) + 9.5H_2\,(g) + H_2O\,(g) \rightarrow 2CO\,(g) + 4CO_2\,(g) + 10.5H_2(g) \quad (3)$$

It can be concluded that the results obtained by plasma steam reforming are consistent with a simplified analysis based on an overall chemical equations.
CO and CO_2 production are not produced from the solid carbon due to electrodes ablation. The carbon consumption has been estimated to be in the range of $1.7\text{x}10^{-7}$ mol.min^{-1}, which means that ablation does not contribute to CO_2 or CO production.

2.2.4. Methanol steam reforming. The above results show the interest to access to a more detailed understanding of the mechanisms involved. We have therefore chosen a molecule that can be considered a model with known kinetics of combustion. This molecule is methanol. Steam reforming using non thermal plasma processes was presented previously [9]. Some authors have also performed plasma reactor experiments in water, alcohol, and hydrocarbons for H_2 production with high H_2 yield at limited energy cost [10, 11].

Some results obtained from Methanol are plotted on figure 4. This graph reports results obtained at a constant injection flow rate for various compositions of the inlet mixture.
The amount of H_2 decreases slightly and the amounts of CO and CO_2 are most sensitive to the initial composition. It is quite difficult to discuss such results; therefore, we propose a simplified analysis based on a combustion model.

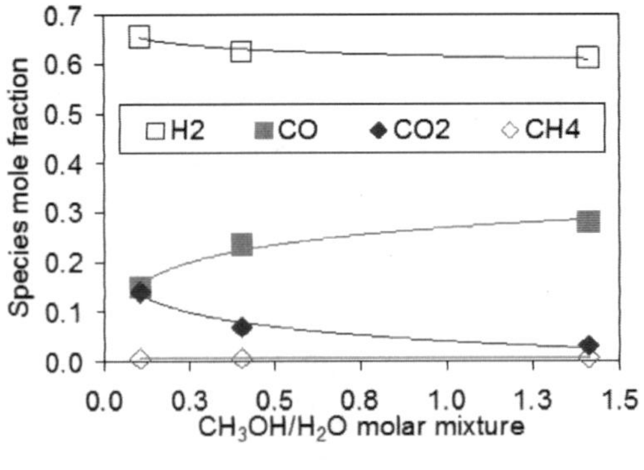

Figure 4. H_2, CO, CO_2 and CH_4 mole fraction as a function of inlet liquid molar Methanol/water ratio (total liquid flow rate = 1.26 sccm).

2.2.5. Modeling and experiment-modeling comparison. Results presented above are obtained from a reactor in which a stationary discharge is produced. Modeling of the discharge involves understanding of various and complex physico-chemical phenomena, especially non thermal discharge modeling, hydrodynamic modeling and kinetics. A simplified way is proposed to describe arc effects with the assumption that the plasma reactor is similar to a cascade of two plug flow reactors (PFR) as shown in figure 5.

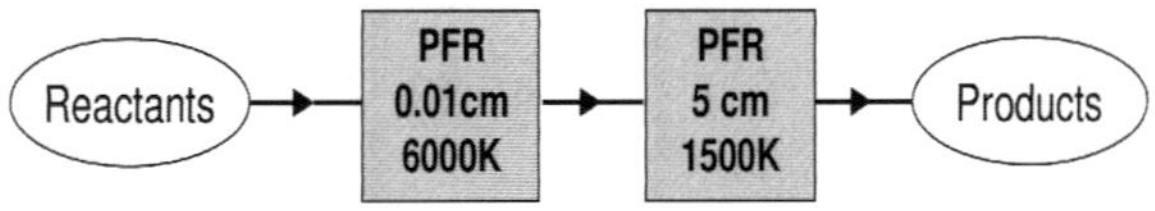

Figure 5. Schematic Plug Flow Reactor (PFR) model.

The first short length (0.01 cm) - high temperature PFR at 6000 K describes the gas mixture injection in the hot plasma zone at the upper electrode level. It is a PFR which use the conventional combustion kinetic mechanisms but does not take into account the electrons and the ions from the plasma. It is supposed to represent a thermal effect equivalent to the plasma effect. This PFR behaves as radical species injector in the surrounding reactive zone. The temperature of 6000 K is an estimation of the average temperature of the ionized channel. It is in the order of magnitude of the temperature measured on the axis of the discharge (6500 < T (K) < 7300) by molecular emission spectroscopy using OH, C_2, and CN bands.

The second 5 cm length – moderate temperature PFR at 1500 K describe the chemical reactions occurring in the reactive zone surrounding the plasma column. Various phenomena such as convection, diffusion, chemical dissociation and recombination occur in that zone. The temperature of 1500 K was obtained by software fitting in order to find equal concentrations of CO and CO_2 in an inlet mixture with 10% mole fraction ethanol.

The mechanisms used in PFR 6000 K and 1500 K modeling is Gri 3.0 [12] taken from literature and the software used is Kintech Chemical Workbench [13]. For simulation dedicated to the methanol mixtures, only O, OH, H, H_2, H_2O, CO and CO_2 concentrations are presented and considered for discussion. Results obtained from an inlet methanol mole fraction of 10% are plotted on figures 6 and 7 for PFR 6000 K and 1500 K, respectively. The percentage (%) corresponds to % mole in vapor phase.

Simulated results of PFR at 6000 K (Figure 6) show that the reactive species (radicals), OH, H, O, can be produced and then injected in the PFR 1500 K.

Composition along PFR length 1500 K is shown on figure 7. From this figure, an increase of the inlet flow corresponds to a decrease of the reactor length. Reduction in the effective length of the reactor leads to a decrease in the H_2 and CO_2 concentrations and an increase in CO. Similar modeling is also applied for methanol inlet concentrations of 5, 15 and 20%.

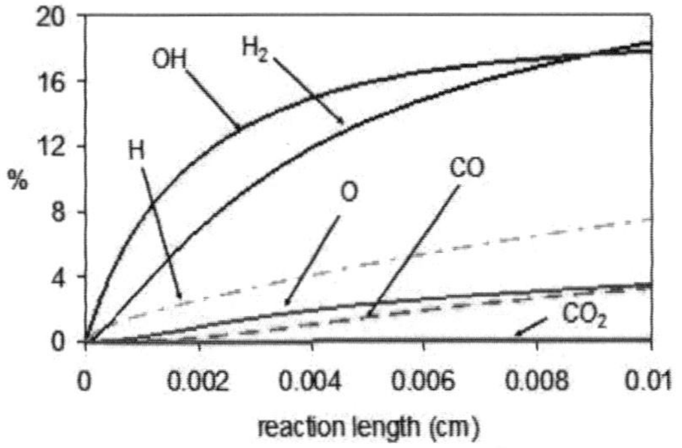

Figure 6. Species concentration as a function of reaction length: PFR 6000 K – 10% methanol.

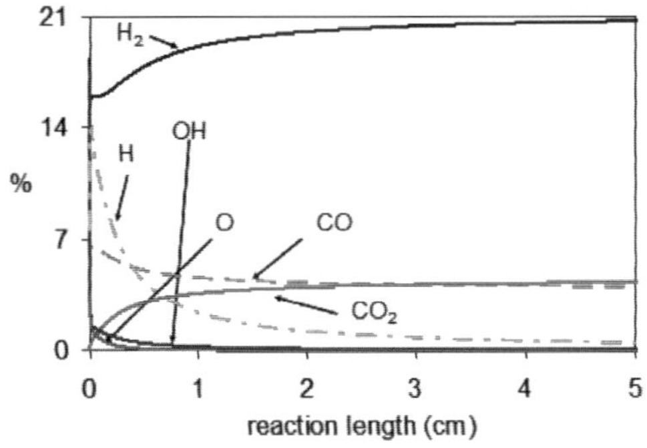

Figure 7. Species concentration as a function of reaction length: PFR 1500 K – 10% methanol.

In figure 8, comparisons between experimental results and modeling data concerning H_2 concentrations H_2/CO, and CO/CO_2 are displayed for various inlet methanol concentrations. It is shown that the general trends in hydrogen production, carbon monoxide and carbon dioxide, are fairly well described by the model. It should be noted that the presence of the first PFR is essential because of the impossibility to find modeling results, compatible with the experiments, using only a single reactor PFR at 1500 K. The best agreements between experimental and modeling results are obtained for CO/CO_2 ratios and H_2 mole fractions. H_2 mole fractions vary slowly. However, a maximum can be observed for a mole fraction of 0.15. The results are not understood today but can be discussed with a parametric simulation. Such a work is underway.

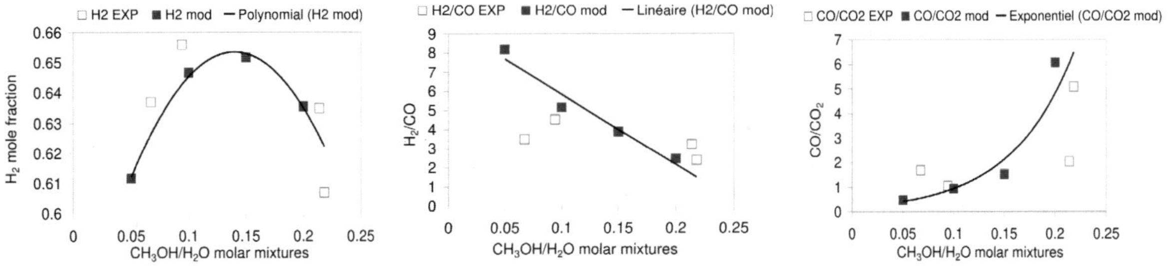

Figure 8. Comparison of model and experimental results.

3. Low current plasma torch and biomass pyro-gasification

3.1. Plasma torch and biomass schematic view of the Pyroplasm arrangement

Started at the end of 2006 and achieved in 2010, the Pyroplasm ANR French project had combined cyclone and a plasma reactor to quantitatively study the abatement efficiency of condensable vapours produced by biomass pyro-gasification [8]. The experimental set-up consists in a laboratory scaled continuous pilot plant [6-8]. It is composed of a biomass feeding section, a heated pyrolysis-cyclone reactor, a plasma reactor and the recovery of all the solids, liquids and gases. Thus, complete mass balances can be established. The plasma reactor can be placed at various locations between the pyrolysis reactor and the gas outlet. Figure 9 shows a schematic view of experimental set-up. Only comments about effects observed in A and B plasma setting are given in the paper.

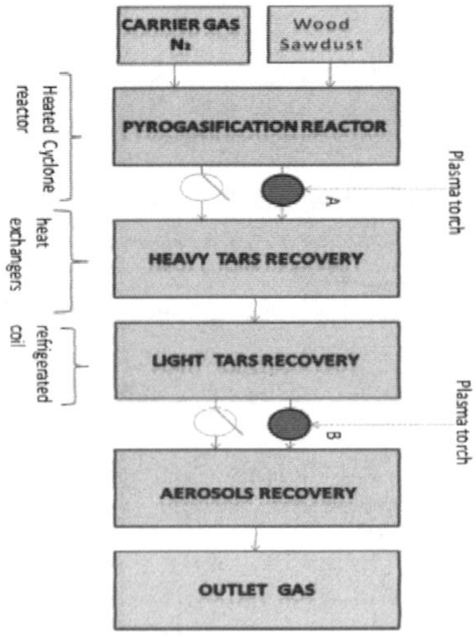

Figure 9. Schematic view of the experimental Pyroplasm set-up.

The raw biomass sawdust is prepared from wood pellets grind (mixture of oak and beech wood). A calibrated screw feeder (placed at the bottom of a stirred hopper) allows delivering a continuous and known mass flow rate in the rage 200 - 380 $g.h^{-1}$, at the entrance of the reactor. Particles are transported from the end of the screw feeder to the top of the cyclone reactor by a nitrogen carrier gas controlled flow ($2\text{-}19\times10^{-4}$ kg.s^{-1}).

The pyro-gasification reactor is a cyclone reactor. The cyclone is made of refractory stainless steel and heated at its walls by induction. The wall temperature can be adjusted from 880 to 1220 K [1, 8]. Char is automatically separated from the gaseous products in a cylindrical collector placed at the bottom of the cyclone reactor. The fluid products include condensed vapours, aerosols and permanent gases. They first cross heat exchangers in order to recover a part of liquids called "Heavy tars". Then, a refrigerated coil is used to condense the residual vapours. The temperature at the exit of the refrigerated coil is regulated close to 258 K. The tars recovered in the coil are called "Light tars". Finally, the aerosols are trapped in an electrostatic precipitator and a membrane filter placed in series. Ultimate vapours are trapped in a silica gel cartridge.

Clean gases are sampled and analysed by gas chromatography. The different concentrations of H_2, CO, CO_2, C_2 and C_3 hydrocarbons are determined. Assuming a complete conversion of the biomass introduced in the reactor, the Mass Balance is given by the following expression and calculated from the outlet masses of char (m_{char}), tars (m_{tars}), gases (m_{gases}), ultimate vapors (m_{uv}) and the inlet biomass (m_{bio}):

$$Mass\ Balance = \frac{m_{char} + m_{tars} + m_{gases} + m_{uv}}{m_{bio}}$$

3.2. The plasma reactor

The reactor, shown in figure 10, is a low current arc plasma torch. The plasma torch has a tip-cylinder configuration. It operates in a non-transferred mode. The inner electrode is a 1.5 mm diameter electrode located on the axis, while the nozzle electrode is a stainless steel cylinder with an axial internal passage. The reactants (i.e. carrier gas, condensable vapours and gases) are injected tangentially in a chamber located right in front of the inner electrode tip.

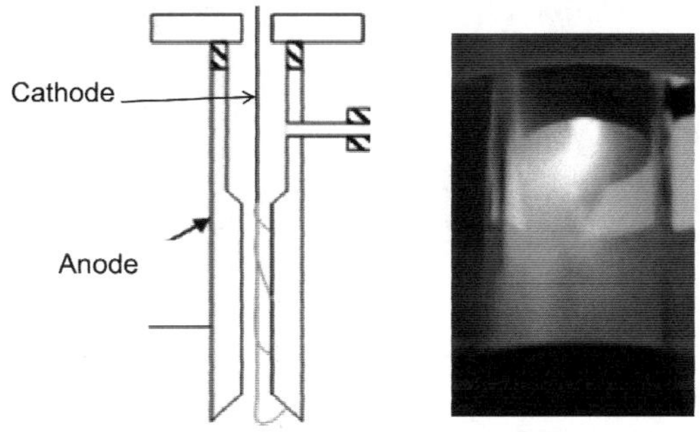

Figure 10. Schematic diagram and photography of low current plasma torch.

The plasma reactor is powered by means of two 50 Hz high voltage transformers (230V/15 kV) with leakage flux which allows sinusoidal current waveforms with a constant RMS value of 155 mA (plasma power 50W) or 0.9 A (plasma power 160 W). A scheme of the high voltage apparatus is displayed figure 11. Voltage is determined by arc properties in ionised gaseous mixture. Examples of voltage and current waveforms are given in figure 12 in case of 160 W power supply.

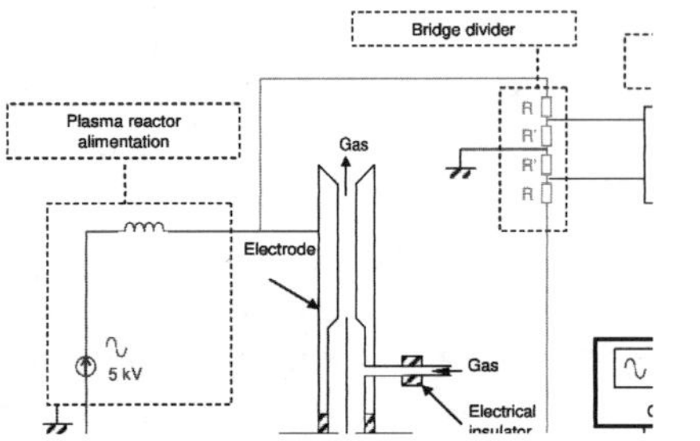

Figure 11. Functional scheme of the plasma torch and power supply.

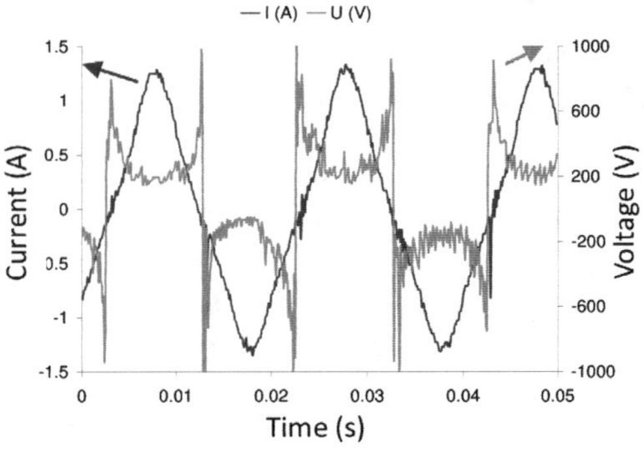

Figure 12. Voltage and current in plasma torch experiments (Average Power: 160W).

3.3. Results and discussion

Experiments are performed with the plasma reactor settled between the outlet of the cyclone reactor and the inlet of the three heat exchangers (A position in figure 9). The mass of all the products (solids, total liquids, permanent gases and ultimate vapours) during wood pyro-gasification at 1108 K is determined. Mass repartition of gases, liquids, solids and ultimate vapours are compared to a previous experiment realized without plasma reactor in which all the operating conditions are identical (cyclone walls temperature, carrier gas flow rate, biomass mass flow rate). Results have been previously published [7] and an example is shown in figure 13.

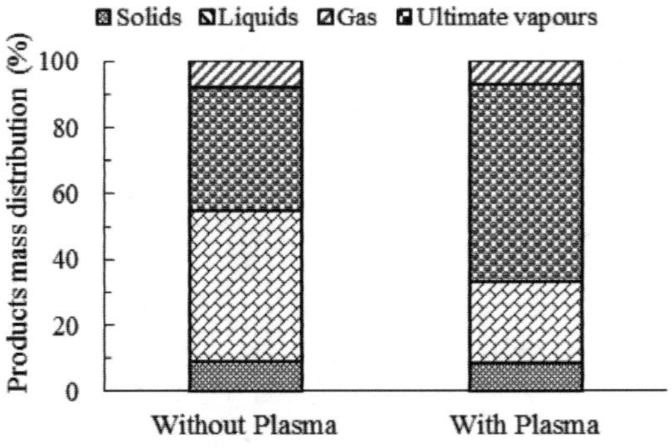

Figure 13. Influence of plasma on products masses distribution [7].

The plasma reactor has a strong influence on the liquid/gas distribution (figure 13). Indeed, the amount of produced permanent gas increases whereas the amount of total liquid decreases. Then, a non-negligible part of condensable vapours seems to be cracked by the plasma reactor, 45% of tars entering the plasma reactor are removed.

Typical composition gas mixtures obtained in case of pyro-gasification at wall temperature of the cyclone of about 1108 K (without plasma) is: 18% H_2, 63% CO, 12.5% CO_2, 2.5% CH4, and 1% C_2-C_3 species.

Experiments with the installation Pyroplasm showed that there is a slight increase of H_2 production, from 18 to 20%, and a decrease of CO production, from 63% to 61%. Interesting results are also observed by setting plasma reactor at the outlet of the refrigerated coil used for light tars trapping (B position in figure 9). Experiments show that 70% of aerosols can be removed by the plasma.

3.4. Short conclusion about Pyroplasm project

The plasma reactor has a strong influence on the liquid/gas products distribution during wood flash pyro-gasification. Indeed, the amount of produced permanent gases increases whereas the amount of total liquid decreases. The aerosol is the fraction most easily destroyed. Destroyed tars generate gases containing more hydrogen than the pyrolysis gases.

The efficiency of a low current plasma torch is significant in the tar conversion process. However, for such a plasma torch, the physical and optical diagnostics are very difficult to perform. To better characterize main physical phenomena, additional studies are developed by using a simplified plasma reactor, called "Statarc" (described in the first part of this paper) and specially design for bio-oil steam reforming.

4. Conclusion

The results presented here show that the combination of plasma reactor with a pyro-gasification reactor is a promising way for a better use of biomass. It is also shown that oils derived from pyrolysis can be used to produce hydrogen rich syngas. The plasma methanol experiments show that the production of reactive species by low energy plasma can generate reactions over a significant volume in comparison to the volume of the plasma filament itself. This study justifies the development of new modeling including the combined effect of kinetic and aerodynamic models. It is shown that the plasma steam reforming of bio-oils generates a syngas with higher hydrogen concentrations compared to those obtained by pyro-gasification. This is due to the steam reforming reaction under the plasma effect.

The mass conversion yields of bio-oils, approximately 25%, are however not sufficient to consider direct industrial applications by using these laboratory scale reactors. The design of new devices has to be performed for industrial scaling.

Acknowledgments

This work was supported by the French National Research Agency (ANR) under Pyroplasm project. The authors gratefully appreciate the technical assistance of G. Coudrat and R. Lainé from GREMI and LRGP laboratories, respectively.

References

[1] French Pyroplasm program 2006-2010 *Pyrolyse de la biomasse à haute température assistée par plasma non thermique* ANR PNRB
[2] Lédé J 2000 *Industrial Engineering Chemistry Research* **39** (4) 893
[3] Broust F, Ferrer M and Lédé J 2002 *12th European biomass conference for energy, industry and climate protection* (Amsterdam The Netherlands) p 545
[4] Broust F 2003 *PhD Thesis INPL-LSGC Nancy University (France).*
[5] Ndiaye FT 2008 *PhD Thesis INPL-LSGC Nancy University (France)*
[6] Lédé J, Broust F, Ndiaye F T and Ferrer M 2007 *Fuel* **86** 1800
[7] Han J and Kim H 2008 *Renewable and Sustainable Energy Reviews* **12** 397
[8] Falcoz Q, Leininger JP, Lédé J, Martin G, Luche J, Khacef A, Aubry O and Cormier J M 2009 *17th conference and exhibition from research to industry and market (*Hamburg Germany)
[9] Aubry O, Met C, Khacef A and Cormier J M 2005 *Chemical Engineering Journal* **106** p 241
[10] Petitpas G, Rollier J D, Darmon A, Gonzalez-Aguilar J, Metkemeijer R and Fulcheri L 2007 *Int. J. of Hydrogen Energy* **32** 2848
[11] Rusu I and Cormier J M 2003 *Chemical Engineering Journal* **91** 23
[12] Smith G, Golden D M, Frenklach M, Moriarty N W, Eiteneer B, Goldenberg M, Bowman CT, Hanson R K, Song S, Gardiner Jr W C, Lissianski V and Qin Z http://www.me.berkeley.edu/gri-mech/
[13] *Chemical Workbench (Kintech Lab.)* http://www.kintechlab.com/

IOP Conf. Series: Materials Science and Engineering **29** (2012) 012012 doi:10.1088/1757-899X/29/1/012012

Sustainable electricity generation by solar pv/diesel hybrid system without storage for off grids areas

Y Azoumah[1,3], D Yamegueu[1,2], X Py[2]

[1]Laboratoire Energie Solaire et Economie d'Energie, Institut International d'Ingénierie de l'Eau et de l'Environnement, 01 BP 594 Ouagadougou 01, Burkina Faso.
[2]Laboratoire Procédés Matériaux et Energie Solaire, Université de Perpignan, Rambla de la thermodynamique, Tecnosud, 66100 Perpignan cedex, France.

E-mail : yao.azoumah@2ie-edu.org, daniel.yamegueu@2ie-edu.org, py@univ-perp.fr

Abstract. Access to energy is known as a key issue for poverty reduction. The electrification rate of sub Saharan countries is one of the lowest among the developing countries. However this part of the world has natural energy resources that could help raising its access to energy, then its economic development. An original "flexy energy" concept of hybrid solar pv/diesel/biofuel power plant, without battery storage, is developed in order to not only make access to energy possible for rural and peri-urban populations in Africa (by reducing the electricity generation cost) but also to make the electricity production sustainable in these areas. Some experimental results conducted on this concept prototype show that the sizing of a pv/diesel hybrid system by taking into account the solar radiation and the load/demand profile of a typical area may lead the diesel generator to operate near its optimal point (70-90 % of its nominal power). Results also show that for a reliability of a PV/diesel hybrid system, the rated power of the diesel generator should be equal to the peak load. By the way, it has been verified through this study that the functioning of a pv/Diesel hybrid system is efficient for higher load and higher solar radiation.

1. Introduction

The electrification rate in sub-saharan Africa is lower than in any other region of the world. Indeed, more than 70 % of the populations of this region are excluded from the benefits linked to the access of electricity. The situation is more catastrophic in rural areas where less than 12% of the populations have access to electricity [1]. In order to contribute to the increasing of the electrification rate in this region, an original concept called "flexy energy", which consists of hybrid solar PV/diesel/biofuel power plant without battery storage, has been developed by Azoumah et al [2]. To assess the technical and economical feasibility of this concept, a PV/diesel hybrid system without storage has been set up on the site of the International Institute for Water and Environmental Engineering (2iE) at Kamboinsé (12°22' N and 1°31'W), located at fifteen kilometers in North of Ouagadougou, capital of Burkina Faso. This prototype consists of a 9.2 kW diesel generator coupled with a PV array of 2.85 kW*p* and an inverter rated at 3.3 kW. The peak Watt (W*p*) is an amount of electrical power that a system is expected to deliver in standard test conditions, STC (AM 1.5, solar cell operating temperature of 25°C

[3] To whom any correspondence should be addressed

and solar radiation of 1000W/m²); the air mass (AM) is the ratio of the mass of the atmosphere through which the beam radiation passes to the mass of the mass it would pass if the sun were at the zenith.

The concern of this paper is to present the technical performances experimentally obtained for the hybrid system under various operating conditions. Four fixed load/demand profiles corresponding to different percentages of the nominal capacity of the diesel generator have been experimented for different solar radiations and consequently various PV contributions. The obtained raw and calculated performances are presented and discussed in this paper. Also, a comparative analysis for three different systems (diesel and solar PV generators only, and hybrid solar pv/diesel system) for powering a same load profile has been done.

2. Experimental

2.1. *Experimental set up*

Hybrid systems can be broadly classified in series hybrid, switched hybrid and parallel hybrid configurations [3]. The present prototype has a parallel hybrid configuration (figure 1). In this scheme, the PV system is connected to the diesel generator by the mean of a DC-AC converter or inverter. The PV and the diesel generators supply a portion of the load demand directly (since they are operating in parallel).

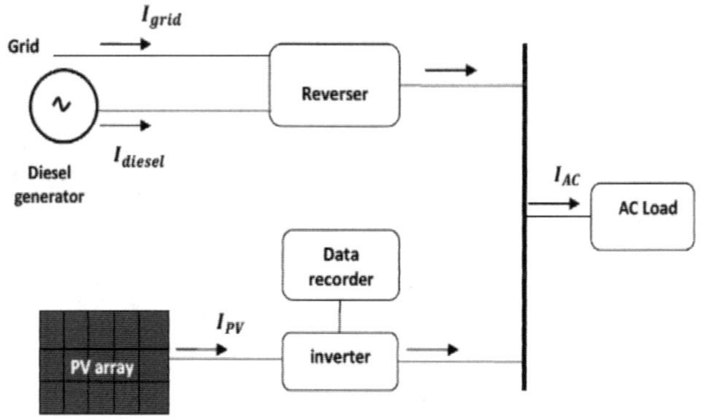

Figure 1. Prototype design

The PV system is composed of 15 modules HIT (Heterojunction with Intrinsic Thin layer), each rated at 190 W_P, totalling 2.85 kW_P DC at Standard Testing Conditions. The PV array covers an area of about 18 m² with PV modules distributed in a parallel/series configuration of three strings connected to one inverter rated at 3.3 kW. The diesel generator rated power is 9.2 kW (11.5 kVA) at 1500 rpm.

2.2. *Results and discussion*

In this section, the behavior of a typical hybrid PV/diesel system without storage which characteristics have been presented in the previous sections is studied. The contribution of the PV and the diesel generators, the specific consumption of the diesel generator for a given load (L) profile have been assessed. For four days, we have considered four constant load profiles (3.7 kW, 5.7 kW, 7.6 kW and 9.7 kW) corresponding respectively to 40%, 62%, 82% and 105% of the diesel generator nominal power.

The behaviors of the PV and the diesel generators in the hybrid system are interdependent. It can be observed from fig. 2 to fig. 5 that for the loads in the order mentioned above, the specific fuel consumption of the diesel generator ranges respectively from: 0.45 to 0.51 l/kWh (see fig.2), 0.35 to 0.41 l/kWh (see fig.3), 0.33 to 0.36 l/kWh (see fig.4) and 0.33 to 0.34 l/kWh (see fig.5).

These results show that for small loads (less than 62% of the rated power of the DG), the contribution of the PV generator (P_{AC} / L) which ranges from 4 to 38% for a load values of 3.7 kW and 5.7 kW does not enable an optimal functioning of the diesel generator. In fact, as mentioned by Ashari et al [4] and also verified by results obtained in the present investigation[5], the maximum fuel efficiency of a diesel generator is about 0.33 l/kW, but for these two rated loads, the specific fuel consumptions are far from this optimal value. This is due to the fact that these loads are already far from the optimal functioning point of the diesel generator and therefore the PV generator contributes rather to decrease the performance of the engine.

However, for high loads (see fig. 4 and fig.5), the contribution of the PV array (which ranges from 1.5% to 33% for a load value representing 82 % of the diesel generator nominal power and from 2.5 % to 22% for a load value representing 105 % of the diesel generator nominal value) does not affect considerably the performance of the diesel engine.

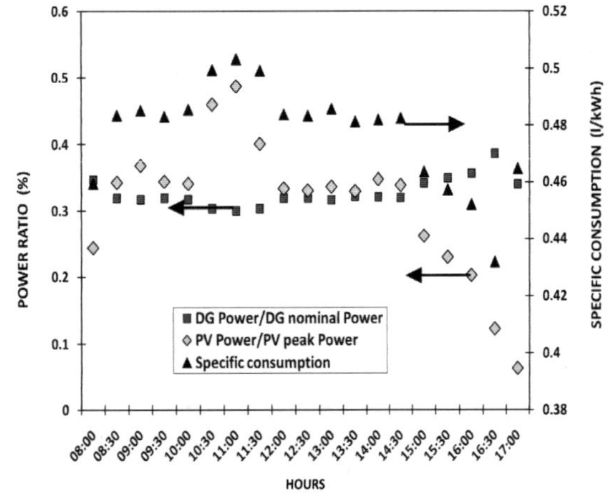

Figure 2. Behaviour of the system for a load equal to 40% of the diesel generator rated power

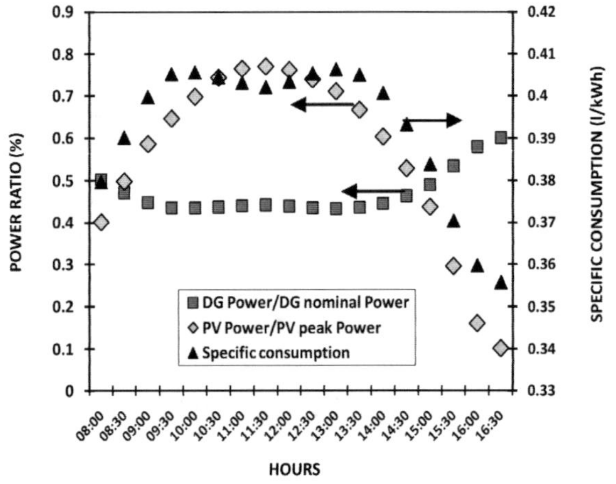

Figure 3. Behaviour of the system for a load equal to 62% of the diesel generator rated power

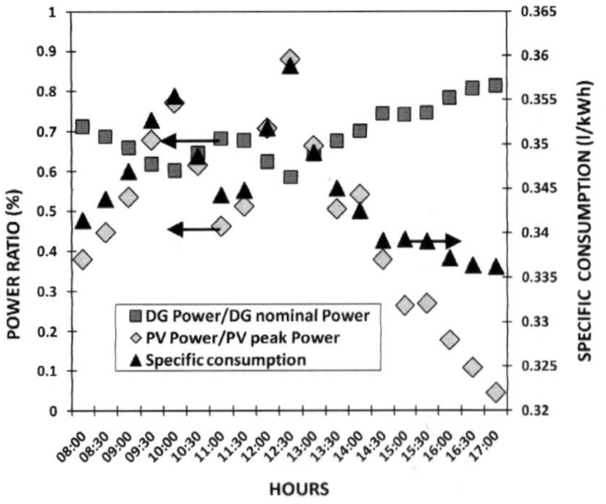

Figure 4. Behaviour of the system for a load equal to 82% of the diesel generator rated power

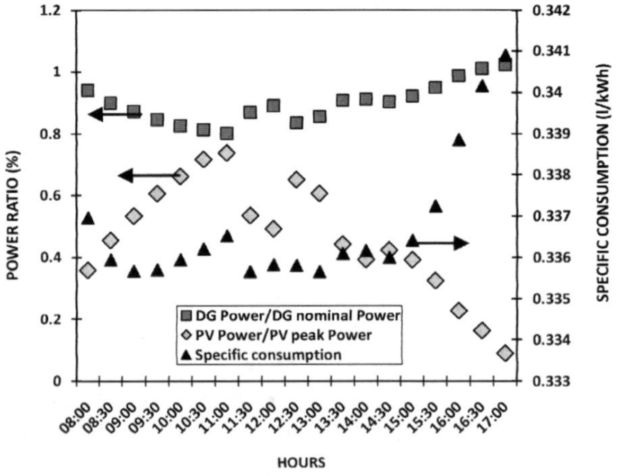

Figure 5. Behaviour of the system for a load equal to 105% of the diesel generator rated power

As a matter of fact, for these loads, the values of the specific fuel consumption vary from 0.33 to 0.36 l/kWh what is close to the optimal functioning point of the diesel generator. As the specific fuel consumption and then the operation cost of the diesel generator increases considerably when the hybrid system operates below the load values representing less than 62% of the diesel generator rated power, the sizing of such a system must be done so that the diesel generator operates near its optimal point (70-90 % of its nominal power).

3. Advantages of using hybrid systems: comparative analysis for three different systems (diesel generator only, solar PV generator only and hybrid solar PV/diesel) for powering a same load profile [2]

3.1. *Economic analysis*

To point out the economic advantages of hybrid systems for remote areas, a comparative analysis is conducted for three types of power generators (DG only, PV generator only and hybrid solar PV/diesel) for powering a same load. For this study a typical daily load of the Sabou village in Burkina Faso is considered (see figure 6). This village has 45,000 inhabitants, thus about 6429 households (if considering seven persons per household).

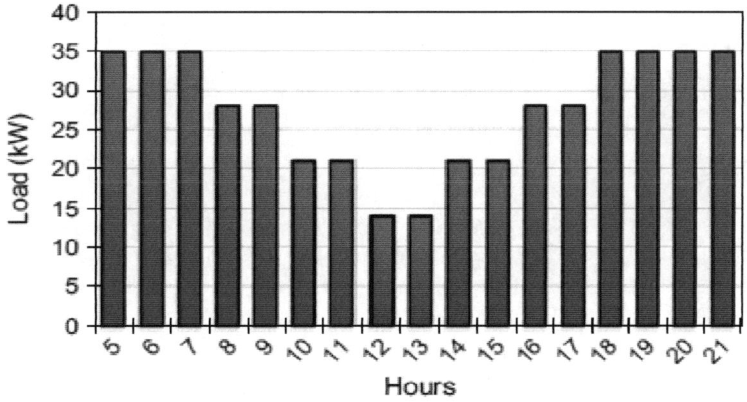

Figure 6. Typical daily load profile of the Sabou village in Burkina Faso.

The economic analysis made in this section is based on the use of the life cycle cost (LCC) method. The life cycle cost is an economic assessment of the cost for a number of alternatives by taking into account all significant costs over the lifetime of each alternative, adding each option's costs for every year and discounting them back to a common base. These costs can be categorized into two types: (i) recurring cost (operation and maintenance costs for the diesel generator, maintenance costs for the PV generator and batteries) and (ii) non-recurring cost (batteries and diesel generator replacement costs).

The life cycle cost is expressed as follows:

$$LCC = C + PW_R + PW_{NR} \tag{1}$$

where C is the initial cost PW_R is the present worth recurring costs and PW_{NR} is the present worth non- recurring costs.

The data of the following table 1 are used for the calculations.

After calculation, one obtained that for the same energy demand estimated at 3,423,700 kWh over 20 years, hybrid solar PV/diesel system Life cycle cost (LCC=983,865 Euros) is lower than the PV generator only (LCC=992,000 Euros) which is at its turn lower than the DG only (LCC=996,000 Euros).

It is important to point out that the hybrid system considered here is not the optimal one in terms of configuration, the load management, and even the whole system management. By considering all these aspects, it is no doubt that the benefits will be more important and it is what we intend to demonstrate through the ''flexy-energy'' concept [2] .

Table 1[a]. Costing data considered in the present study

	Diesel generator system (38.5 kW)	**PV**	**Battery**
Initial cost	15,400 €	700€/kWp	0.07€/Wh
Annual operation cost	0.9€/l of fuel	0	0
Annual maintenance cost	0.18€*RT	3% of annualized initial cost	15% of annualized initial cost
Salvage value	10% of the diesel generator initial cost	10% of the PV generator initial cost	0
Replacement cost	Equal to the initial cost	0	Equal to the initial cost
Life span (years)	10	20	5
Interest rate, i (%)	8	8	8
Escalation rate, e (%)	4	4	4

[a] In the table 1, RT is the running time of the diesel generator.

3.2. *Environmental analysis*

3.2.1. *Environmental evaluation of the DG only.* Petroleum-based products are one of the main causes of anthropogenic carbon dioxide (CO_2) emissions to the atmosphere. As a matter of fact, the combustion of each liter of diesel produced 2.7 kg of CO_2. During the life cycle of the diesel generator, 1, 244, 286.28 L of diesel oil will be consumed; this will lead to about 3360 tons of CO_2 released in the nature.

3.2.2. *Environmental evaluation of PV generator only.* The mayor drawback with this scenario is the recycling of batteries at the end of their lifetime. In fact, there is no policy concerning recycling of batteries in sub-Saharan African region and this particular aspect represents a considerable environmental concern. In the case studied, over the lifetime of the system, for a total capacity of 41,037.5 Ah, it is 411 batteries of 100 Ah (12 V) which will be probably thrown in the nature.

3.2.3. *Environmental evaluation of PV/diesel hybrid system.* By considering in the first time that the diesel generator in this scenario is fueled by fossil fuels, about 2915 tons of CO_2 are released in the nature. Therefore, it is 445 tons of CO_2 emission which is saved over the life cycle of the hybrid system compared to the diesel generator system only.

From the present case studied, one can conclude on the environmental aspect that, the hybrid system enabled to save CO_2 emission of about 445 tons compared to the diesel generator system only. Furthermore, for the PV system only, more than 411 batteries of 100 Ah (12 V) will be thrown in the nature against zero battery released when considering the hybrid system scenario.

4. Conclusion

First, this paper presents the experimental results of a hybrid pv/diesel system without storage. These results are very useful for the design optimization and the reliability enhancement of PV/Diesel hybrid systems. Second, the economical and environmental advantages of the hybrid system compared to Diesel and PV systems only are displayed and show clearly that the PV/diesel hybrid systems can be an appropriate technological solution for the electrification of remote areas.

References

[1] World Energy Outlook (WEO) www.iea.org/weo/electricity.asp
[2] Azoumah Y, Yamegueu D, Ginies P, Coulibaly Y and Girard P 2011 Sustainable electricity generation for rural and peri-urban populations of sub-Saharan Africa: the "flexy-energy" concept *Energy Policy* **39** 131-41
[3] Nayar C V 1995 Recent developments in decentralised mini-grid diesel power systems in Australia *Applied Energy* **52** 229- 42
[4] Ashari M and Nayar CV 1999 An optimum dispatch strategy using set points for a Photovoltaic (pv)–diesel–battery hybrid power system *Solar Energy* **66**(1) 1-9
[5] Yamegueu D, Azoumah Y, Py X and Zongo N 2011Experimental study of electricity generation by Solar pv/diesel hybrid systems without battery storage for off-grid areas *Renewable Energy* **36** 1780-87

IOP Conf. Series: Materials Science and Engineering **29** (2012) 012013 doi:10.1088/1757-899X/29/1/012013

Investigation of building energy autonomy in the sahelian environment

O Coulibaly[1], A Ouedraogo[2], F Kuznik[3], D Baillis[3] and J Koulidiati[1]

[1]UO-LPCE, Université de Ouagadougou, Laboratoire de Physique et de Chimie de l'Environnement, 03 BP 7021 Ouagadougou 03, Burkina Faso
[2] UO-UFR/SEA - Unité de Formation et de Recherche en Science Exacte et Appliquée de l'Université de Ouagadougou, 03 BP 7021 Ouagadougou 03, Burkina Faso
[3]CETHIL. – Centre de Thermique de Lyon ; INSA Lyon : 20, Av des Arts, 69621 Villeurbanne, France

Email : ousmane.coulibaly@univ-ouaga.bf, abdoulay@univ-ouaga.bf

Abstract. In this study, the energy generation of a set of photovoltaic panels is compared with the energy load of a building in order to analyse its autonomy in the sahelian environment when taking into account, the orientation, the insulation and the energy transfer optimisation of its windows. The Type 56 TRNSYS multizone building model is utilized for the energy load simulation and the Type 94 model of the same code enables the coupling of photovoltaic (PV) panels with the building. Without insulation, the PV energy generation represents 73.52 and 111.79% of the building electric energy load, respectively for poly-crystalline and mono-crystalline panels. For the same PV characteristics and when we insulate the roof and the floor, the energy generation increases to represent successively 121.09 and 184.13%. In the meantime, for building without insulation and with insulate the roof, the floor and 2 cm insulated walls, the energy consumption ratios decrease respectively from 201.13 to 105.20 kWh/m^2/year. The investigations finally show that it is even possible to generate excess energy (positive energy building) and reduce the number and incident surface area of the PV panels if we conjugate the previous model with building passive architectural design mode (orientation, solar protection …).

1. Introduction

Due both to the cost and time of completion, numerical computation is more and more perceived as an efficient tool especially for building thermal load investigation and transient analysis of renewable energy systems. Because of the lack of thermal regulation, building thermal efficiency is very poor in Sub-Saharan Africa in general and in Burkina Faso in particular. The consequences are excessive energy consumption for air conditioned buildings on one hand and very unconformable leaving conditions on the other hand for natural convection houses. As a matter of fact, the climate is rather mild during the cooling season which is also very short, so heating is not needed. Estimations put energy consumption in public and private buildings and in the tertiary sector to somewhere between 250 and 450 kWh/m²conditioned/year, depending on the number of stories [1]. To promote renewable energy, the government of Burkina Faso seeks to sensitize the population for better energy conservation and substitution to solar energy, which is abundant throughout the country. For fact, the daily solar irradiation amounts to 5 kWh/m^2/day corresponding to an available energy of about $5\cdot10^8$

GWh/year in the whole country[2]. Therefore, the investigation of building energy autonomy by coupling PV panels and optimizing the losses through its envelope is a very relevant research topic in these climatic conditions. Inspiration for the actual study comes from the work of Al-Ajmi et al. [3], Annabi et al. [4] and Diez-Webster U. et al. [5]. First of all, we model an individual house utilizing TRNSYS software, next we integrate a set of PV panels and finally, we compare their energy generation with the house electric energy load when taking into account, the insulation, the orientation and the energy transfer optimisation of its windows.

2. Modeling with TRNSYS

2.1. The TRNSYS code

TRNSYS for « TRaNsient SYstem Simulation program » is a modular environment developed by a network of institutions [6]: The University of Wisconsin Madison, TESS, a private enterprise of the same city, the group TransSolar of Germany and the CSTB (Technique Scientific Center of Building) in France. TRNSYS SIMULATION STUDIO is the home office which allows on one hand to deal with the existing models in dynamic simulation and on the other hand, to design new models which must be compatible with a predefined protocol in order to interface with the heart of the computations [7]. A TRNSYS simulation project consists of the following:

- choose a set of mathematical models of physical components based on existing models in TRNSYS library or designed new models;
- describe the interactions between these models.

2. 2. TRNSYS model for building

The heat transfer through the building envelop is treated utilizing TRNSYS ''TYPE 56'' model. All three heat transfer modes are concerned:

- conduction through the envelop;
- convection on the envelop by the adjacent fluids;
- short and long radiation wave lengths.

Balanced energy equations for each wall surface are written as:

$$\phi_{cond} + \phi_{conv} + \phi_{LWL} + \phi_{SWL} = 0 \quad (1)$$

where:

ϕ_{LWL} : Flux density of long radiation wave length (W/m^2)

ϕ_{SWL}: Flux density of short radiation wave length (W/m^2)

ϕ_{conv} Convection flux density (W/m^2)

ϕ_{cond}: Conduction flux density (W/m^2)

The Type 56 TRNSYS model treats the heat conduction equations using the Z- transfer function technique. When studying wall heat transfer, Berges [8] compares the computation time of various numerical methods. She inferred that the Z- transfer function technique performs well compared to the finite - difference method.

$$\phi_{Cond,e}(t) = \sum_{n=0}^{Na} a_n T_{s,e}(t - n\Delta t) - \sum_{n=0}^{Nb} b_n T_{S,i}(t - n\Delta t) - \sum_{n=1}^{Nd} d_n \phi_{cond,e}(t - n\Delta t) \quad (2)$$

d_n are coefficients of the Z - transform functions related to the heat flux density

Na, Nb, Nc et Nd are number of coefficients a_n, b_n, c_n *et* d_n

Δ t : space time for the computation of the transfer coefficients

$T_{s,e}(t - n\Delta t)$: Wall outer surface temperature at time (t –n Δt) (K)

$T_{s,i}(t - n\Delta t)$: Wall inner surface temperature at time (t –n Δt) (K)

IOP Conf. Series: Materials Science and Engineering **29** (2012) 012013 doi:10.1088/1757-899X/29/1/012013

$$\phi_{Cond,i}(t) = \sum_{n=0}^{Nb} b_n T_{s,e}(t - n\Delta t) - \sum_{n=0}^{Nc} c_n T_{S,i}(t - n\Delta t) - \sum_{n=1}^{Nd} d_n \phi_{cond,i}(t - n\Delta t) \tag{3}$$

where:

a_n , b_n et c_n are coefficients of the Z - transfer functions related to the surface temperatures (W/ m^2 K)

✓ Conduction heat transfer

All glazing surfaces are supposed to be inertial. Hence, one generally utilizes steady state equations to model the conduction heat transfer.

$$\phi_{cond} = K(T_{si} - T_{se}) \tag{4}$$

where:

K is the conductance 'surface – surface' glazing (W/m^2 K)

✓ Convective heat transfer on the inner and outer surfaces

$$\phi_{conv,e} = h_{c,e}(T_{s,i} - T_e) \tag{5}$$

$$\phi_{conv,i} = h_{c,i}(T_{s,i} - T_i) \tag{6}$$

In these relations:

$h_{c,e}$ and $h_{c,i}$ are respectively the convective heat transfer coefficients on the outer and inner surfaces (W/m^2.K). T_e and T_i are inner and outer ambient air temperatures.

✓ Radiation heat transfer

The radiation transfer makes a distinction between short wave length (SWL), less than 2.5 μm and long wave length (LWL), greater than 2.5 μm.

$$\phi_{rd} = \phi_{SWL} + \phi_{LWL} \tag{7}$$

For an external wall, the equations of the heat transfer by radiation are written as:

$$\phi_{rd,e} = \phi_{SWL,e} + \phi_{LWL,e} \tag{8}$$

$$\phi_{LWL,e} = h_{r,c}(T_{s,e} - T_c) + h_{r,e}(T_{s,e} - T_e) \tag{9}$$

where:

$h_{r,e}$ is the LWL radiation transfer coefficient with the environment.

$h_{r,c}$ is the LWL radiation transfer coefficient with the sky.

For the SWL radiation coefficient, the solar irradiation flux is computed as far as the ''Type 16''model is concerned, based on meteorological data (height H_s and solar azimuth Az_s, direct and diffuse flux density, $\phi_{dir,HZ}$, ϕ_{dif} on an horizontal surface), which are inputs for the model.

✓ *Zone heat balance*

The house convective heat balance is treated assuming that the variation of the air temperature is due to the sum of the radiation flux related to the inward new air, radiation fluxes exchange between the walls and inner air, in addition to other inner radiation and convection.

$$\sum_{i=1}^{Ns} S_i h_{ci}(T_{si} - T_i) + m_a C_a(T_a - T_i) + A_p = \rho_a C_a V \frac{dT_i}{dt} \tag{10}$$

IOP Conf. Series: Materials Science and Engineering **29** (2012) 012013 doi:10.1088/1757-899X/29/1/012013

3. Investigation of the house

3.1. Description

The investigation concerns a type of low income house built in Ouagadougou by a government of Burkina Faso project. It has a surface area of 50 m², south oriented front door, keeping with the conclusion of a study which finds reduced energy consumption when a building is south oriented [9]. The configuration of the house is shown in figure 1. It has two bedrooms, a living room (stay) and toilets. The walls are made of 25 cm void bricks, cement insulated with white inner painted and thick yellow outer painted. It has also a 15 cm thick concrete floor, metal-frame double planes wooden doors, while the windows are made of ordinary 4 mm metal-frame thick glazing, having pure conductance equal to 5.7 W/(m².K) and a solar factor of 0.85. The roof is made of 20 cm thick concrete and the enclosures are in 15 cm chipboard, inner and outer cement insulated and both sides white painted. To optimize the windows heat transfer, they are rather made of plane framed double glazing of 1.4 W/(m².K) conductance and a solar factor of 0.59. They are also protected from the solar radiation by exterior shades. The surface ratio of the windows to the walls is about 15%.

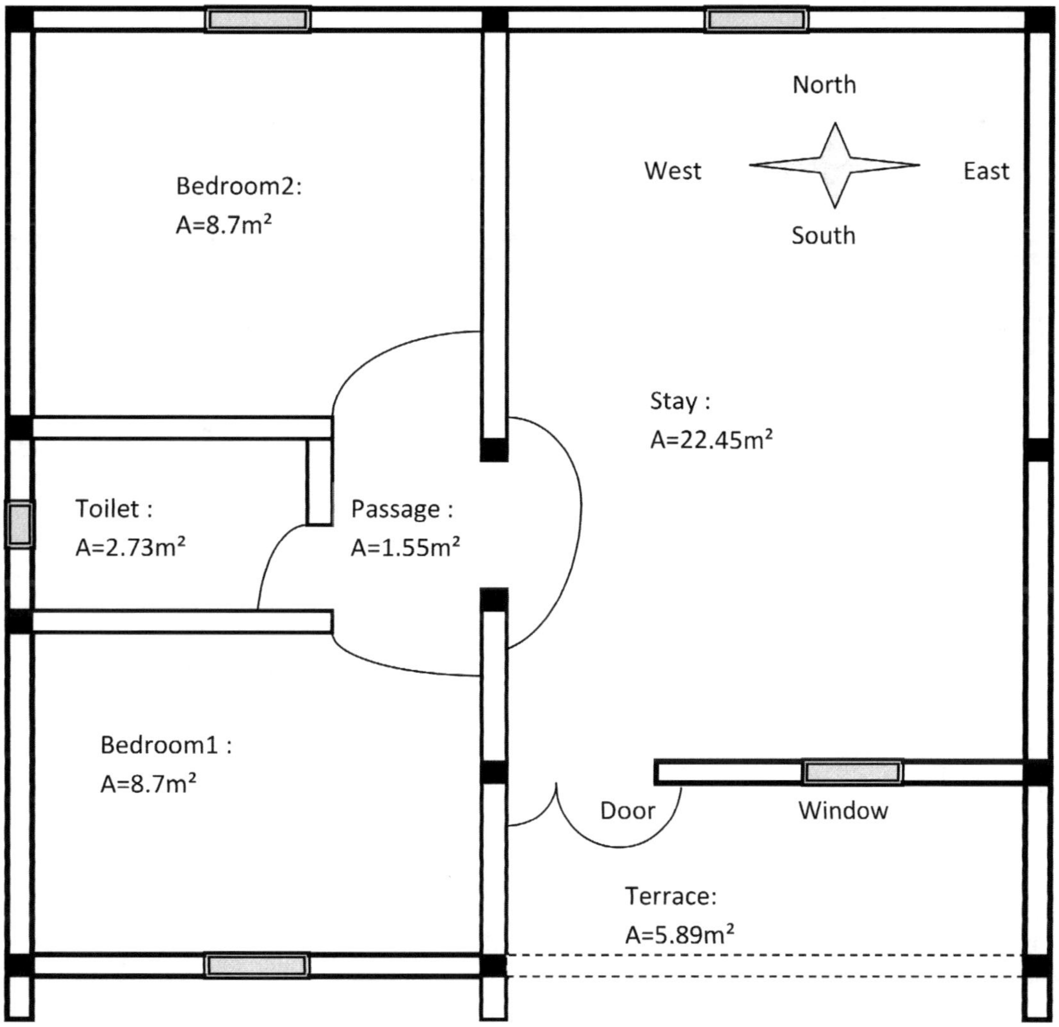

Figure 1. Configuration of the apartment

3.2. Multizone modeling
To have a refined modeling, function of its usage, we divide the house into six (6) zones which will serve as input for TRNSYS sofwares: Zone1 living room (stay), Zone2 terrace, Zone3 joint passage, Zone4 bedroom1, Zone5 bedroom2 and Zone6 toilets.

3.3. Meteorological data utilized in the model
To keep with the sahelian environment, *METEONORM* data of the city of Ouagadougou, delivered by TRNSY*S* version 16.1 have been utilized as inputs.

3.4. Internal heat load and regulation equipments

3.4.1 Indoor comfort conditions. **Previous investigations have set the comfort zone in tropical climate in the temperature interval 24°C - 27°C with the relative humidity fluctuating between 50 and 55%, [10]. We assume that the air conditioning unit starts running when the indoor temperature becomes upper than 27 °C with a relative humidity of 50%. Weeks and weekends occupancy scenarios have been designed for the bedrooms and the living room [11]. Table 1 shows the internal heat load and the regulation equipments utilized for the simulation.**

Table 1. Indoor heat load and regulation equipments.

Unit involved	Number	Occupancy[11]	Nominal power	Consign	Observations
Air conditioner	1	Schedule	100 kW	27°C	-
Dehumidification set	1	Schedule	100 kW	50%	-
Fluorescent Lamp	1	Schedule	8 W/m²	-	Convective/swl/lwl 50/25/25 %
Fan	1	Schedule	1 V/h	-	-
Occupants	4/ living room 2/bedroom	Schedule	131 W	-	Latent/convective/swl/lwl 45/28/0/27 %
Television	1	Schedule	60 W	-	UC=100%
DVD	1	Schedule	150 W	-	UC=40%
Refrigerator	1	Constant occupation	70 W	-	UC=100%

UC : Utilization Coefficient

3.4.2. Construction materials and walls. Table 2 displays the thermophysical properties of the void bricks wall and other surfaces of the building.

Table 2. Thermophysical Properties of construction materials [1].

Building components	Construction materials	Thickness (m)	Heat Conductivity (W/ m.K)	Density (kg/m^3)	Heat capacity (J/ kg. K)
Wall	- Cement finish	0.025	0.87	2200	1050
	- Void bricks	0.200	0.67	1250	880
	-Polystyrene (insulation)	0.020	0.04	18	1450
Floor	- Concrete	0.150	1.75	2100	653
	- Polyurethane (insulation)	0.050	0.03	34	1450
Roof	- Wood wool (insulation)	0.100	0.04	140	1700
	- Concrete	0.220	1.75	2100	653
	- Adhesive (asphalt)	0.005	0.17	1050	1000

4. Coupling the PV system with the building

The Type 94 TRNSYS model developed by Duffie et al. is utilized for coupling PV system with the building of concern [12]. The selected panels are made of thirty six (36) crystalline PV cells. Twenty heights (28) panels with a total surface area of 24.92 m^2, which is half that of the roof (50 m^2), are been installed. They are south oriented with 15° slope which corresponds to the best tilt for optimizing the solar energy gain in Ouagadougou (Burkina Faso). The panels' efficiency is measured by the ratio of the energy generated to the incident energy received. Estimation puts its value to about 10.52% (average annual value obtained with the Type 94 model) for the poly-crystalline panels, the most available. The efficiency increases up to 16% for mono-crystalline panels which are unfortunately more expensive [5]. The assembly shown in figure 3 is utilized for the building description in TRNSYS. It comprises two parts: the upper set leads to the computation of the total solar radiation and the PV panels' generation, model 94 [12]. Whereas the lower set is for the computation of the multizone building energy load utilizing the 56 model type.

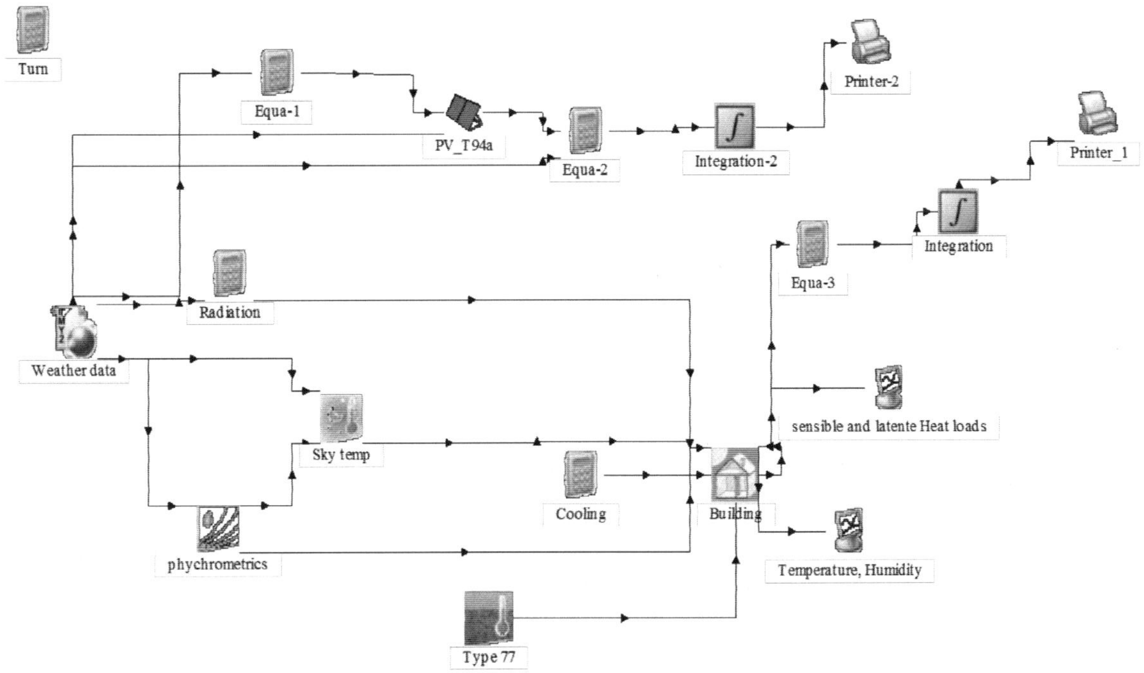

Figure 2. Modules assembly for the building description in TRNSYS

5. Results and discussions

5.1. Optimisation of the windows and equipments specific consumption

Figure 3 shows the yearly evolution of the sensible and latent energy loads. As expected, the months of April and May, the hottest of the year, (average maximum temperature of 41.8 °C and 41.6 °C respectively) are those for which the overall gains heat and, consequently, the cooling loads are greatest. The raison is that the solar heat flux is at its maximum during these months as shown below:

- April : global flux = 5698 W/(m².day), direct flux = 3448 W/(m².day) and diffuse flux = 2250 W/(m².day)
- May: global flux = 5674 W/(m².day), direct flux = 3313 W/(m².day), and diffuse flux = 2361 W/(m².day).

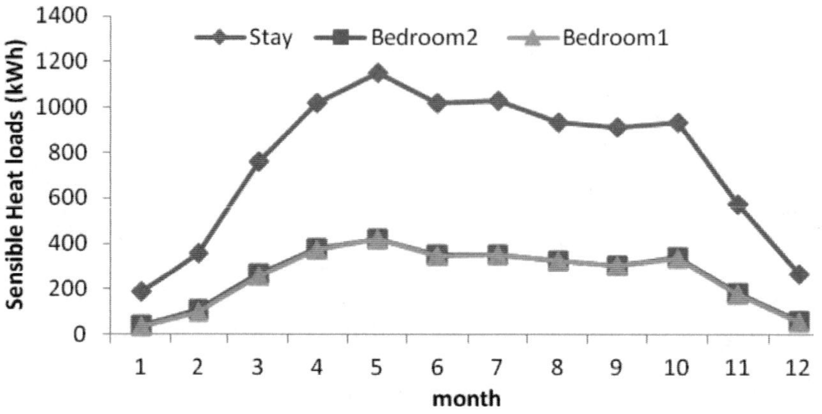

Figure 3a. Yearly evolution of the sensible energy load

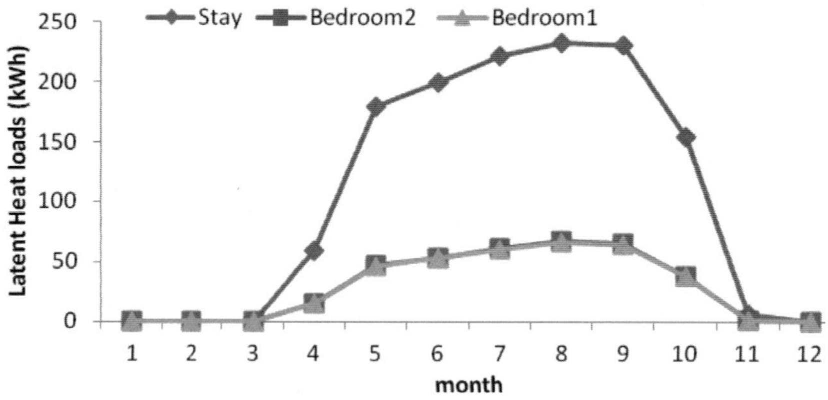

Figure 3b. Yearly evolution of the latent energy load

Table 3 gives the sensible and latent energy load in each bedroom, function of the type of glazing. One sees as expected that the latent energy varies little as a function of the type of glazing. In the meantime, the sensible energy load decreases especially when we add window shades.

Table 3. Yearly sensible and latent energy loads in the bedrooms.

	Sensible load (kWh)			Latent load (kWh)		
Unit	Stay	Bedroom 2	Bedroom1	Stay	Bedroom 2	Bedroom1
Simple_glazing	9148.84	3137.62	3093.30	1283.46	350.18	344.34
Double _glazing	8922.91	3109.80	3085.69	1281.81	351.85	346.17
Double_glazing + Window shades	8623.25	2964.77	2972.39	1285.21	357.83	353.92

The yearly energy load differences between the simple and double glazing are about 1.50%; the differences increase to about 4.61% when we compare the simple and the double glazing with window shades, figure 4.

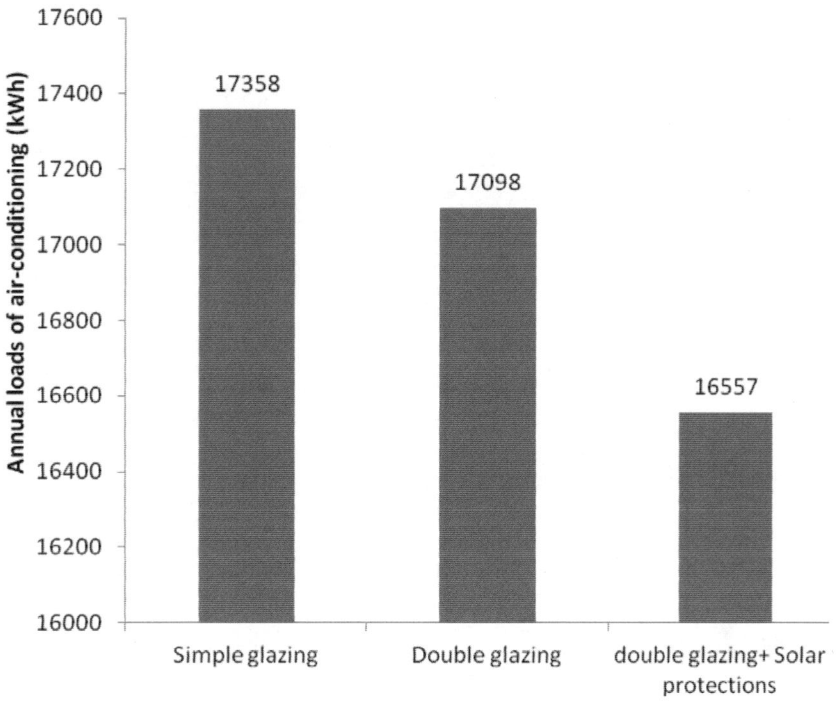

Figure 4. Window energy load optimisation and air conditioning load

The yearly energy consumptions of the equipments and fluorescent lamps are estimated when we take into account the occupancy time and their energy characteristics. The results are presented in table 4 below.

Table 4. Specific energy consumption.

	Electric consumption (kWh)		
Lighting	Equipments	Fans	Total
151.84	540.20	409.92	1101.96

5.2. Insulation, building consumption and PV panels energy generation

In this section, we compare the electric generation of the PV panels with the building total electric consumption. Figure 5 proves that the maximum of annual total radiation in Ouagadougou is obtained for 15° slope.

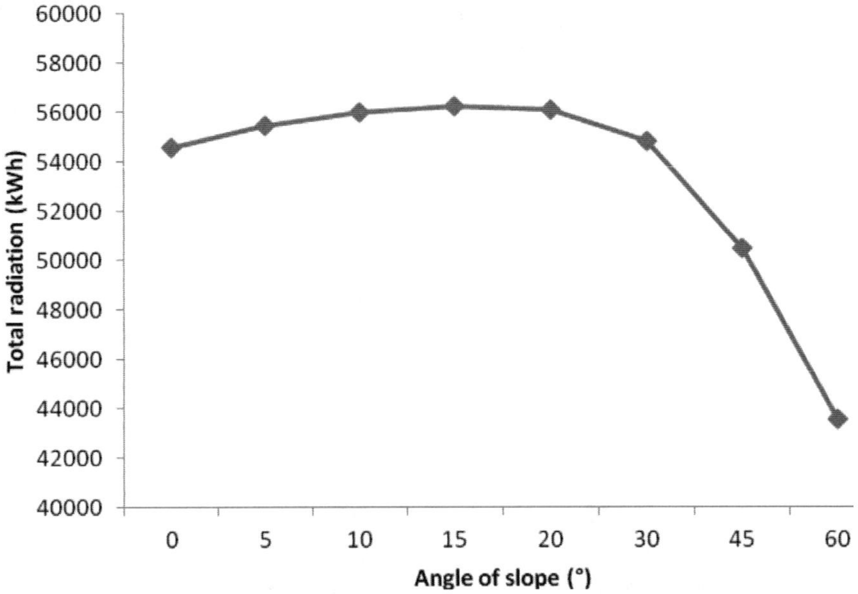

Figure 5. Variation of total radiation as a function of angle of slope

The table 5 shows the PV panel's generation and total radiation. The production of the polycrystalline panel's is obtained with the Type 94 of TRNSYS code. Total radiation being the same and knowing that the monocristalline panel's efficiency is 16.00% [5]; we calculated the production of monocrystalline panel's.

Table 5. PV panels generation and total solar radiation.

Module Type	Total radiation	PV generation (kWh)	Efficiency
Polycristalline	56209.91	5914.49	10.52%
Monocristalline	56209.91	8993.59	16.00%

The electric consumption ratio R_0 defined by reference [1] is computed by the following equation:

$$R_0 = \frac{Electric\ \text{consumption}}{Air\ conditioned\ surface} \qquad (11)$$

The air conditioned electric consumption load is computed assuming that the average COefficient of Performance (COP) of the air conditioner is equal to 2.5. We chose the value 2.5 to show that even if an air-conditioner which is not very powerful was utilized, we can obtain surpluses of PV productions through the optimization of the building envelope.

$$Electric\ \ consumption\ load = \frac{Thermal\ \text{energie}\ \ load}{COP} \tag{12}$$

Table 6 shows the insulation effects on the poly and mono crystalline PV panels' generation and on the building total electric consumption.

Table 6. Insulation effects on the contribution of PV panels' energy generation.

Some insulation cases considered	case1 : no insulation	case2 : Window optimization	Case3 : Roof insulation	Case4 : Roof insulation+floor	Case4+2cm insulation (wall)
Energy load (kWh)	17 357.73	16 555.37	10 826.83	9 456.31	7 764.92
Air conditioned electric consumption load (kWh)	6 943.09	6 622.15	4 330.73	3 782.52	3 105.97
electric consumption load +specifics (kWh)	8 045.05	7 724.11	5 432.69	4 884.48	4 207.93
Ratio R_0 (kWh/m²/year)	201.13	193.10	135.82	122.11	105.20
Polycristalline Contribution (%)	73.52	76.57	108.87	121.09	140.56
Monocristalline Contribution (%)	111.79	116.44	165.55	184.13	213.73

Without insulation, the electric consumption amounts to 8045.05 kWh. Hence, the PV generation represents 73.52% and 111.79% of the total electric consumption respectively for the poly and mono crystalline panels. Now, if we insulate the building by doing the following: optimizing the windows heat transfer, insulating the roof with 10 cm wood wool and insulating the floor with 5 cm polyurethane foam, the building electric consumption will amount to 4884.48 kWh. This time, the PV generation will represent 121.09 and 184.13% of the total electric consumption due to air conditioning, respectively for the same PV characteristics. Adding 2 cm inner insulate to the wall will increase these proportions to respectively 140.56 and 213.73%. We can then infer that with the abundant solar radiation in hot tropical climate, one can reduce both the number and incident surface area of PV panels if we conjugate the precedent practices with the building passive architectural design mode (orientation, window shades, etc.). Of course, it is necessary to proceed with some economical studies of the all process. Moreover, for building without insulation and with 2 cm

IOP Conf. Series: Materials Science and Engineering **29** (2012) 012013 doi:10.1088/1757-899X/29/1/012013

insulated wall, the energy consumption ratios decreases respectively from 201.13 to 105.20 kWh/m^2/year, which is an indication for better building efficiency. The investigation of author [1] which is so far the most accepted literature data in the topic, has concluded that the energy consumption reference of an apartment in tropical humid climate amounts to 160 kWh/m²/year. The results of the actual study are in the same range.

Figures 6 and 7 compare the PV panels' energy generation with the monthly electric consumption respectively for a no insulated building and a Case4+2cm wall insulated building. In the first case, we see that during the hot months and the raining season (increase of latent heat of dehumidification), the energy consumptions are greater than the energy generation depending on the PV panel type, figure 6. With insulation, the monthly electric generation is greater than the consumption regardless of the month, for monocristalline panels, figure 7.

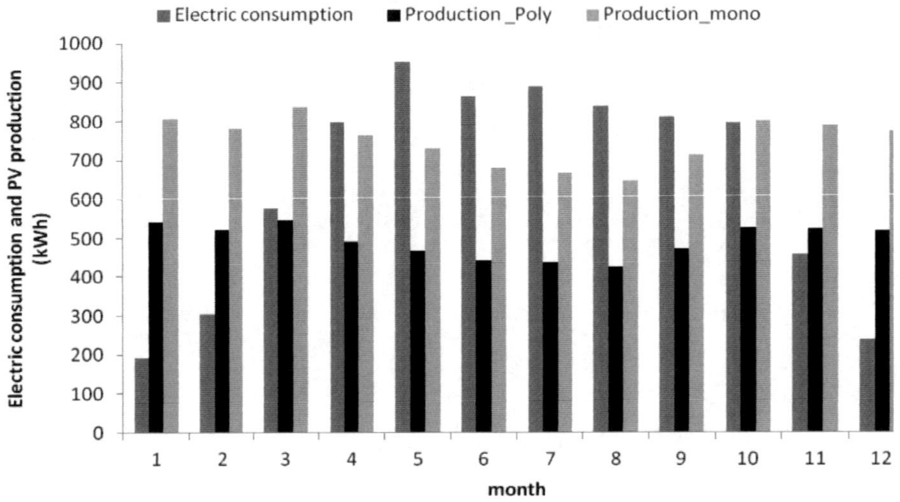

Figure 6. PV panels monthly energy generation and building consumption with no insulation

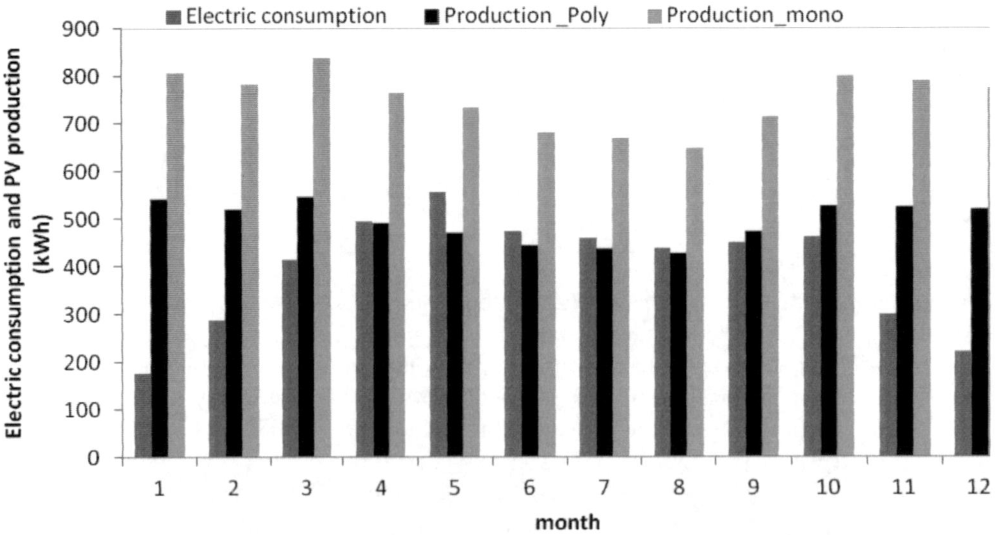

Figure 7. PV panels monthly energy generation and building consumption with case4+2 cm insulated wall

6. Conclusion

The actual work is part of a much larger effort to gather necessary and enough scientific data for designing the country's energy regulation code. As a matter of fact, it is well known that Sub-Saharan countries are allocating huge amounts of their hard gained currencies for the payment of the price tag of their ever increasing building energy consumption. Hence, they need to formulate in a near future a coherent building energy policy where renewable energy especially solar energy must play a bigger role. The results of the simulation, which are quite similar to the only accepted reference data in the field, show that energy savings can be made when we simply optimize our windows. Furthermore, energy autonomy might be reached and we can even generate excess energy if we couple the building with PV panels and insulate the roof, the floor, the wall and finally implement the passive designed mode. Of course, care must be taken in the choice of the construction materials and meteorological data. To complete the investigation, however, economical analysis of the all process must be undertaken to make sure that the entire operation is feasible. If so, besides being energy efficient, the excess energy generated by these types of buildings might be injected into the utility electric grid, reducing both the need for fossil fuel import and air pollution.

References

[1] Claessens J et al 2003 *Efficacité énergétique de la climatisation en pays tropicale Tome 1 : Conception des nouveaux bâtiments* IEPF
[2] IRSAT-CNRST 1995 *Plan stratégique de l'énergie au Burkina Faso*
[3] Al-ajmi F F and Hambi V I 2008 *Energy and buildings* **40** 1101-1109
[4] Annabi M, Mokhtari A and Hafrad T A 2006 *revue des énergies renouvelables* **vol. 9 N°2** 99-106.
[5] Diez-Webster U. et al. 2006 *Modélisation d'une maison à énergie positive* Projet de fin d'études INSA Lyon France
[6] Klein S A et al. 2006 *TRNSYS version 16 User manual* (solar energy laboratory University of Wisconsin USA)
[7] Keilholz W and Sette P 2005 Les évolutions de TRNSYS - La version 16 *Actes de Journée thématique SFT-IBPSA La Rochelle* http://perso.univ-lr.fr/fcherqui/IBPSAmars/textes/AM1-05-Abadie.pdf
[8] BERGES B 1986 Analyse d'un Comportement Conductif de l'Enveloppe d'un Bâtiment - Recherche de Modèles de Transfert Adaptés au Couplage Enveloppe – Équipement Régulation *Thèse de Doctorat (*Université de Nice France)
[9] Coulibaly O, kuznik F, Baillis D and Ouedraogo A 2010 Étude de l'influence des matériaux de construction sur les charges de climatisation d'une maison indivuduelle en climat tropical sec *IBPSA 09 au 10 Novembre* (Moret sur Loing France)
[10] Givoni B 1994 *renewable Energy* **5** Part II 908-916
[11] Coulibaly O 2011 Contribution à l'élaboration d'une réglementation thermique et énergétique des bâtiments au Burkina Faso, Données de base multiparamétriques et modélisation thermo-aéraulique sous CoDyBa et TRNSYS *thése de doctorat* (Université de Ouagadougou Burkina Faso)
[12] Duffie John A and William A B 1991 Solar Engineering of Thermal Processes (New York: John Wiley & Sons, Inc)

IOP Conf. Series: Materials Science and Engineering **29** (2012) 012014 doi:10.1088/1757-899X/29/1/012014

A guideline for sizing Photovoltaic panels across different climatic zones in Burkina Faso

M Waongo[1,2], Z Koalaga[1], F Zougmore[1]

[1]LAME-UO, Laboratoire de Matériaux et Environnement, Département de Physique, UFR/SEA, Université de Ouagadougou, 03 BP 7021 Ouagadougou 03, Burkina Faso
[2]Karlsruhe Institute of Technology (KIT), Institute for Meteorology and Climate Research Atmospheric Environmental Research (IMK-IFU), 82467 Garmisch-Partenkirchen, Germany

Email: moussa.waongo@imk.fzk.de, koalaga@univ-ouaga.bf

Abstract. In many developing countries a long time series of solar radiation measurements are not often available. This is due to the cost, maintenance and calibration requirements of measuring devices. Consequently, the use of solar energy by Photovoltaic (PV) conversion copes with the choice and the optimization of the PV system. This work concerned the analysis of climate parameters strongly influencing the Photovoltaic (PV) systems energy production and the simulation of an ideal system based on a single PV module. Estimation and analysis of time series of climate parameters covered a set of six weather stations with respect to the three climatic zones in Burkina Faso (BF), over 38 years. The analysis showed that the solar irradiation in BF lies between 3 kWh/m^2/day and 7.5 kWh/m^2/day. The highest values of the solar irradiation are measured in the Northern part of the country while lowest values are measured in the Southern part. Daily mean temperature for all weather stations was greater than the Standard Test Condition (STC) temperature (25°C) over a long period of the year. Information on solar irradiation and temperature is fundamental for PV systems sizing process. For PV performance evaluation, a simulation is carried out using an ideal system composed of a single PV module from TENESOL Company. This simulation is performed for three classes of climatic conditions "Mean situation", "Adverse situation", and "Beneficial situation", and evaluated for six sitesacross BF. The results revealed intra-annual and spatial variability of Maximum Power (MP). Across BF, MP varied between 60 W/day and 190 W/day in Sahelian zone, between 65 W/day and 185 W/day in soudano-sahelian zone, and between 67 W/day and 208 W/day in Soudanian zone. MP intra-annual variability is higher during the period July-August, mainly for "Beneficial situation". The negative effect of temperature on PV energy production is specially amplified in Sahelian zone due to its highest temperatures. This study allowed to estimate and evaluate the role of climatic parameters on the energy production of PV and can be used as a guide in PV panels sizing.

2. Introduction

Energy plays a key role in many aspects of human development, such as poverty and hunger reduction, and the improvement of health services. The relationship between energy availability and life conditions improvement has been recognized and largely mentioned in many publications [1-5]. Therefore, face to the crisis of fossil energies and negative impacts on environment of the latter, renewable energies are considered as suitable energies for the future over the world. This is also due to the fact that renewable energy resources have some advantages if compared to fossil ones. One major advantage with the use of renewable energies is that as they are renewable they are therefore sustainable and so will never run out. In addition, renewable energies are unable to produce carbon-based warming and polluting agents into the atmosphere. Thus, they can be considered as one of solutions for climate change issue.

The use of renewable energies has experienced a significant growth in the world [6]. Among them, solar energy might be a good candidate to substitute fossil energy at least in tropical zone, which is characterized by highest global solar radiation values. In developing countries, locate in tropical zone, energy is a scarce resource. For this reason, the use of solar energy by Photovoltaic (PV) conversion might be an alternative to supply electricity for lighting and water pumping to remote or rural areas [7-9]. This will contribute to improve the life of concerned populations. In urban areas with a repetitive black out due to high demand for energy, PV energy might be attractive for electricity companies [10-11]. However, the choice and the optimization of a Photovoltaic system depend enormously on climatic conditions such as solar radiation and temperature at the location of interest and financial capacity of the interested person. Unfortunately, for many developing countries as Burkina Faso, solar radiation measurements are rarely available due to the cost, maintenance and calibration requirements of measuring devices. Thus, it is important to elaborate methods to estimate the solar radiation based on readily available meteorological data. Furthermore, for PV panels sizing, it is essential to evaluate the impact of climate conditions over the year on energy production from PV [12].

The first objective of this study is to estimate the potential solar radiation over Burkina Faso using available meteorological data. The second aim is to analyze intra-annual variability of output energy of a single Photovoltaic module using two major PV energy production drivers such as radiation and temperature.

Nomenclature	
h	Measured daily sunshine duration (hours)
H	maximal daily sunshine duration (hours)
I	current (A)
I_{ph}	photo- current (A)
I_S	saturation current of the diode (A)
K	constant of Boltzmann
MP	maximum power (W)
n	ideality factor of the diode
N_P	number of cells in parallel
N_S	number of cells in series

q	electron charge (C)
R_a	Extraterrestrial solar irradiation ($J/m^2/day$)
R_G	global solar irradiation ($J/m^2/day$)
R_P	shunt resistance (Ω)
R_S	seriesresistance (Ω)
T	Absolute temperature (K)
V	voltage of cells (V)

3. Materials and Methods

3.1. Photovoltaic system drivers

For a given site and a specific PV module, the converted energy from solar radiation depends on several parameters and variables such as the earth-sun movement, sensors orientation and especially weather and climate conditions. Latter include solar radiation, temperature, wind, cloud covering and dust. In fact, incident solar radiation depends strongly on the day and the latitude of the location. Solar radiation and temperature are the most represented variables in PV system efficiency study.

The measured solar radiation so-called solar irradiation is the part of solar radiation, which reaches the surface of earth after many physical processes, or not. This incident radiation is composed by two partsof radiation mainly made up visible radiation (0.38 µm - 0.78 µm):

- direct radiation (D) which represents the part of the solar radiation that reach the earth without any transformation;
- diffuse radiation (S) which represents the radiation that reaches the surface of the earth after some physical transformation such as absorption, reflexion and diffusion, by atmospheric clouds and gases.

Global solar irradiation (R_G) is the sum of both parts of radiation which can be written in this form:

$$R_G = D + S \tag{1}$$

At the top of the earth atmosphere, incident solar radiation flux so-called extra-terrestrial global irradiation or solar constant is estimate to be equal to 1367 Wm^{-2} [13]. The daily R_G is measured in $J/cm^2/day$ using a pyranometer. Some differences can be observed in R_G daily value depending on the landscape. Sunshine duration usually measured by a Campbell-Stock heliograph is also use to estimate R_G in many regions over the world. Sunshine duration during a given period is defined as the sum of that sub-period for which the direct solar irradiance exceeds 120 Wm^{-2}.

The operating temperature of a module is determined by the equilibrium between the heat produced by the PV module, the heat lost to the environment and the ambient operating temperature. The heat produced by the module depends on the operating point of the module, the optical properties of the module and solar cells, and the packing density of the solar cells in the PV module. The heat lost to the environment can proceed via one of three mechanisms; conduction, convection and radiation. Thus, thermal condition during the day plays a crucial role in PV efficiency.

3.2. Study sites and data sets

For our study we used data sets by National Meteorological Services (NMS) of BF. They provided daily data for sunshine duration, mean temperature from 1971 to 2008 and solar global irradiation from 1971 to 2004. Sunshine duration and mean temperature data sets had no missing data contrary to solar global irradiation data set. The complete time series of solar global irradiation are less than fifteen years. For years with missing values, the rate of daily missing value is more than 25%.

These data sets cover six out of ten principal meteorological stations. These stations are chosen to be representative for each of the three agro-climatological zones in Burkina Faso (figure 1). In each agro-climatological zone, two principal meteorological stations are chosen to be reflect sub-climate patterns and also for the quality of recorded data.

For the PV simulation under Burkina Faso climate conditions, we used characteristics of a single polycrystalline PV module from TENESOL company so-called TE 2000.

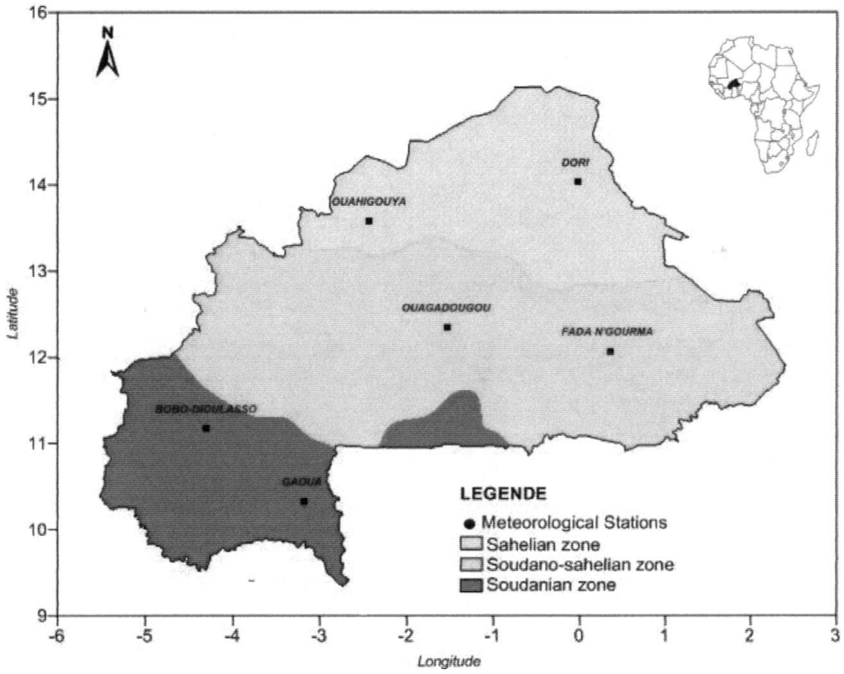

Figure 1. Location of six principal meteorological stations in Burkina Faso

3.3. Global solar irradiation estimation

Several approaches have been developed for estimating the surface global solar irradiation, such as remote sensing retrievals and a number of empirical models based on surface meteorological data [14-16]. Among the different categories of empirical models, sunshine based models are most popular in PV design, hydrological and agricultural studies [17], because of its low computational cost and accessible inputs. In this study, Ångström-Prescott model [18] is used to compute global solar irradiation. This model was proposed by Ångströmin 1924 and modified by Prescott in 1940. It has been widely calibrated for many different climatic zones:

$$R_G = R_a \left(a + b \times \frac{h}{H}\right) \qquad (2)$$

Where $\mathbf{R_G}$ is daily actual global radiation (J/m^2/day), $\mathbf{R_a}$ is daily extra atmosphere global solar radiation (J/m^2/day), **h** is measured daily sunshine duration (hours), **H** is maximal daily sunshine duration (hours) and **a** and **b** are empirical parameters.

The steps of computation of $\mathbf{R_a}$ and **H** are given by literature [17]. Values of **a** and **b** for different climatic areas over the world are available in literature [18]. For our study **a** and **b** are set to 0.25 and 0.45 respectively.

To analyze accuracy and relevance of estimated global solar irradiation using Ångström-Prescott model, Pearson correlation coefficient **r** and the root mean square error **(RMSE)** were computed for corresponding period where measured global irradiation (R_O) and estimated global solar irradiation ($\mathbf{R_{Ge}}$) were available.

3.4. Modeling of solar global radiation and mean temperature

A statistical analysis of empirical distribution of $\mathbf{R_{Ge}}$ and mean temperature (Tm) was done on daily time scale in order to determine three subsets of data for each climatic variable. The definition of each subset data reflects the potential impact of solar radiation in PV system for energy production:

- the first subset of data so called "adverse situation" correspond to the lower quartile (Q1) for daily $\mathbf{R_{Ge}}$ and daily Tm,
- the second subset of data so called "mean situation" correspond to the daily mean value for $\mathbf{R_{Ge}}$ and daily Tm,
- the third subset of data so called "beneficial situation" correspond to the upper quartile (Q3) for daily $\mathbf{R_{Ge}}$ and daily Tm.

Using data of daily sunshine duration, global insolation expressed in kWh/m^2/day was computed for each subset. Then, the three subsets of temperature and daily global insolation were fitted using polynomial approach. MATLAB software was used for this process.

3.5. Simulation of a single Photovoltaic module

A Photovoltaic module is the basic element of each Photovoltaic system. It consists of many jointly connected solar cells. A solar cell can be described by an electrical model (figure 2a) [19-21]. In our study we assume the electric model to be ideal (figure 2b). The electrical model equation of a module composed with Ns cell in serial and Np cells in parallel is given by:

$$I = N_p I_{ph} - N_p I_s \left(e^{\left[\frac{q\left(\frac{V}{N_S}+\frac{IR_S}{N_P}\right)}{n\,k\,T}\right]} - 1 \right) - N_p \left(\frac{\frac{V}{N_S}+\frac{IR_S}{N_P}}{R_p} \right) \quad (3)$$

The following non linear equation represents the ideal characteristic of electric voltage-current (I-V) for a single PV module:

$$I = N_p I_{ph} - N_p I_s \left(e^{\left[\frac{q\,V}{N_S\,n\,k\,T}\right]} - 1 \right) \quad (4)$$

where I is the module electric current (A), I_{ph} is the cell photocurrent (A), I_s is the reverse saturation current (A), V is a cell voltage (V), q is the charge of electron (=1.6×10^{-19}C), n is the diode ideality factor (1 for ideal diode), T is the thermal condition (K), k is Boltzmann constant, N_p and N_s are the number of parallel and series cells in the PV module respectively.

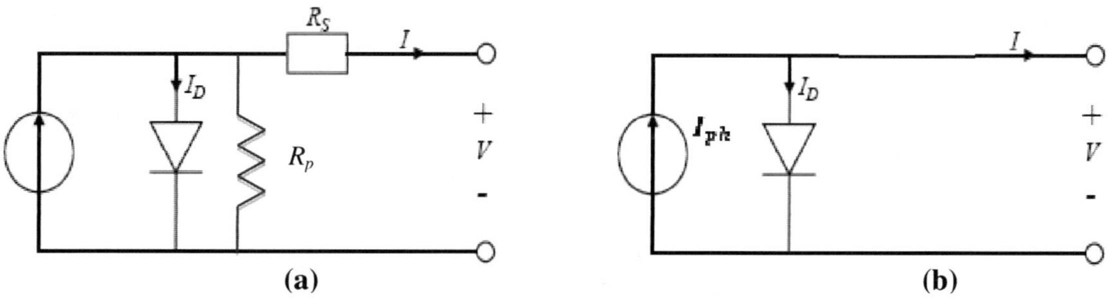

Figure 2. Simplified solar cell electric model (a) and an ideal solar cell electric model (b).

The electrical model for one PV module was computed using MATLAB software [22-23]. Then, characteristics of TE 2000 module and the three subsets of climate data were used as inputs in our model in order to simulate the potential impact of climate conditions on the module energy production assuming no waste of energy due to material resistances. In sum, three simulations were carried out according to scenarios "adverse situation", "mean situation" and "beneficial situation".

4. Results and discussion

4.1. Daily based global irradiation correlation analysis

Figure 3 shows that the determination coefficient (R^2) between measured global irradiation (R_O) and estimated global irradiation (R_{Ge}) varied between 0.47 and 0.84 with an average of 0.68. In general, all the scatters are located above line 1:1. This shows that R_{Ge} was overestimated R_O. This overestimation is clearer for Ouahigouya, Ouagadougou, Fada N'Gourma and Gaoua stations. For all stations RMSE is less than 3.85 MJ/m^2/day. Highest value was reached in Ouahigouya station and lowest one in Dori station (table 1). Correlation coefficient (table 1) for linear trend varied between 0.70 (Ouahigouya) and 0.92 (Ouagadougou). The linear Trend equation (figure 3) can be a good approximation for R_O mainly for Ouagadougou and Gaoua to minimize RMSE. The empirical parameter a and b proposed by Frere and Popov (1979) and used in this study is more global and need to be calibrate for each site or at least to identify more accurate values for a and b for entire Burkina Faso. A robust empirical relationship between R_O and R_{Ge} can be help for PV panel sizing and PV performance study as R_O is often not available.

4.2. Daily global insolation and temperature fitting

Daily global irradiation plots for all six stations showed that global irradiation varied between 3 kWh/m^2/day and 7.5 kWh/m^2/day (figures 4, 5, 6). So, the mean value of global irradiation decreases from North to South. The highest values were reached in Dori (figure 4) and the lowest values in Gaoua (not shown here). The period from April to May correspond to a peak of global irradiation for all stations. A second peak was observed from September to October. The latter peak is relatively low in term of value and duration if compared to the former. From December to January, global irradiation drops to reach the lowest values.

Mean temperature in all sites varied between 20°C and 36°C with an average greater than 25°C (figures 4, 5, 6). The highest mean temperature was measured in Dori, which is located in the Northern area and the lowest mean temperature in the southern area of BF. A bimodal shape was observed for mean temperature. The first peak was reached on April-May and the second one, less than the former, on October. Based on this analysis, temperature represents a limiting factor for all stations. The highest values of global radiation on April match with the highest values of mean temperature (more than 30°C). It is admitted that the Sahel zone countries have very important incidental solar energy; this study shows that solar energy is very variable according to the site and period of the year.

For PV applications, a polynomial fitting using a polynomial function of degree 10 for global irradiation and mean temperature (figure 4, 5, 6) was done for all stations. The computed polynomial function was able to reproduce a reliable daily time series of global irradiation and temperature. The fitting function was more accurate for temperature if compare to global irradiation. The coefficients of polynomial functions (not shown here) can be used for daily based PV simulations or other applications in case of missing data or data accessibility problems.

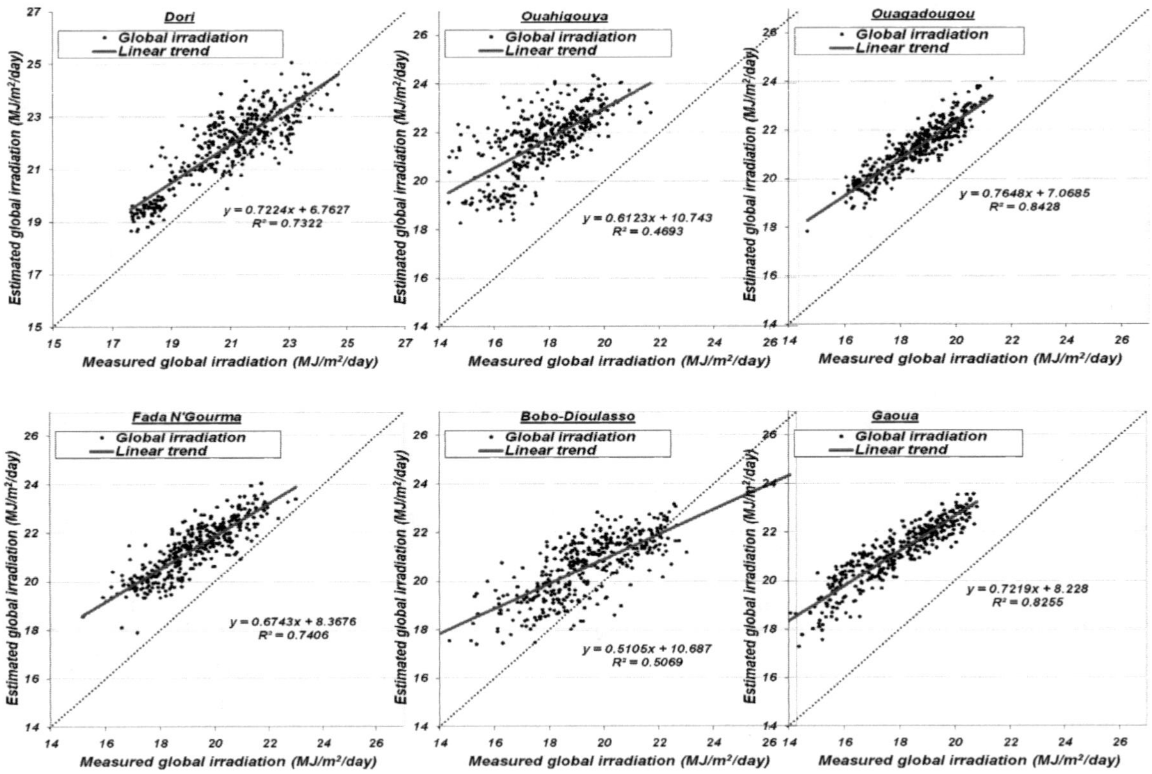

Figure 3. Correlation results between R_O and R_{Ge} for all six sites.

Table 1. Correlation coefficient and root mean square error between R_O and R_{Ge} for all six sites.

	Dori	Ouahigouya	Ouagadougou	Fada N'Gourma	Bobo-Dioulasso	Gaoua
R	0.86	0.69	0.92	0.86	0.71	0.91
RMSE	0.96	3.85	2.72	2.11	1.22	3.27

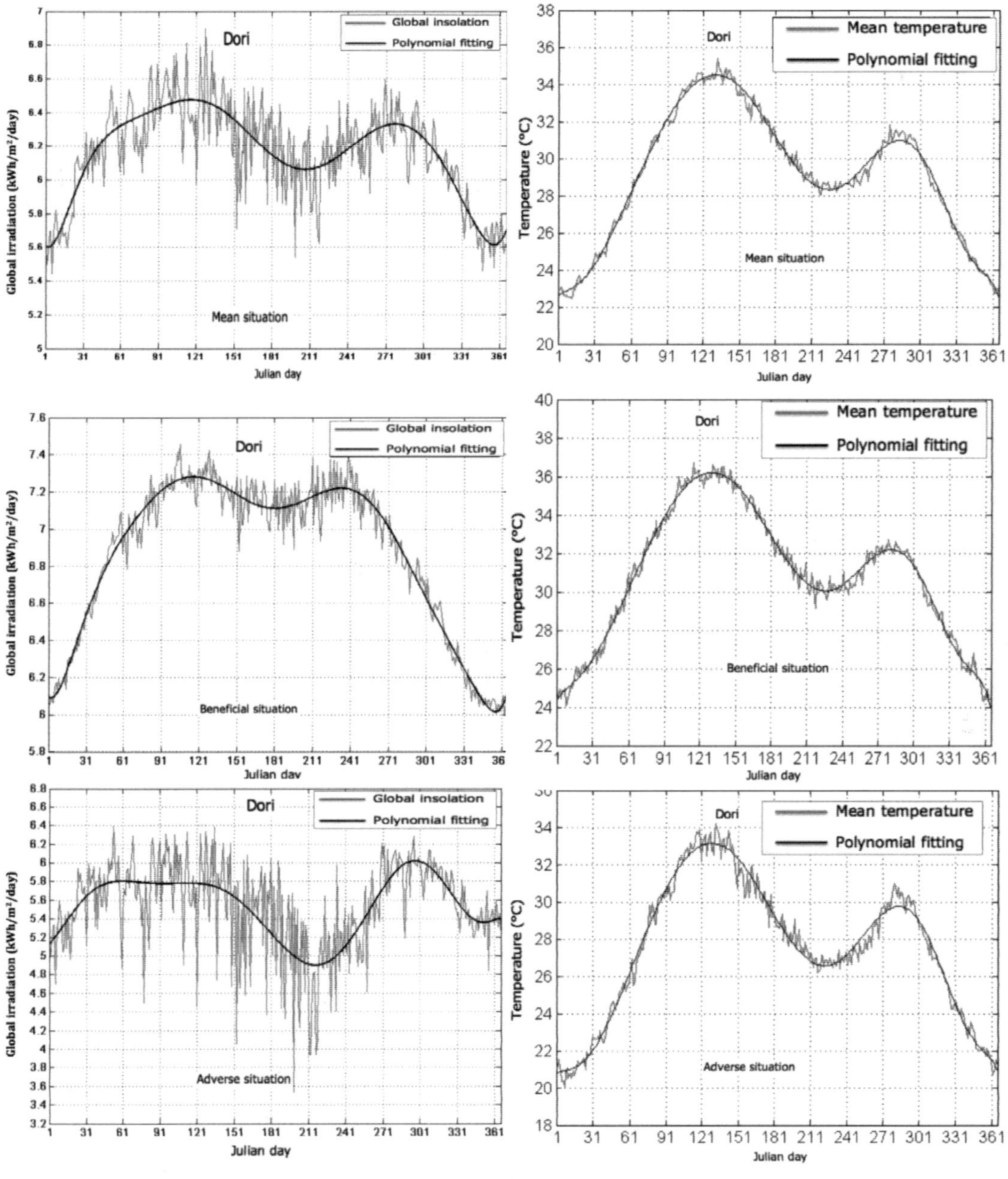

Figure 4. Daily global irradiation and daily mean temperature for three climatic conditions at station Dori.

IOP Conf. Series: Materials Science and Engineering **29** (2012) 012014 doi:10.1088/1757-899X/29/1/012014

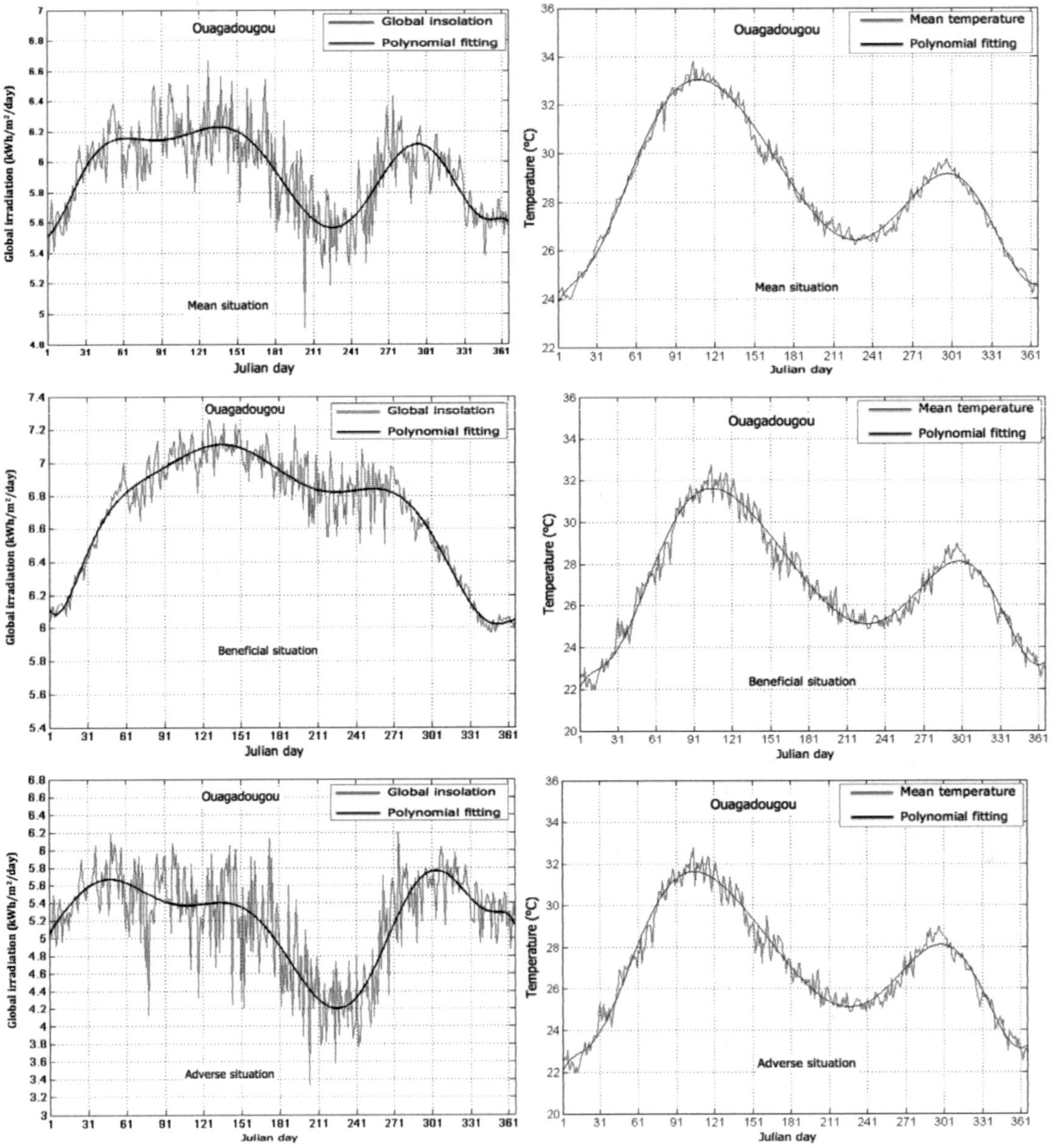

Figure 5. Daily global irradiation and daily mean temperature for three climatic conditionsat station Ouagadougou.

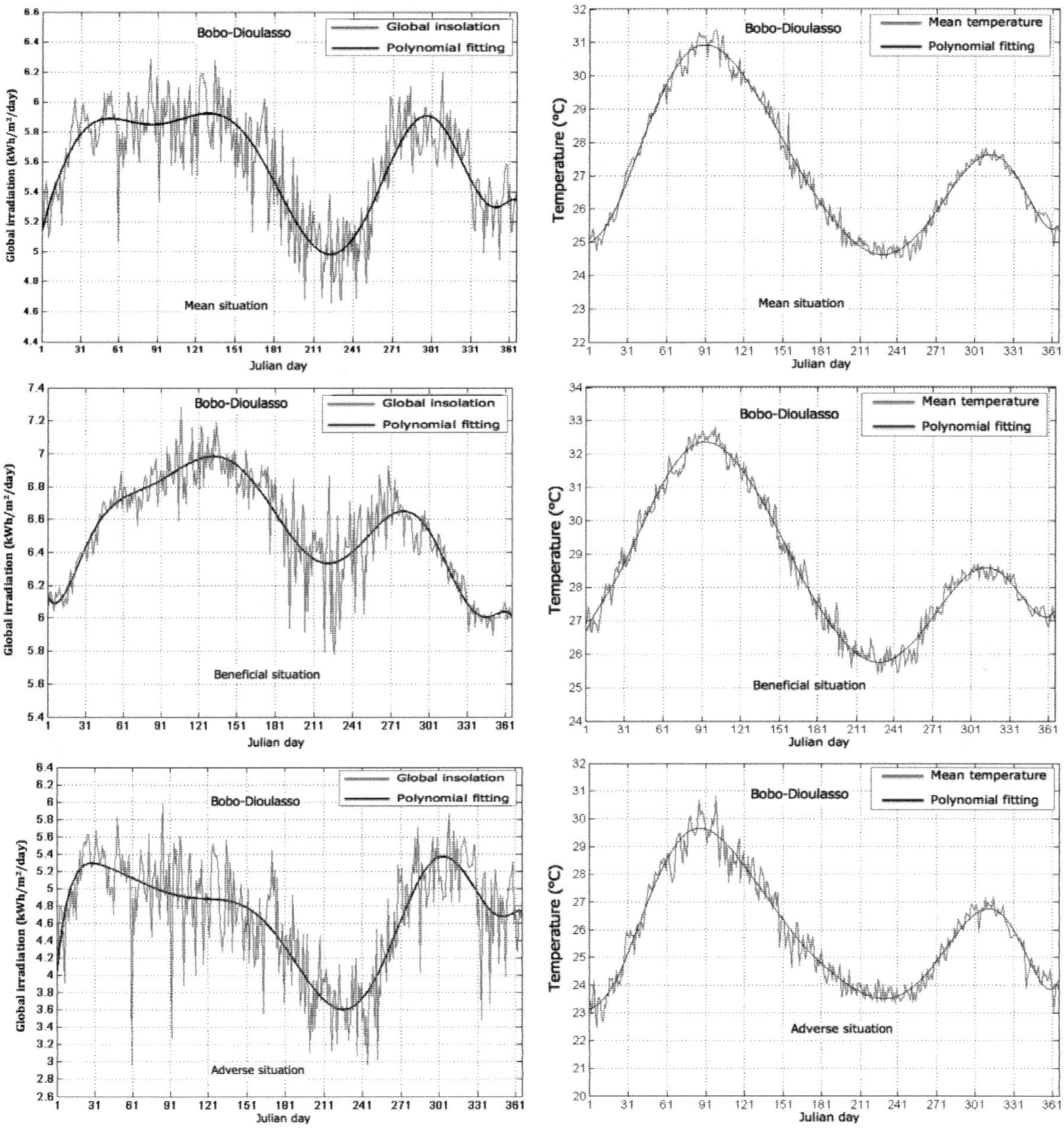

Figure 6. Daily global irradiation and daily mean temperature for three climatic conditions at station Bobo-Dioulasso.

4.3. Single PV module simulation

4.3.1. TENESOL PV module TE 2000 characteristics. Most of existing softwares for PV system modelling [24] are commercial or sometimes not adapted for many areas. To face this problem we developed under MATLAB an open source program to simulate a PV module TE 2000. The inputs for this model are TE 2000 characteristics [25], solar radiation and mean temperature. The model successfully reproduce Current-Voltage (I-V) characteristics (figure 7), the influence of temperature in reducing PV module output power (figures 8b, 8d) and the energy conversion efficiency from different incident solar radiations (figures 8a, 8c). In Standard Test conditions (STC), TE 2000 module produces a power of 210 W at maximum power point (MPP). This MP is reduced to 50% when incident solar radiation is set to 700 W. MP at STC is reduced to 200 W when temperature is set to 30°C. So, the model was able to track the MP for TE 2000 module as mention by TENESOL Company [25].

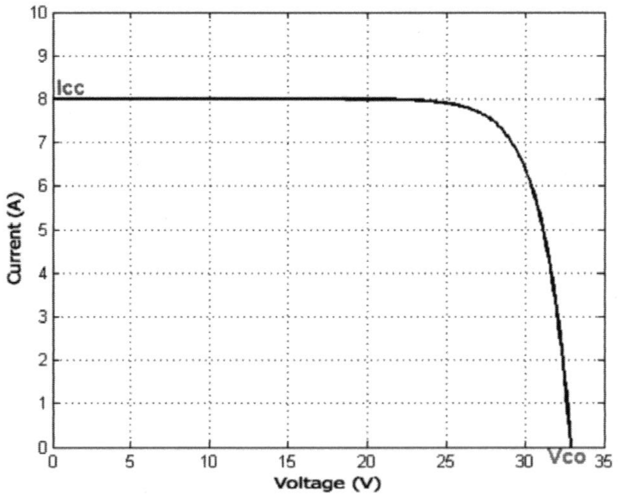

Figure 7. PV module TE 2000 characteristic I-V.

4.3.2. Simulation under different climatic conditions. Using three different climatic conditions of solar radiation and temperature, the power yield of a TE 2000 PV module is simulated. For each station and for each climate condition, simulated MP (power generated by the electrical model of PV TE 2000 module) is tracked in daily time scale [24].

In Dori and Ouahigouya, which are located in Sahelian zone the MP of TE 2000 module varied between 60 W (adverse situation) and 190 W (beneficial situation) (figure 9). The mean of the MP in this zone is estimated to be 95 W with a standard deviation of 12 W. Further, the PV module MP varied within this zone. For the same climatic zone, the MP was drove by temperature and the station of Ouahigouya got the highest MP compared to the station Dori (figure 9). During January, February, October, November and December, the maximum power provided by module is lower than 100 W in *beneficial situation* and lower to 70 W during December.

In Ouagadougou and Fada N'Gourma in Soudano-sahelian zone, the average of the MP was 90 W with a standard deviation of 14 W. In *beneficial situation* the maximum power is about 120 W and can reach 180 W during the periods of September and August. In this climatic zone, the periods of

IOP Conf. Series: Materials Science and Engineering **29** (2012) 012014 doi:10.1088/1757-899X/29/1/012014

lowest MP were in January, February, November and December (figure 9). An average of 82 W was estimated during these periods. For the *adverse situation*, the MP drop to 63 W during these periods.

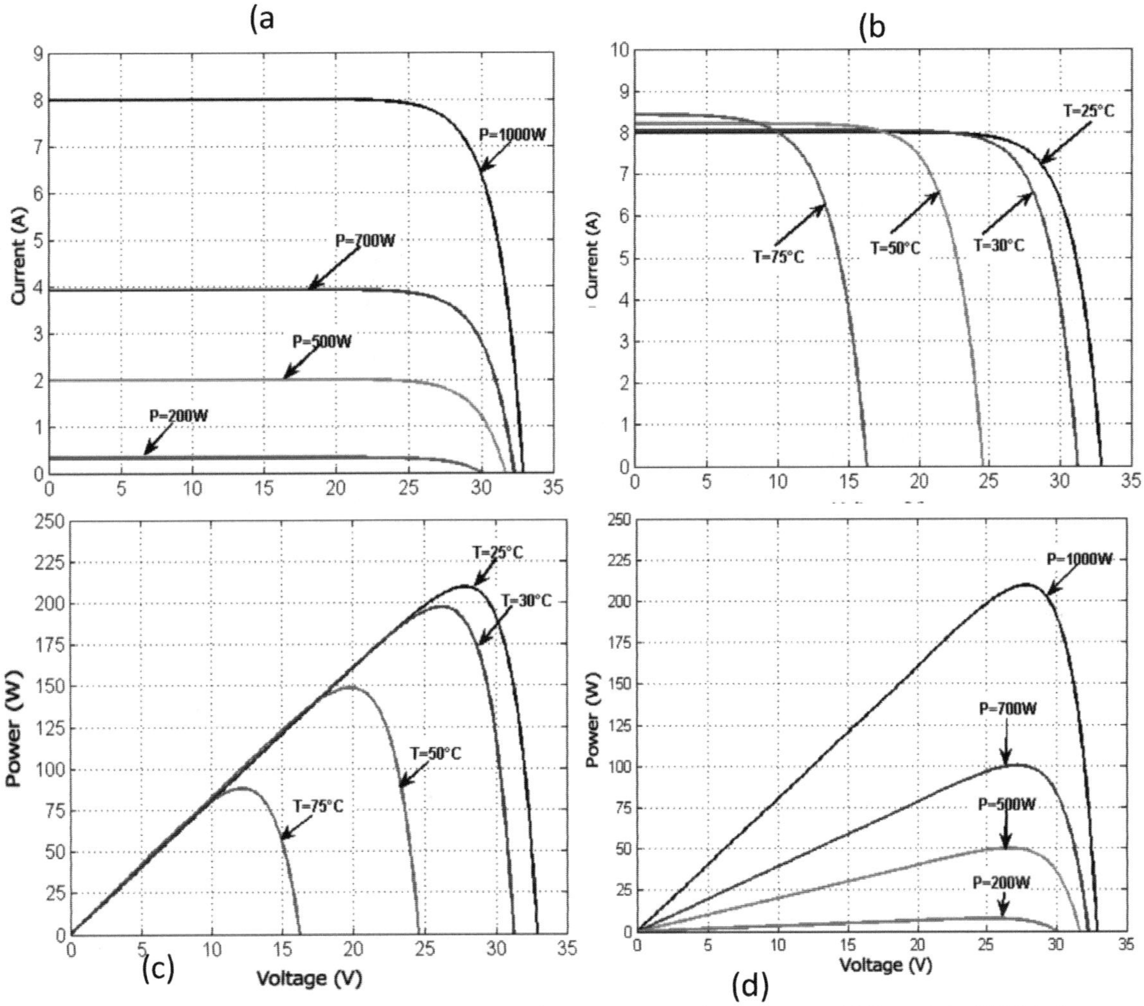

Figure 8. Standard global irradiation and temperatures influence on PV module TE 2000.

The MP varied between 67 W and 208 W in Bobo-Dioulasso and Gaoua in Soudanian zone (figure 9). The mean was 120 W with a standard deviation of 18 W. In *beneficial situation*, MP was 190 W during the period from July to September. In *adverse situation* the mean value of MP in these areas was 90 W. The lowest values of MP were got on November, December and January. The highest MP in Soudanian zone was mainly due to lowest temperatures in this area.

IOP Conf. Series: Materials Science and Engineering **29** (2012) 012014 doi:10.1088/1757-899X/29/1/012014

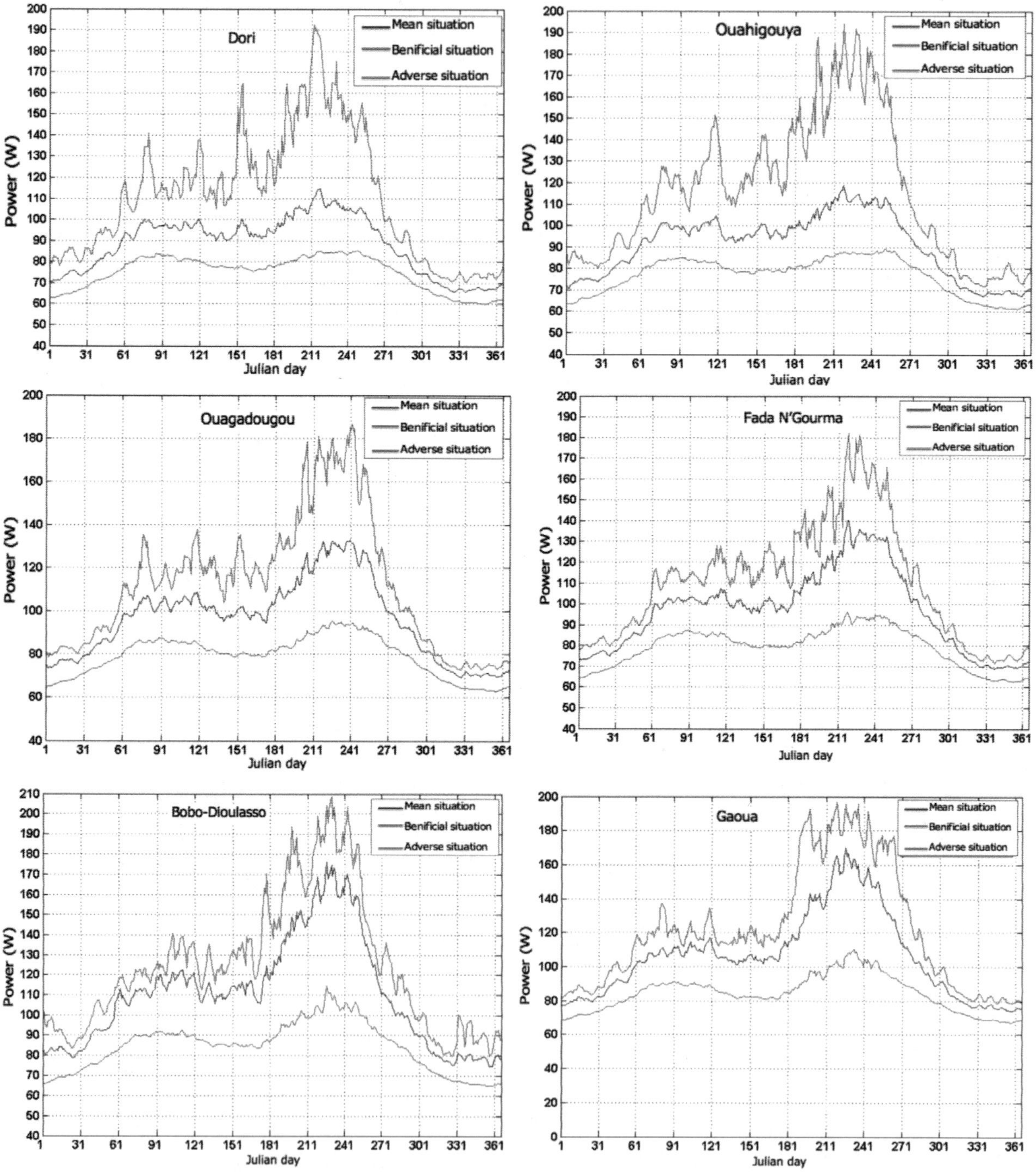

Figure 9. PV module TE 2000 model output for different climatic conditions.

5. Conclusion

This study concerned the analysis of climatic variables, which strongly influence Photovoltaic systems energy production and the performance analysis of PV system using a single PV module. The analysis of climate parameters was done for six (6) stations in Burkina Faso. It results that solar irradiation in BF lies between 3 kWh/m^2/day and 7.5 kWh/m^2/day. This study revealed that solar irradiation availability varied remarkably within the sub-climatic zones. Between sub-climatic zones, the highest values of the solar irradiation are located in the Northern part of the country while lowest values are found in the South.

PV system simulation was conducted to measure the impact of three classes of conditions (mean situation, beneficial situation and adverse situation) on energy production of PV system using solar radiation and temperature. The combination effect of radiation and temperature is more favorable for Soudanian zone than Sahelian zone. In fact, the negative effect of temperature on PV energy production is amplified in Sahelian zone due to its highest temperatures. It comes out that in BF, temperature greatly affects the efficiency of PV energy production in BF.

The daily MP was greater than 60 W for all the three classes of climatic conditions with the highest values and high variability during the rainy season in BF, from July to September. The high variability is due to the fact that a cloudy day reduces the incoming solar energy (direct solar irradiation) while the later is high during a clear sky day. During the rainy season, the low level of aerosol and dust in the atmosphere contribute to high values of incoming solar energy if compared to the dry season from November to April. In addition, the rainy season is also characterized by low temperatures. Those reasons can justify the high value of MP from July to September.

The results achieved in this study can be helpful for solar panel sizing or PV energy production optimization.

Acknowledgements

The authors are grateful to P. Laux and E. Bleher from Karlruhe Institute of Technology, the Institute for Meteorology and Climate Research, Atmospheric and Environmental Research (IMK-IFU) in Germany for their careful reading and insighful comments.The help of the National Meteorological Services (NMS) of BF which provided the climate data is highly appreciated.

References

[1] Connor H and Williamson L 2006 Ultimate Challenge: Energy for Global Ecodevelopment *HELIO International Global Report* available at www.helio-international.org
[2] Minvielle J P 1999 *La question énergétique au Sahel* Karthala (Collection Economie et developpement Paris)
[3] Martin J-L 2002 The G8 at the bedside of developing countries : renewable energy as part of sustainable development to break the vicious circle of poverty *Revue de l'énergie* **540** 585-593
[4] Karekezi S 2002 Poverty and energy in Africa - A brief review *Energy Policy* **30** 915-919
[5] Akella A K, Saini R P and Sharma M P 2009 Social, economical and environmental impacts of renewable energy systems *Renewable Energy* **34** 390-396
[6] Multon B 1999 Production d'énergie électrique par sources renouvelables *Techniques de l'Ingénieur, traité D4005*
[7] Karekezi S and Ranja T 1997 *Renewable energy technologies in Africa* pp. xvi + 269
[8] Foley G 1995 Photovoltaic applications in rural areas of the developing world *World Bank Technical paper number 304 energy serie*

[9] Ramakumar R, Abouzahr I, Krishnan K and Ashenayi K 1995 Design scenarios for integrated renewable energy systems *Energy Conversion, IEEE Transactions on* **10** 736-46
[10] Kaundinya D P, Balachandra P and Ravindranath N H 2009 Grid-connected versus stand-alone energy systems for decentralized power - A review of literature *Renewable and Sustainable Energy Reviews* **13** 2041-50
[11] PERSEUS 2007 *Guide des installations photovoltaïques raccordées au réseau électrique, destiné aux particuliers* Edition ADEME, HESPUL et al - pp 7- 50
[12] Buresh M 1983 *Photovoltaic Energy Systems (*McGraw-Hill Book Company New-York)
[13] Sanroma E, Palle E and Sanchez-Lorenzo A 2010 Long-term changes in insolation and temperature at different altitudes *Environ. Res. Lett.* **5** 024006
[14] Paulescu M and Schlett Z 2003 A simplified but accurate spectral solar irradiance model *Theoretical and Applied Climatology* **75** Numbers 3-4 DOI:10.1007/s00704-003-0731-y
[15] Paulescu M, Fara L and Tulcan-Paulescu E 2006 Models for obtaining daily global solar irradiation from air temperature data *Atmospheric Research* **79** 227-40
[16] Bourges B 1991 *European simplified methods for active solar system design* (Dordrecht The Netherlands Boston : Kluwer Academic Publishers)
[17] Allen R G, Pereira L S, Raes D and Smith M 1998 Crop evapotranspiration: Guidelines for computing crop requirements *FAO Irrigationand Drainage Paper* No. 56
[18] Frere M and Popov G F 1979 Agrometeorological crop monitoring and forecasting *FAO plant production and protection paper* (Rome Italy) p 64
[19] Zhou W, Yang H and Fang Z 2007 A novel model for Photovoltaic array performance prediction *Applied Energy* **84** (12)1187-119
[20] Francisco M 2005 Model of Photovoltaic Module in Matlab *2Do Congreso Iberoamericano de Estudiantes de Ingenieria Electrica, Electronica Ycomputacion* (II CIBELEC Espagna)
[21] TsaiH-L, Tu C-S and Su Y-J 2008 Development of Generalized Photovoltaic Model using MATLAB/ SIMULINK Proc. *of the World Congress on Engineering and Computer ScienceWCECS 2008* (San Francisco USA)
[22] TsaiH-L 2010 Insolation-oriented model of Photovoltaic module using Matlab/Simulink *Solar Energy* **84** 1318-1326
[23] Altas I H and Sharaf A M 2007 A Photovoltaic Array Simulation Model for Matlab-Simulink GUI Environment *Int. Conf. on Clean Electrical Power ICCEP 07 21-23 May* (Capri Italy)
[24] Benatlallah A and Mostefaoui R 2002 Logiciel de simulation PV *Revue des Energies Renouvelables* : Zones Arides
[25] Tenesol *TE 2000 Module characteristics* http://www.tenesol.com

IOP Conf. Series: Materials Science and Engineering **29** (2012) 012015 doi:10.1088/1757-899X/29/1/012015

Grid-connected photovoltaic (PV) systems with batteries storage as solution to electrical grid outages in Burkina Faso

D Abdoulaye[1], Z Koalaga[1], F Zougmore[1]

[1]LAME-UO, Laboratoire de Matériaux et Environnement, Département de Physique, UFR/SEA, Université de Ouagadougou, 03 BP 7021Ouagadougou 03, Burkina Faso

Email: abdoulayedaouda2004@yahoo.fr, koalaga@univ-ouaga.bf

Abstract: This paper deals with a key solution for power outages problem experienced by many African countries and this through grid-connected photovoltaic (PV) systems with batteries storage. African grids are characterized by an insufficient power supply and frequent interruptions. Due to this fact, users who especially use classical grid-connected photovoltaic systems are unable to profit from their installation even if there is sun. In this study, we suggest the using of a grid-connected photovoltaic system with batteries storage as a solution to these problems. This photovoltaic system works by injecting the surplus of electricity production into grid and can also deliver electricity as a stand-alone system with all security needed. To achieve our study objectives, firstly we conducted a survey of a real situation of one African electrical grid, the case of Burkina Faso (SONABEL: National Electricity Company of Burkina). Secondly, as study case, we undertake a sizing, a modeling and a simulation of a grid-connected PV system with batteries storage for the LAME laboratory at the University of Ouagadougou. The simulation shows that the proposed grid-connected system allows users to profit from their photovoltaic installation at any time even if the public electrical grid has some failures either during the day or at night.

1. Introduction

The electricity problem in Africa particularly in West Africa is getting more and more crucial these last years. So, in addition to the incapacity of the installed powers, we observe recently an increase in power cuts frequencies especially during period of severe hot season (in March, April, May) for many of these countries.

This problem may be explained by the fact that he electricity demand is higher than the capacity of electricity networks. In these conditions, the electricity companies are obliged to plan the electricity distribution. They cut the electricity in certain districts to feed the others [1]. In addition, all West African countries grids experience some instability. So the imported part of energy production from neighboring countries is also subject to this instability. This situation represents a real handicap to the economic and social development of these countries. We point out this fact from a survey that we conducted on Burkina-Faso electrical network.

It becomes thus urgent to find a solution which can at the same time increase the electrical production and resolve the problem of power cuts and grids instability. The photovoltaic systems represent a real asset for many African countries, especially those which have permanent sunshine all year long such as Burkina Faso. Some users have installed classical grid-connected photovoltaic but they are unable to profit from their installation even if there is sun when the grid experiences some outages.

In this paper, we propose a grid-connected photovoltaic system with battery storage as solution. The system will work in the injection mode of the surplus of production. We conceive it to feed directly the electrical loads and the electricity network is considered as a secondary source of supply and storage.

Firstly, we make a survey of a real situation of one African electrical grid, the case of Burkina Faso. Secondly, we undertake a sizing, a modeling and a simulation of a grid-connected PV system with storage for one physics laboratory at the University of Ouagadougou as study case.

2. Survey of one African electrical grid (case of Burkina Faso)

As for all the African countries, the electrification of Burkina-Faso territory is mainly ensured by the National Company of Electricity (SONABEL). SONABEL has thus in charge of the production, the transport, distribution and marketing of electricity mainly in urban areas. Another institution called Electrification Development Fund (FDE) intervenes only in small villages for rural electrification. The rate of electricity coverage in the national territory is 19% for year 2010 [2]. SONABEL grid presents some problems related to production, transportation and distribution. Here we focus our study on one crucial issue, namely power deficit for users.

2.1. *Evolution in energy production [1]*

SONABEL energy production is characterized by a very important local production from thermal power plants and hydraulic, and a production imported from Ivory Coast, Ghana and Togo. The local production represents 83% of the total production and the 17% correspond to the imported part.

2.1.1. *Thermal production.* The thermal production remains the most important part. It represents 67% of the total production. It is a sure production, because it does not depend on random parameters as the wind and the rainfall. On the other hand, it requires huge recurrent expenses because of fuel purchase.

The figure 1 gives the thermal production evolution between 2000 and 2009. We notice an increasing evolution in this type of production. Over 10 years, it almost doubled, with 94, 2% increase. This situation is explained by an increase in electricity demand, which requires new investments for new materials for power plants and recurrent expenses in fuel purchase. This has impacts on kWh price. So, we noticed an increase in kWh price during these last years (on 2004 and 2006).

2.1.2. *Hydraulic production.* The hydraulic production is made by hydroelectric power plants installed in the regions with large quantities of water. This production represents 16% of the total production. But it is not constant because it depends on the rainfall.

The figure 1 gives the hydraulic production evolution between 2000 and 2009. We notice that the hydraulic production knew an increase less important than the thermal production. Between 2000 and 2009, it is increased from 98 GWH to 132 GWH, i.e 34.7% increase. This increase is also relative, because for 2001, 2002 and 2006 the production decreased probably due to rainfall. Indeed, hydraulic production can only increase in years with heavy rainfall.

2.1.3. *Imported production.* The imported energy represents 17% of the total production (in 2009). It comes from Côte d'Ivoire, Ghana and Togo. These connections were made to reinforce SONABEL energy supply. The interconnection Côte d'Ivoire – Burkina Faso dates back to 2001, the connection with Ghana is done in 2003 and the Togo connection is established in 2006.

The figure 1 gives the imported production evolution. Between 2001 and 2009, the imported part increased by 116.4%, 67 GWh in 2001 and 145 GWh in 2009. We notice that this production which increases much more as compare to the others. We can explain this fact by the policy adopted by

SONABEL to face energy need. This policy has to favor the exchange of sub-regional energy to secure the reserve of electricity [1].

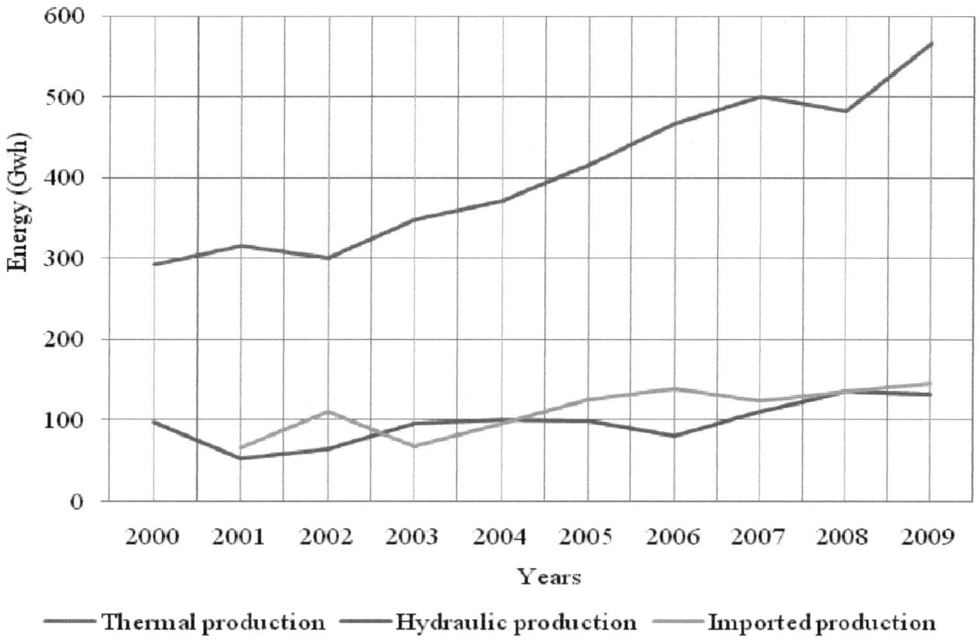

Figure 1. Production curves

2.1.4. *Impact on total production.* From the production curves components evolution over years above mentioned, we can notice that between 2001 and 2009 the total production increased from 435 GWh to 844 GWh (thermal production + hydraulics + imported), i.e 94% increase. It is explained by a strong increase in energy demand and consequently expenses. Among these three types of production, the thermal production incurs more expenses because of fuel purchase and maintenance of power plants. The hydraulic production cannot increase considerably because of low rainfall, and the imported energy can increase within the limits of supplier capacities. This study allows us to make the following remarks:

- a strong increase in energy demand, due to population and economic growth;
- an incapacity of available power to meet the actual demand;
- a possible increase of kWh price because it depends on oil price.

2.2. *Electric interruptions Problems in SONABEL network [1]*
The main causes of the electric interruptions are:

- the unballasting due to the deficit of power;
- cuts due to works on the electricity network or technical incidents.

2.2.1. *Average electric interruptions registered in the regional consumption center of Ouagadougou (CRCO) and Bobo-Dioulasso (CRCB).* We consider the two main centers of consumption in Ouagadougou and Bobo-Dioulasso. They consist of several electrical departure points. The Tables 1 below give the registered average interruptions, their duration, and the corresponding energy not distributed in 2008 and 2009.

By analyzing this table we notice that the electric interruptions and the average time of cut increased during these two years. This fact is explained particularly by a deficit of power. So:

- for the CRCO, the accentuation in power deficit is translated in the average time of cut which increased from 41, 3 minutes in 2008, to 112, 3 minutes in 2009.
- for the CRCB, this average time also doubled for power deficit because of the disconnection of Côte d'Ivoire – Burkina Faso electric line during the period of intensification of pylons on the territory of Côte d'Ivoire.

We can also notice the big quantity of energy lost by the company because of interruptions representing therefore an income lost.

Table 1. Electric interruption in CRCO and CRCB

Center	CRCO		CRCB	
Years	2008	2009	2008	2009
Incidents	723	886	550	414
Works	174	236	150	143
Unballasting	103	268	52	50
Total	1000	1390	752	607
Energy Not Distributed (MWh)	4 952	15263	443	467
Average time of cut (minute)	41.3	112.3	101	202.5

2.2.2. *Monthly interruptions evolution in the CRCO [3].* The electric interruptions registered in this center concern all the electrical departure points. We are going to make a study of the monthly interruptions over the last three years namely 2008, 2009 and 2010. The histograms given on figures 2, 3 and 4 show the number of monthly interruptions.

We notice that electric interruptions during each year are frequent from March up to October. It is explained by the fact that from this month, the climate is characterized by a strong heat and electricity demand (for ventilators, air conditioners) over this period is higher than the installed powers. So, it is necessary to proceed to momentary unballasting.

In June, begin the rainy season, characterized by incidents provoked by a rise of transformers tension and other damages caused by rainwater. On the other hand between November and February, we notice a decrease in interruptions, which can be explained by the beginning of a cold season. In this period, electricity demand is limited.

Now that we observed the monthly evolution of the interruptions, we will see the interruptions annual evolution.

IOP Conf. Series: Materials Science and Engineering **29** (2012) 012015 doi:10.1088/1757-899X/29/1/012015

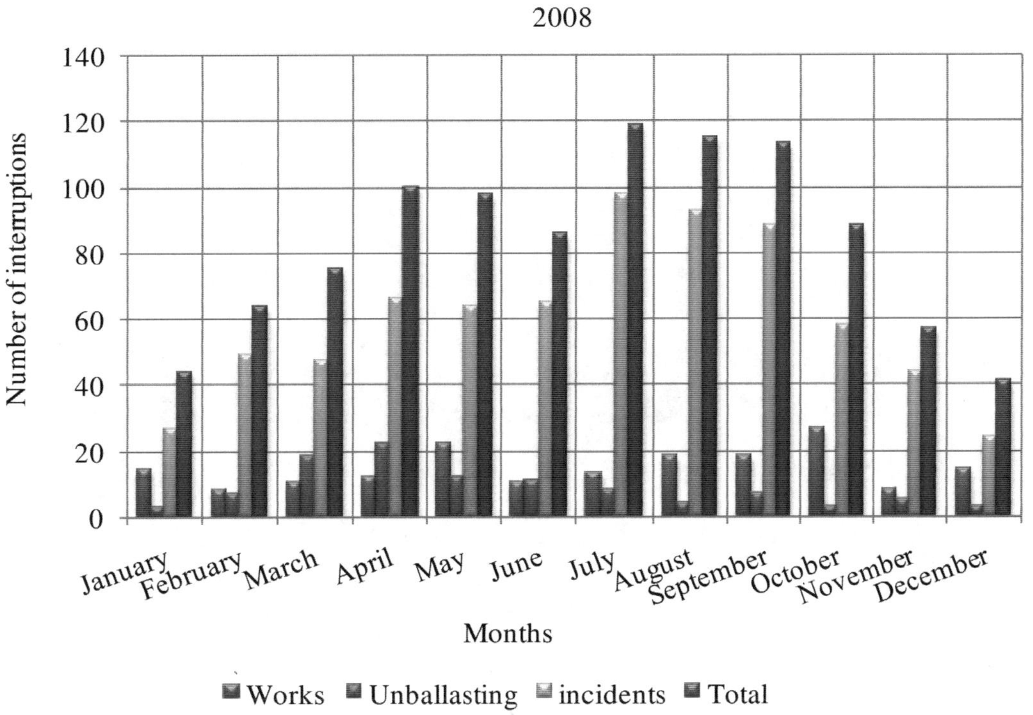

Figure 2. Evolution of electric interruptions (2008)

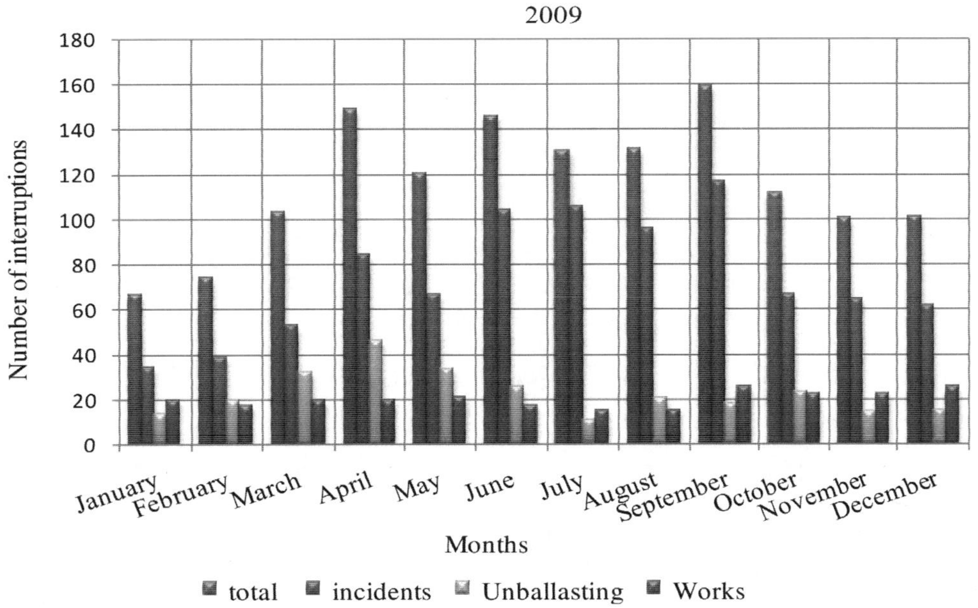

Figure 3. Evolution of electric interruptions (2009)

IOP Conf. Series: Materials Science and Engineering **29** (2012) 012015 doi:10.1088/1757-899X/29/1/012015

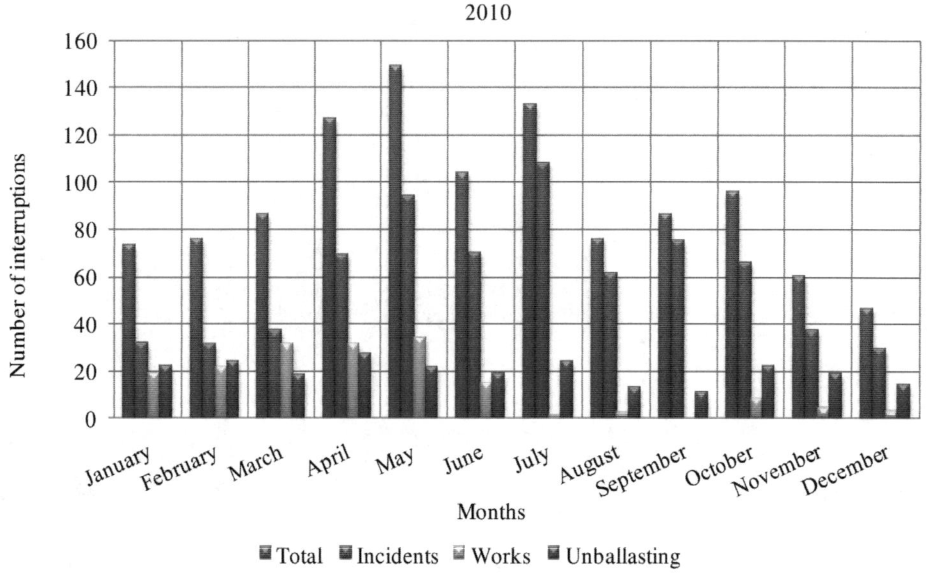

Figure 4. Evolution of electric interruptions (2010)

2.2.3. *Annual evolution in electric interruptions [3].* The following curve (figure 5) shows the evolution of the annual interruptions, between 2002 and 2010.

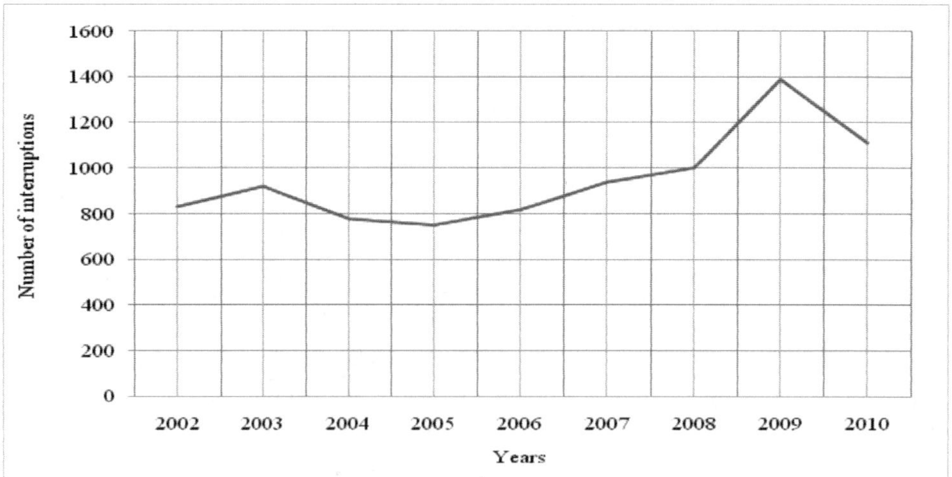

Figure 5. Curve of electric interruptions evolution.

Although we notice a relative decline in electric interruptions in 2004, 2005 and 2010, in general, we can say that annual electric interruptions increased considerably and will continue with the same evolution. This annual interruptions increase is explained by the strong increase in energy demand. As in the future, energy demand will continue to rise, it is therefore necessary to look for solutions at user's level through the use of photovoltaic systems.

3. Sizing, Modeling and simulation of an example of grid-connected PV system

To face problems detailed in the survey part, the photovoltaic systems are to be considered as serious candidates specially the grid-connected PV system with batteries storage. As study case, we take as loads, the energy needs of the LAME laboratory of Ouagadougou University and size a pilot PV system to be implemented. The photovoltaic production is completely intended for the loads of the laboratory and the possible surplus is injected on the University electricity network. To have an idea of the proposed system performances, we undertake a simulation of the produced energy availability of the pilot system after designing and modeling it.

3.1. *Pilot PV system design*

For the plan, two scenarios are possible as indicated on figure 6 below.

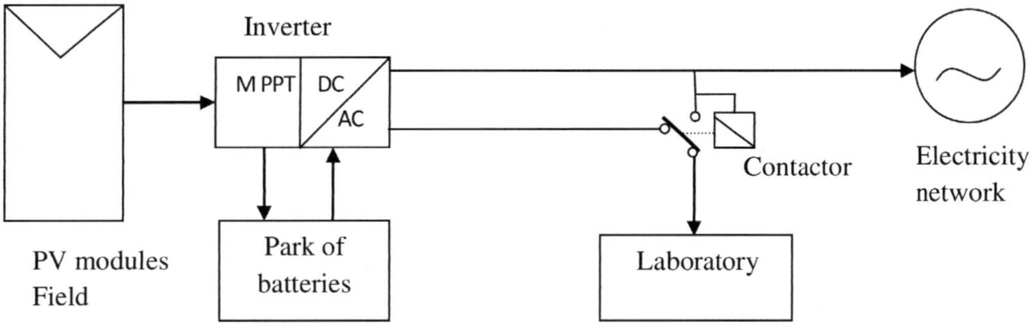

(a): Grid-connected PV system with storage based on one inverter [4].

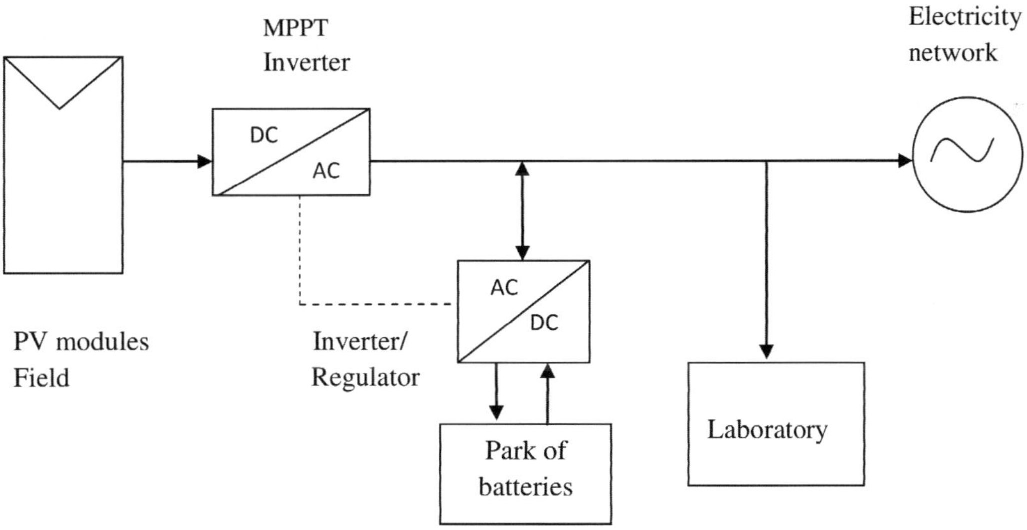

(b) : Grid-connected PV system with storage based on two inverters [5].

Figure 6. Grid-connected PV system with storage.

For our system, we choose the architecture a), because it presents some advantages compared to configuration b). As the number of components is reduced, the system efficiency is better, and also its cost is lower. Moreover, it is simpler to realize and maintain.

The adopted system comprises:

- a field of PV modules;
- an inverter with build-in MPPT regulator;

- a park of batteries;
- the LAME laboratory as charge;
- the University electricity network.

The inverter is the main component of this system. It must ensure tree functions: DC/AC conversion, MPPT regulation and islanding function. The role of the MPPT regulator is to put the modules in their maximal power point for a better production and to ensure an optimal charge of the batteries. The DC/AC converter transforms the direct current (DC) into alternative current (AC) synchronized with the network to feed the loads of the laboratory. In standard mode, the alternative power produced by the PV generator is used by loads and possibly injected in the electricity network. In the case of grid failure, the inverter switches automatically in an islanding configuration and works in autonomous mode [4] with a guaranteed safe operation. Thus, the interest of the storage is to guarantee the continuity of power supply as long as necessary. The primary source is the grid-connected photovoltaic system and the electricity network is considered a secondary source of supply and an additional means of energy storage.

3.2. Sizing of the pilot PV system

In this section we determine the characteristics of the PV generator composed of PV modules field, batteries park and inverter necessary to feed the loads of LAME laboratory. The other components of the system are the loads that we consider constant and the electrical network.

From the table 2 giving energy needs of the laboratory, we use classical sizing method [5, 6] to determine the following characteristics for the PV generator of the pilot PV system:

- Modules field: it is made of 16 modules grouped in 8 strings with 2 modules in each. The main characteristics of the used modules in standard test conditions (STC) are given by the manufactory: V_{max} = 26.7 V, I_{max} = 7.5 A, P_{max} = 200 W [7]. The peak power of the modules field is 3.2 kWc.
- Batteries park: The batteries must supply power for a given number of grid failure. By taking into account the average number of interruption per day and their average time given in the survey part, the cumulative time of grid failure can reach 12 hours per day. As the highest operation duration of equipments is 6 hours per day, we choose an autonomy duration of 12 hours. We then need to provide a storage capacity of 200 Ah in order to ensure the continuity of services.
- Inverter: It must be able to deliver the power needed by loads and also to support the maximum power supplied by the modules field as it act as a regulator. So, we choose an inverter having a nominal power of 3.0 kW.

The characteristics of the electricity network are: 230 V AC / 50 Hz.

Table 2. Loads of the LAME Laboratory

Equipment	Number	Power (W)	operation duration (Hours/day)
Desktop computers	5	300	4
Laptop computers	3	300	2
Fluorescent lamps	4	36	4
Fans	1	75	6

For these calculations, we have used a value of 5.1 kW/m^2/day for solar irradiation corresponding to December, the unfavorable situation as highlighted above.

3.3. *Modeling of the pilot PV system*

The modeling of the pilot PV system allows us to have an idea on power availability which will be supplied to the laboratory by simulation. To make this work, we have to take into account some parameters such as solar radiation and temperature of the installation site. This can be done with the following diagram that gives an overview (figure 7) [8]:

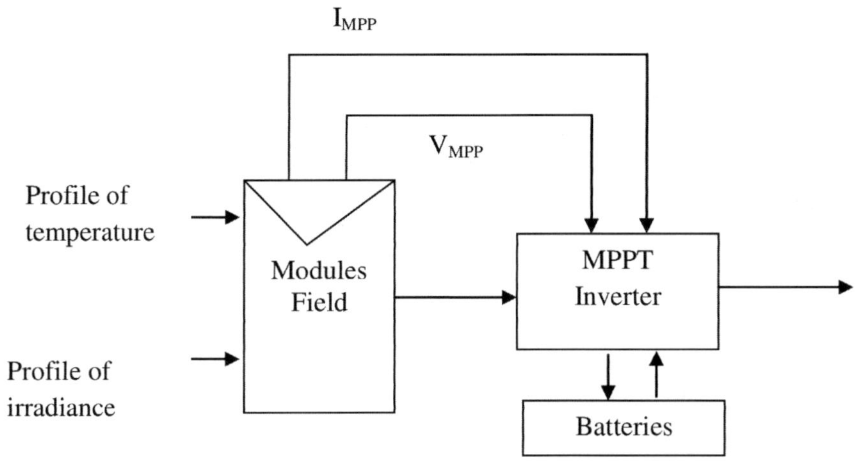

Figure 7. Diagram of the PV generator model.

3.3.1. *Mathematical model of the PV modules field.* The PV modules field is composed of N_p modules strings with N_s modules in each string. The equation describing the relation between the current I supplied by a module and its voltage V is derived from the cell model [6] and can be written as:

$$I = n_p I_{ph} - n_p I_{sat}(T)\left(\exp\left(\frac{q\left(\frac{V}{n_s} + R_S \frac{I}{n_p}\right)}{nkT_c}\right) - 1\right) - n_p \frac{\frac{V}{n_s} + R_S \frac{I}{n_p}}{R_p} \quad (1)$$

where I_{ph} is the cell photocurrent, I_{sat} is the diode reverse saturation current. n_s and n_p are respectively the series and parallel cell number. R_s and R_p are the intrinsic series and shunt resistance of the cell respectively. n is the diode ideality factor varying between 1 and 2. q is the electron charge and k is Boltzman's constant. T_c the cell temperature in Kelvin is affected by external factors such as global irradiance, ambient temperature and wind speed. It is determined by the following formula [8]:

$$T_c = T_a + \left[\frac{NOCT - 20}{800}\right] \times \phi \quad (2)$$

where Ta is the ambient temperature, NOCT is the Nominal Operating Cell Temperature and ϕ is the irradiance of the site.

As a maximum power point tracking (MPPT) technique is used to track the peak power to maximize the produced energy, some authors [8 - 11] derived the MPP coordinates (V_{mpp}, I_{mpp}) from equations (1) and manufacturer data. For the TENESOL modules, the manufacturer gives the dependence of V_{mpp} and I_{mpp} to the irradiation. By taking this fact into account, we rewrite the equations usually used [11, 12] to calculate the MPP coordinates in any condition of irradiance and temperature as following:

$$\mathrm{I}_{mpp} = \frac{\phi}{\phi_r} I_{mppr} + \left(\frac{d\mathrm{I}}{dT}\right)(T_c - T_r) \tag{3}$$

$$V_{mpp} = \mathrm{k}(\phi) V_{mppr} + \left(\frac{dV}{dT}\right)(T_c - T_r) \tag{4}$$

where I_{mppr} and V_{mppr} are the maximal current and voltage corresponding to the reference (standard) irradiance respectively, ϕ_r is the reference irradiance, (dI/dT) and (dV/dT) are respectively current and voltage temperature coefficients. T_r is the reference temperature in Celsius degree. From the data of the manufacturer [7], the coefficient k(ϕ) that gives the Voltage dependence to irradiance can be expressed by:

$$k(\phi) = a_4\phi^4 + a_3\phi^3 + a_2\phi^2 + a_1\phi + a_0 \tag{5}$$

where $a_4 = -1.27\text{x}10^{-12}$, $a_3 = 3.48*10^{-9}$,$a_2 = 3.49*10^{-6}$, $a_1 = 1.538*10^{-3}$, $a_0 = 0.7404$.

So, for the module field, the MPP coordinates (I_{mppf},V_{mppf}) are given by:

$$\mathrm{I}_{mppf} = N_P I_{mpp} \tag{6}$$

$$V_{mppf} = N_s V_{mpp} \tag{7}$$

3.3.2. *Mathematical model of the inverter.* The inverter acts as an interface between the PV generator and the electricity network. It has a build-in MPPT regulator which controls all the time the maximal power point (MPP). This point is characterized by the coordinates (I_{mppf}, V_{mppf}) defined above. The power at inverter input is given by:

$$P_{DC} = \mathrm{I}_{mppf} V_{mppf} \tag{8}$$

The output power of the inverter is then done by:

$$P_{AC} = \eta_{inv} I_{mppf} V_{mppf} \tag{9}$$

where η_{inv} is the inverter efficiency.

3.3.3. *Mathematical model of the Batteries.* In general, the battery model describes the relationship between its voltage, current and state of charge (SOC) [9]. As our study is based on the availability of the pilot system power for loads at any time, we only consider the battery discharging mode. The worst case is when the battery must totally supply power if the modules are in darkness and the grid fails. We model the battery as a switch controlled by the build-in regulator of the inverter using the value of Low Battery Voltage (LBV) as control parameter. As long as the battery voltage is higher than the LBV, the switch is on and the power can be delivered by the battery when it must supply power.

3.3.4. *Mathematical model of the electrical network.* We also consider the electricity network as a simple switch allowing the energy exchange between the PV generator, loads and electrical network.

If electricity network is available and stable, the switch is closed. When the grid experiences some outages, the switch is open and loads are isolated from the grid. The switch operating is controlled by the inverter that measures the electricity network state by using the network voltage and frequency. At this stage we have all formula necessary to undertake the simulation of the pilot PV system availability. The loads are considered to be constant at any time.

3.4. *Simulation of the pilot PV system availability*

The main objective of the study is to show the interest of the proposed system as compared to classical grid-connected systems in African countries. So, we focus the simulation on the availability of power provided by the pilot system for the user loads in critical situations in day as at night. To do the simulation, some data such as the irradiance and some coefficients from the Manufacturer database [7] are to be provided.

3.4.1. *Irradiance database on the site of Ouagadougou.* The irradiance data of Ouagadougou were obtained by the use of a solar recorder [13] and a processing with Excel software. The figure 8 gives some daily irradiance data recorded in April, August and December. We have the highest irradiance value in April. As this month also corresponds to an important number of grid failures, we use its irradiance values as entries for the simulation model.

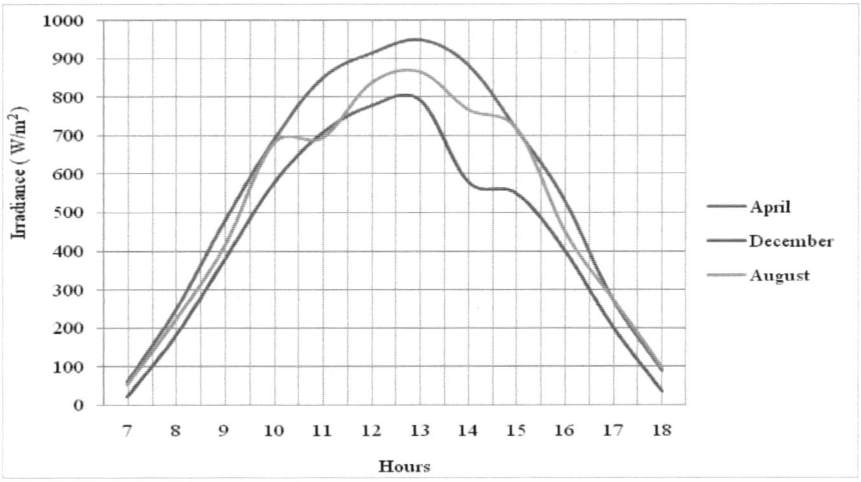

Figure 8. Sample of daily irradiance.

3.4.2. *Simulation algorithm.* The system model, consisting of the mathematical equations listed above in this paper, is implemented using Excel Software. The simulation data-flow diagram is shown in figure 9. Using the meteorological data (irradiance and temperature), the manufacturer data of modules and the state of electrical network, the program calculates the MPP coordinates (I_{mppf}, V_{mppf}), output power produced by photovoltaic generator or energy availability as well in day as at night. To test the simulation program, we use the MPP values (I_{MPPf}, V_{MPPf}) at STC and the NOCT conditions given by the manufacturer.

3.5. *Results and discussion*

In order to investigate the ability of the pilot PV system to supply power at any time or whenever it is necessary, we consider some significant operating conditions to highlight our system performance compared to classical grid-connected system having the same peak power. In this view, we simulate the power produced by the pilot system in normal case and the power availability for loads in two scenarios described below.

IOP Conf. Series: Materials Science and Engineering **29** (2012) 012015 doi:10.1088/1757-899X/29/1/012015

The power provided by the pilot system for loads in day in normal case is given in figure 10. The profile of the daily maximum power follows the hourly irradiance. We notice that it varies between 170 W at 7 am and 2530 W at 1 pm.

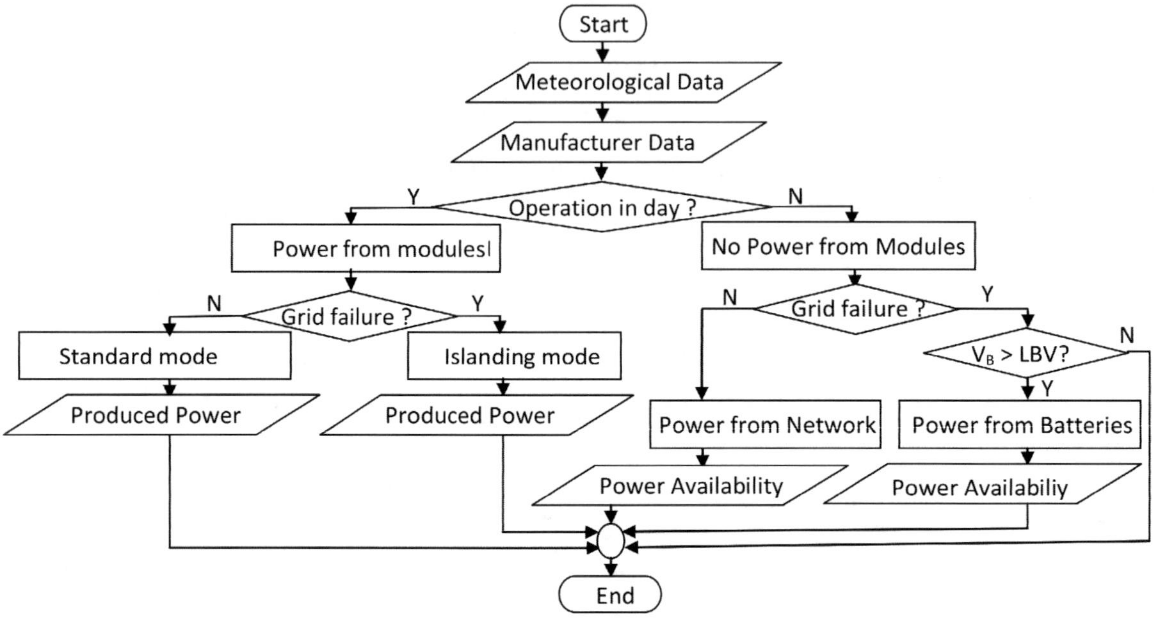

Figure 9. Simulation data-flow diagram

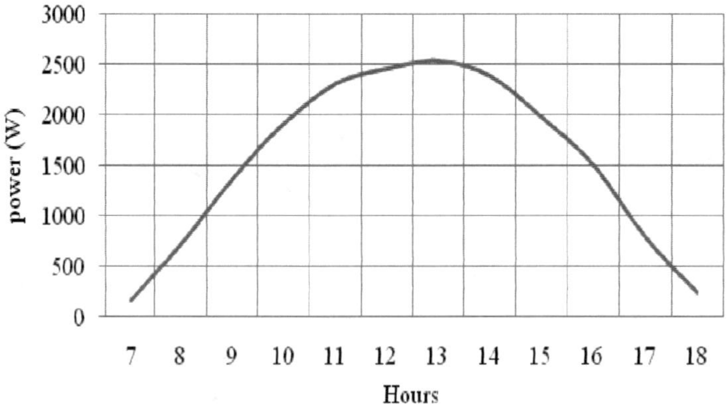

Figure 10. Daily evolution of the power P_{MPP}

To see what happens in case several interruptions occur in the day, we simulate the availability of the electricity network, the power provided by a photovoltaic system without storage and that given by our system. Results obtained are reported on figures 11. By observing the three given curves, we notice that in case of network cut, the classical grid-connected system cannot supply power although the solar energy is available. Its operation depends on the electricity network. On the other hand, no electric interruption is observed for the grid-connected system with storage.

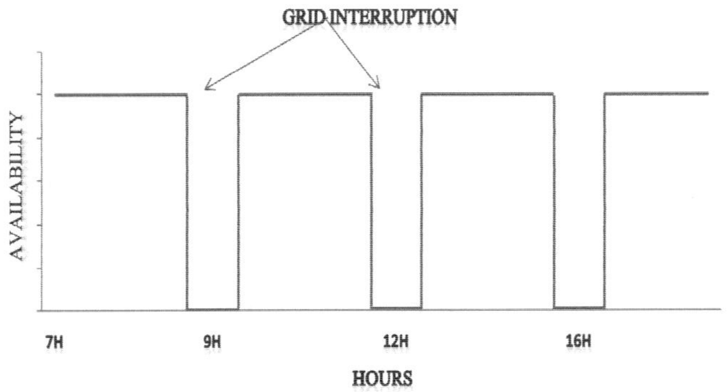

(a): Network operation

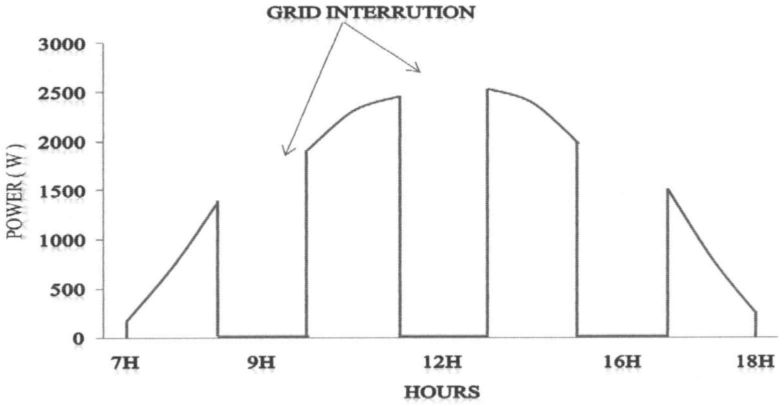

(b): Production of a grid-connected PV system without storage

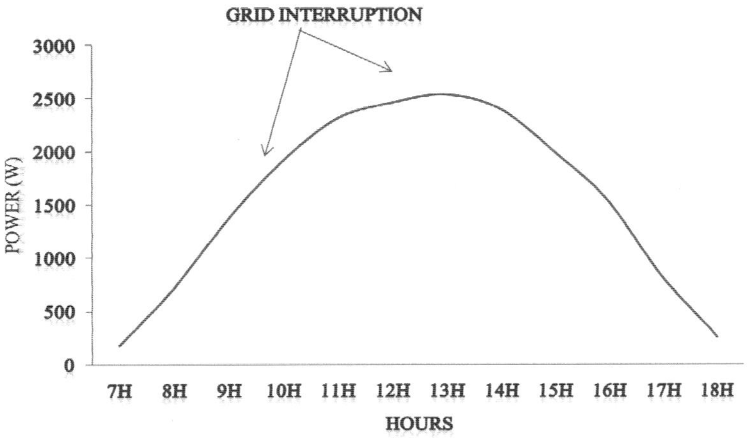

(c): Production of a grid-connected PV system with storage connected

Figure 11. Grid-connected PV system operation in the day

Now we take the case of many interruptions occurring at night or a cloudy day and we simulate the availability of the electricity network, the classical grid-connected system and the studied system. The figure 12 gives the results obtained. First of all, for the two grid-connected systems, the modules are in darkness and they provide no energy. The electricity network supplies power until an outage takes place. When a failure appears, we observe that the classical grid-connected system cannot provide energy for the loads. The availability of energy in that case only depends on the electricity network capacity to be stable. On the other hand, for the proposed grid-connected system, the batteries provide the necessary energy. So, no electric interruption is observed at the output of the grid-connected system with storage. We see that the proposed grid-connected system allows users to profit from their photovoltaic installation at any time even if the public grid has some failures. By considering other situations we get to the same conclusion.

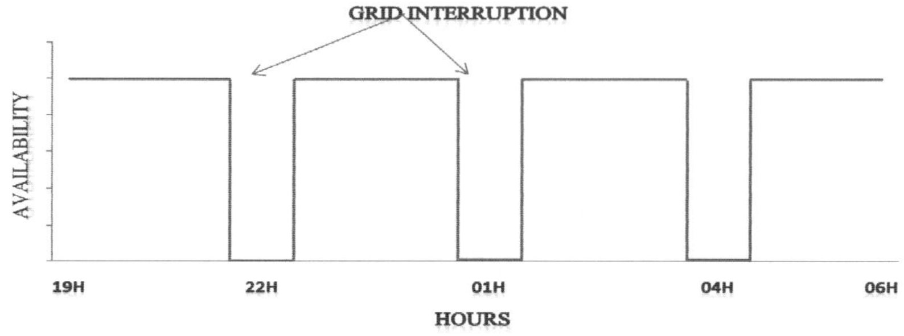

(a): Network operation

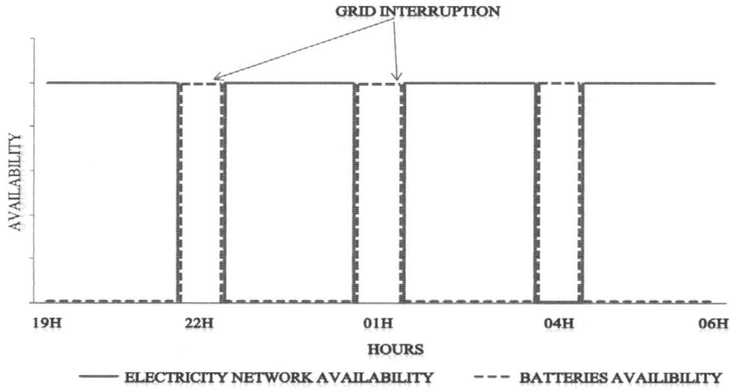

(b): Storage batteries operation

Figure 12. Grid-connected PV system operation in the night

4. Conclusion

In this paper, we have taught about a key solution for power outages problem experienced by many African countries. We have noticed that the electricity distribution in Africa is impeded by the grid failures through a survey of a real situation of Burkina Faso electrical grid. For example, during year 2009, we registered 1390 power cuts in the CRCO with an average of 4 cuts per day. The cumulative time of grid failure can reach 12 hours per day. This is a real problem for the economy, and the prosperity of African countries. To resolve this problem of failure at user level, the grid-connected photovoltaic systems with batteries storage must be considered as one major candidate. Indeed, the simulation of a pilot system, consisting of a grid-connected photovoltaic system with batteries storage, shows that the overall system allows answering user energy demand permanently. Generally, the periods of frequent cuts correspond to the periods of strong irradiation. So, users can profit from their photovoltaic installation at any time even if the public grid has some failures either in day or at night.

Appendix
The table A.1 gives the adopted nomenclature.

Table A.1. Nomenclature

Tc	Functioning cellule Temperature	°C
Ta	Ambient Temperature	°C
I_{MPP}	Current at the point of maximal power	A
V_{MPP}	Voltage at the point of maximal power	V
ϕ	Irradiance	W
ϕ_r :	Standard irradiance, 1000 W/m^2	W
Imr :	Maximal Current corresponding to the standard irradiance	A
$\frac{dIcc}{dT} = It$	Temperature Coefficient of the court-circuit current.	A
V	Voltage of the solar module	V
P_{MPP}	Power at the maximal power point	Wh
P_{DC}	Continuous power	kWh/j
I_{ph}	Photo-current	A
I_{sh}	Current in the shunt resistance	A
q	Charge of the electron	C
k	Constant of Boltzmann	
n_p	Number of modules in parallel	
n_s	Number of modules in series	
R_s	Series resistance	Ω
R_{sh}	Shunt resistance	Ω
I_{sat}	Saturation current of photovoltaic cell	A
I_{max}	Maximal current	A
V_{max}	Maximal voltage	V
P_{max}	Maximal power	W
DC	Direct Current	A
AC	Alternative Current	A
NOCT	Nominal operating cell temperature	°C

References

[1] SONABEL 2009 *Annual activities Report* avalaible at www.sonabel.bf

[2] Minister of mines and energy of Burkina Faso 2010 *Internal report*

[3] Dispatching of the regional consumption center of Ouagadougou (CRCO) 2010 *Sonabel Internal Report*

[4] Ralf B 2010 Inverter for autonomous photovoltaic system and connected to the electricity network (in French) *Master I thesis* Unversity of Reunion

[5] Riffonneau Y, Barruel F and Bacha S 2008 Problematic of associate storage with PV systems connected to the electricity network *Revue des Energies Renouvelables* **11** 407-422

[6] Buresh M 1983 *Photovoltaic energy system* (McGraw-Hill Inc New York)

[7] TENESOL catalogue choice of solar module *data sheet 200* W at www.tenesol.com

[8] Chouder A , Silvestre S and Malek A 2006 Simulation of photovoltaic grid connected inverter in case of grid-failure *Revue des Energies Renouvelables* **9** 285-296

[9] Sukamongkol Y, Chungpaibulpatana S, Ongsakul W 2001 A simulation model for predicting the performance of a solar photovoltaic system with alternating current loads *Renewable Energy* **27** 237–258

[10] Cherfa F, Chouder A, Hadj Arab A, Oussaïd R, Chenlo F and Sylverter S 2007 Modeling and simulation of the CDER Photovoltaic mini-central components connected to the electricity network *(in french) Revue des Energies Renouvelables ICRESD-07 Tlemcen* pp 29-34

[11] Sridhar R, Jeevananathan Dr, Thamizh Selvan N and Saikat Banerjee 2010 Modeling of PV Array and Performance Enhancement by MPPT Algorithm *Int. Journal of Computer Applications* **7** 35-39

[12] Tsai H-L, Tu C-S, and Su Y-J 2008 Development of Generalized Photovoltaic Model Using MATLAB/SIMULINK *Proc. of the World Congress on Engineering and Computer Science October 22 - 24* (San Francisco USA)

[13] INERA (Institut de l'Environnement et de la Recherches Agricoles) 2010 *Internal Report*

IOP Conf. Series: Materials Science and Engineering **29** (2012) 012016 doi:10.1088/1757-899X/29/1/012016

Combined receiver of solar radiation with holographic planar concentrator

O Iurevych[1], S Gubin[1] and M Dudeck[2]

[1]National Aerospace University "Kharkiv Aviation Institute"
17, Tchkalova st, Kharkiv, 61070, Ukraine
[2]Jean le Rond d'Alembert Institut, Pierre and Marie Curie University
4 place Jussieu 75005 Paris, France

E-mail: yurevi.elena@yandex.ru, michel.dudeck@upmc.fr

Abstract. In the present work a combined solar radiation receiver is described. This combined system uses a holographic film to concentrate the solar radiation and to optimize the efficiency of the sensor. A mathematical model is used to calculate the Volt-Ampere behavior and the thermal and photovoltaic efficiencies of a test case to present the advantages of the suggested system.

1. Introduction

Today and for future, the conversion of the sunlight into energy is the one of the most challenging solutions as source of heat and electricity. The solar energy is available in any place of the world and a density of energy from 150 W/m^2 to 300 W/m^2 is received on most of places at the Earth surface. The solar energy allows an environmental protection with an inexhaustible energy source. Solar sensors can be installed on building or associated to allow a large energy distribution. However, a large use of solar captors comes up against current difficulties as size, price of sensors, price of energy and level of energetic efficiency. A large effort is carried out in laboratories in order to increase the efficiency of solar cells and an efficiency of around 40% is now obtained. Another way to improve the efficiency of the device and to decrease the size is a global optimization of the sensor.

At the present time, photoelectrical method for solar radiation transformation makes electricity production possible, and the heat sun spectrum is used for solar collectors for hot-water supply and heating. They two technologies are extensively used.

The solar radiation spectrum arriving on the photoelectrical converter is not completely transformed in energy and then the obtained performances are bounded and the efficiency of transformation is decreasing with temperature rise. The cost of the photoelectric converter (PEC) is significant.

It is possible to partially compensate these disadvantages by concentration of solar radiation and reduction of photoactive space of PEC. With these objectives, we present a combined solar system with a holographic film allowing a separation of the sun radiative spectrum with a first part directed to a photovoltaic device to product electricity and with a second part directed to a heat exchanger (water circuit). In this device, the energy transfer to the heat exchanger allows to decrease the temperature of the solar collector (PEC).

In the section 2, the suggested combined device and its sub-elements are described and the technological choices constitute a first analysis for a future device which will be manufactured at the

University of Kharkov. In the section 3, the set of relations describing the behavior of the combined system are shortly presented. They correspond to a stationary state of the solar system and to global balances for each sub-system (no time and local space variations). In the section 4, the numerical obtained results are shown for a choosen test case in order to illustrate the performances and the advantages of the combined device (current-voltage characteristics and efficiencies). In the conclusion, the possible applications of the suggested combined solar device are commented.

2. Description of the combined solar receiver

A combined solar receiver with photovoltaic conversion and solar collector heating of water is proposed to improve the solar conversion efficiency.

A schematic view of the combined solar receiver is presented on figure 1. The principle of such a solar receiver is based on the fact that visible solar radiation is reflected from the holographic film (2). This radiation arrives on the photoconverter (3) to be converted into electricity after succesive reflexions on the inner side of a waveguide made of polycarbonate (1). The plan holographic film is inserted in the solid polycarbonate material. The visible solar radiation which is typically from near UV to near IR (maximum in yellow around 570 nm) is used to generate electricity. The thermal solar radiations (IR) arrive to an absorber (5) after crossing a heat conductive compound (4).

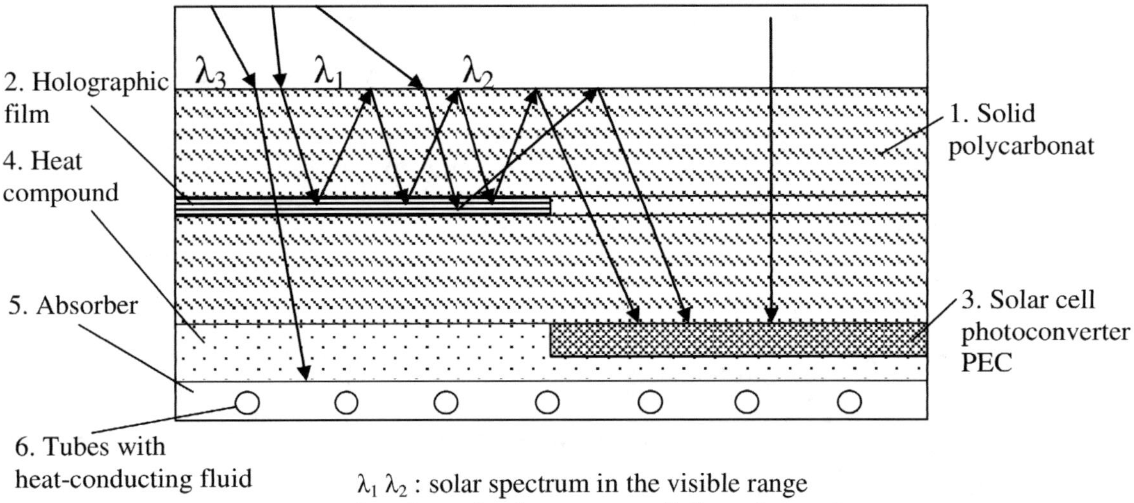

λ_1 λ_2 : solar spectrum in the visible range
λ_3 : thermal solar spectrum (infra-red range)

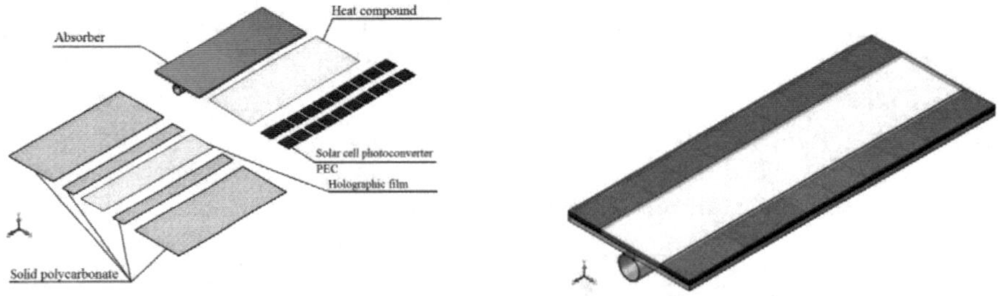

Figure 1. Schematic view of the combined receiver

The holographic film is crossed by the long-wavelength (IR) solar radiations, which go to the absorber (5). A set of tubes with a liquid flow crosses this absorber (5) in order to be heated by the IR sun radiation and to permit to obtain a heated liquid flow. The compound (4) is required to decrease the temperature of the photoelectric converter temperature and its heat resistance. A thermal insulation is used to decrease energy losses of absorber to the environment.

Such three-dimensional hologram operates in following manner: the recording of three-dimensional volume hologram in the mass of a photomaterial is related to a number of strata structures identical to the number of used wave-lengths λ_1, λ_2, λ_3..., of the incident light. As result a set of stratas forms λ_1, λ_2, λ_3..., interlacing of which corresponds to local "object waves" phase shift (Bragg-effect hologram).

The independence of strata's elements interaction with light as selecting or compounding filters leads to occurrence of mutually superimposed monochrome images in the process of hologram exposure to light with wave-length λ_1, λ_2, λ_3.... In a process of exposure such a hologram to white light each wave by certain hade (Bragg angle) will reflect from "it is" stratas system. In such a manner, the problem can be stated as a mutual mathematical description of concentration, photovoltaic conversion and heat-removing processes.

The main advantage of this concentrating system is found in the increase the global system efficient and more economic energetic transformation. Most of present solar radiation concentrators allows to concentrate almost whole emission band resulting in getting visible part, which can be converted into electricity and heat part (infrared - IR), which increases the operating temperature of a solar cell alike. That is exactly why it is recommended to use a 3-D hologram in the form of thin-film material as a concentrator.

3. Modelling

The different relations describing the behaviour of all the sub-elements are briefly presented the efficiency of the coupled thermal-voltaic solar device has been calculated to illustrate its advantages with a set of parameters of holographic concentrator, photovoltaic and thermal device retained as an example.

3.1 Solar radiation

The solar radiation E_0 (W/m^2) cross the solid polycarbonate layer (waveguide) and arrives on the holographic film with only a density of $k_{op} E_0$ (k_{op} , k'_{op} are the optical coefficients of transmission of the polycarbonate layers). Then, the holographic film re-emits the density of energy $k_h k_{op} E_0$ (k_h is the coefficient of emission of the holographic film). Taking into account the concentration coefficient k_c related to the ratio of surfaces of the holographic film and the photoconverter, $k_c = \frac{S_{hf}}{S_{PEC}}$ where S_{hf} is the surface of the holographic film and S_{PEC} is the surface of the photoconverter, the solar radiation E_{PEC} (W/m^2) influencing the surface of the photoconverter (PEC) is:

$$E_{PEC}=k'_{op}.E_0+k'_{op}.k_h.k_c.E_0 \quad (1)$$

The electric energy obtained from the photoconverter is $\eta_{PEC}.E_{PEC}$, if η_{PEC} is its efficiency. Then, the part of energy coming form the photoconvecter to the solar converter is $(1-\eta_{PEC}).E_{PEC}$ and the solar energy crossing the holographic film is $k'_{op}.(1-k_h)E_0$. Finally, the density E_{col} (W/m^2) of the solar energy arriving on the solar collector to heat the calorific fluid is (see figure 2):

$$E_{col}=(1-\eta_{PEC}).E_{PEC}+(1-k_h)k'_{op}E_0=k'_{op}E_0(1+k_h.k_c)(1-\eta_{PEC})+(1-k_h)k'_{op}E_0$$

This relation is improved by the addition of a parameter k_s associated to the ration of the surface of the photovoltaic cells and the solar panel. The previous relation becomes:

$$E_{col}=k_s.(1-\eta_{PEC}).E_{PEC}+(1-k_h)k'_{op}E_0=k_s k'_{op}E_0(1+k_h.k_c)(1-\eta_{PEC})+(1-k_s)(1-k_h)k'_{op}E_0 \quad (2)$$

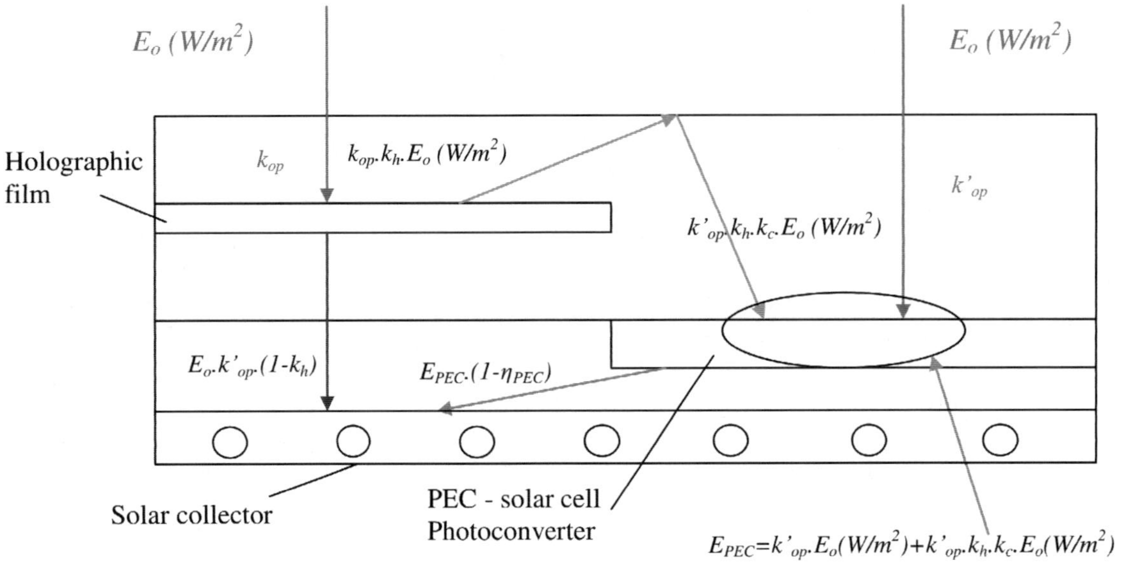

Figure 2. Energy transfer

3.2 Photovoltaic current-voltage

The photovoltaic system is described with the use of an equivalent schematic circuit which includes a resistance R_s *(Ω)* in series, a resistance R_{sh} *(Ω)* in parallel (shunt), a diode crossed by a current I_d, and a photo-current I_{pht} *(A)*. Then, the current-voltage I_{ch}-U_{ch} curve of photovoltaic system is written as [4],

$$I_{ch}=I_{pht}-I_d-\frac{U_{ch}+I_{ch}.R_s}{R_{sh}} \tag{3}$$

where I_d *(A)* is the diode current which is expressed as $I_d=I_0(e^{\frac{U_d}{V_0}}-1)$, U_d *(V)* is the diode voltage as $U_d=U_{ch}+R_s.I_{ch}$, I_0 *(A)* is the reverse saturation current which is temperature dependent and V_0 is a diode parameter expressed as a function of the diode parameter A by $V_0=\frac{AkT}{e}$ (k is the constant of Boltzmann, T is the temperature and $e>0$ is the charge of electron). If one assumes that the shunt resistance (see relation 9) is enough large to neglect the last term, the relation (3) is redacted to,

$$I_{ch}=I_{pht}-I_0(e^{\frac{U_d}{V_0}}-1) \tag{4}$$

with

$$U_d=U_{ch}+I_{ch}R_s \tag{5}$$

The photovoltaic current I_{pht} is

$$I_{pht}=I_{cc}(\frac{E_c}{E_0})-dI_t(\frac{E_c}{E_0})(T_0-T) \tag{6}$$

where the current I_{cc} *(A)* corresponding to nominal conditions with a solar flux E_0 *(W/m²)* at the temperature T_0 *(K)* (short-circuit current), E_c *(W/m²)* is the received solar energy by PEC, dI_t is a thermal parameter for the current and T *(K)* is the temperature.

The relation between photovoltaic current and photovoltaic voltage is function of the solar radiation and temperature, and a set of parameters describing the behaviour of the photovoltaic system (A, R, I_{cc}, $E_0(T_0)$, dI_t).

Shockley pattern with ideal diode conduction formula with predetermined conditions of photovoltage [4] and temperature at the heart is used for photovoltaic cell's specifications description. In this ideal model, the diode parameters can be determined as on a basis of semiconductor structure calculations, as on a basis of experimental research.

So as to calculate the current-voltage I_{ch}-U_{ch} photovoltaic curve, the Idling voltage U_{it} (which is the voltage when the photovoltaic circuit is opened $I_{ch}=0$, $U_{ch} = U_i$) as a function of temperature is expressed as:

$$U_{it} = U_i + dUt.(T_0 - T) \tag{7}$$

where dUt is a voltage temperature coefficient and $U_i\ (V)$ is the floating voltage under the standard temperature $T_0\ (K)$ and illumination changing. $U_i\ (V)$ is depending of $I_{opt}\ (A)$ and $I_o\ (A)$,

$$U_i = \frac{AkT}{e}.\ln\left(\frac{I_{opt}}{I_o} + 1\right) \tag{8}$$

This relation (7) allows us to calculate A from the values of parameters U_i, I_{opt}, I_o given by the manufacturer. From the physical characteristic (p-n junction), the resistance R_s (Ω) in series is given by:

$$R_s = \frac{U_{opt} + dU_t(T_0 - T)}{[I_{opt}(\frac{E_{PEC}}{E_0}) - dI_t.(T_0 - T)].r_d} \tag{9}$$

where $r_d = dU/dI$ is a differential resistance at the beginning of illumination and U_{opt} is a voltage at the optimal point. The reverse saturation current is temperature-dependent and deduced from the short-circuit condition.

$$I_0 = \frac{I^{pht}}{\exp(\frac{e.U_{it}}{A.k.T})} \tag{10}$$

where $T\ (K)$ is the environmental temperature.

The power $P_u\ (W)$ delivered by solar energy plant by unit of surface is $P_u = \eta.E_{PEC}.N_{PEC}.S_p$ where E_{PEC} is a month-average, daily-average value of the energy of the solar radiation influencing the PEC's surface density (W\m^2), η is the efficiency of the photoconverter, N_{PEC} is the number of photoconverters per unit of surface, $S_p\ (m^2)$ is the area of a photoconverter. The maximum power is given by Pmax = $U_{opt}.I_{opt}$.

3.3 Thermal balance of the absorber

A schematic description is used to describe the balance energy of the absorber. The received energy by the fluid in the collector is proportional to the radiative energy $E_{col}\ (W/m^2)$ influencing the collector per unit of surface with an optical efficiency η_0. A lost of energy is due to the difference between the fluid at a mean temperature $T_{hc}\ (K)$ and ambient air at the temperature $T_{air}\ (K)$. A linear law is introduced with a coefficient of exchange K_{ht}. The heat productivity of the solar collector:

$$Q = F_i.A \quad [\eta_0 E_{col} - K_{ht}(T_{hc} - T_{air})] \tag{11}$$

where F_i is the heat collector's absorber's isothermal coefficient. It shows the identity or eveness of temperature's distribution over the heat collector's absorber's surface. $A\ (m^2)$ is the area of the heat-receiving surface.

IOP Conf. Series: Materials Science and Engineering **29** (2012) 012016 doi:10.1088/1757-899X/29/1/012016

4. Results for a test case

In the process of pattern creation it was accepted that at the initiation stage the concentrator must be described by concentration coefficient fixed value, which parameters may vary for different engineering solutions within known wave-lengths of solar radiation $k_c(\lambda_{rp})$.

As exemple, numerical results are presented for a test case in order to illustrate the advantages combined receiver of solar radiation.

The surface of the combined receiver is 0.735 m^2 and the surface of the photoconverter is 0.245 m^2. The incident solar radiation E_0 is 1353 W/m^2 giving a density of energy E_{PEC} on the photoconverter of 2494 W/m^2 and a density of energy E_{col} for the thermal conversion of 1123 W/m^2. The used data for the photoconverter are the table 1.

Table 1. Used data for the photoconverter

size	U_i	I_{cc}	U_{opt}	I_{opt}	dI_t	dU_t	dP_t	η
m	V	A	V	A	A/ degree	V/ degree		%
0.07x0.07	1.025	1.4945	0.9	1.4014	0.00098	0.0018	0.48	19

The electric power delivered by the photoconverter is 106 W and the heat productivity of the solar collector is 572 W.

As a result of mathematical simulation, we got voltage-current U_{ch}-I_{ch} (figure 3.a) and voltage-power U_{ch}-P_u (figure 3.b) curves for different environmental temperatures in the range 28°C- 68°C. The current is almost constant up to a threshold at 35V-40V (this value depends of the ambient temperature). The current is lightly increasing with the ambient temperature. An important fall of current is appearing after the threshold and he reaches a value equal to zero for a voltage around 50V. They are ordinary results for photoconvertor.

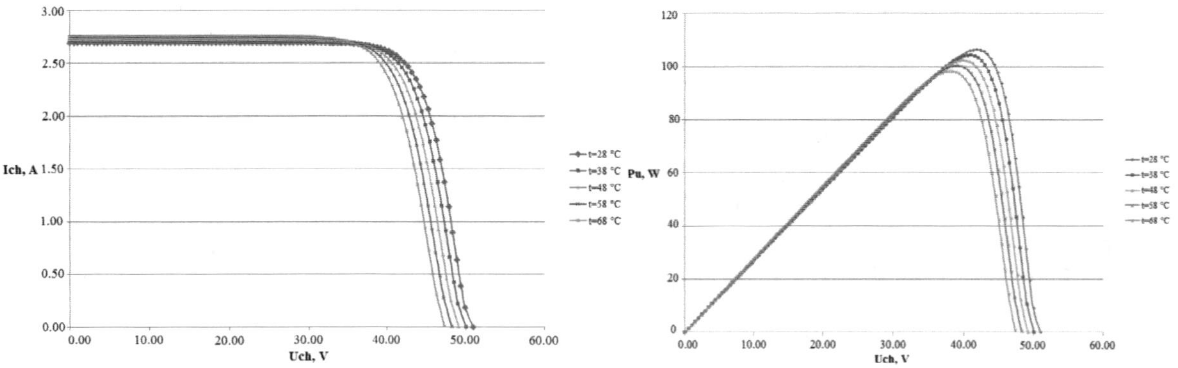

a) b)

Figure 3. Characteristics under different environmental temperature: a) voltage-current b) voltage-power

The diagram of dependence of heat, produced by the collector to environmental temperature is a linear dependence (figure 4). In the process of environmental temperature changing within 28°C to 68°C, the heat amount, produced by the collector varies from 572 W to 625 W.

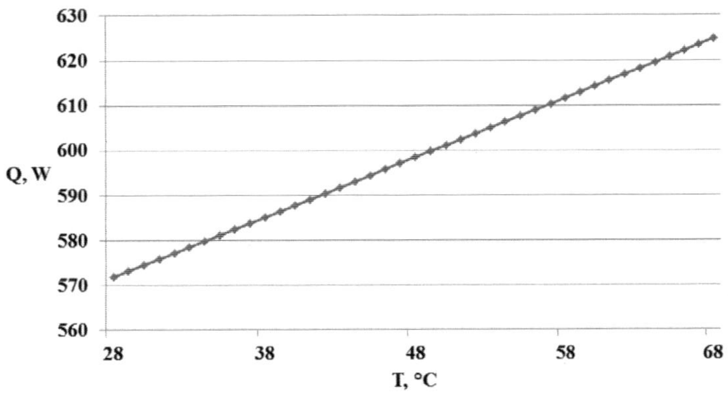

Figure 4. Thermal energy as a function of ambient temperature

The efficiency of the photoconverter $\eta_{PEC}=\dfrac{P_u}{A.E_0}$ is *10.7%* and the efficiency of the thermal collector $\eta_{TC}=\dfrac{Q}{A.E_{col}}$ is *70%* at a temperature of *28°C* and for *A=0.735 m²*. The efficiencies are presented on figure 4 for different values of the environmental temperature. The efficiency of the photoconverter is lightly decreasing with the environmental temperature from 10.7% to 9.9% when the temperature is increasing from 28°C to 68°C. It is due to a weak decrease of power P_u that is caused by a weak increase of the number of the bearers of electric charges and a decrease of the photoelectric voltage U_{ch} .

The efficiency of the solar converter is lightly increasing with the temperature due to an increase of the collected power (from 70 % to 75.7 %).

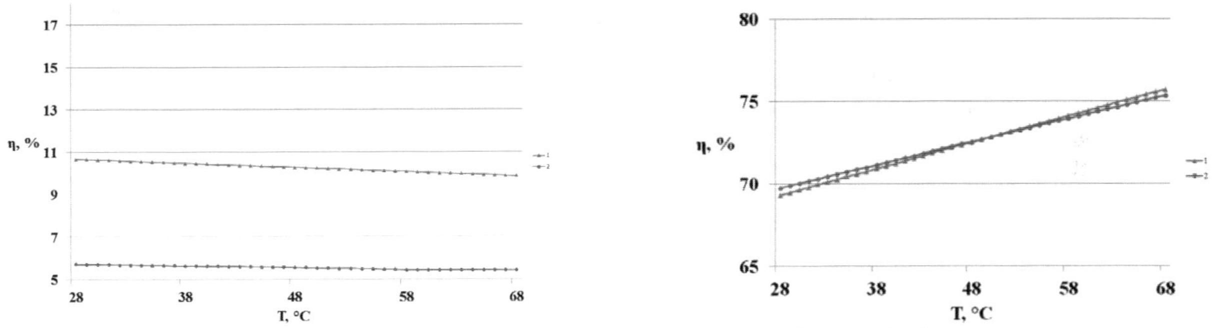

a) Electric photoconverter *b) Thermal solar convertor*

Figure 5. Efficiency: 1- with concentrator, 2- without concentrator

The figure 5 clearly indicates that the suggested combined receiver of solar radiation with holographic planar concentrator allows a more efficient use of the solar radiation.

5. Conclusion

Combined solar receiver possesses a variety of advantages vs. others: more efficient utilization of solar radiation influencing the 1m² surface, no need for a solar tracking system. However, the complexity of it is production is in the combination of contradictory specifications: in solar energy concentration, PC's equilibrium temperature increasing, necessity of it is decreasing by solar collector cooling, and in a necessity of simultaneous increasing of solar collector operational temperature for more efficient heating of operational fluid. It means the optimal combination of parameters of concentration level and operational temperature with maximal PC's energy efficiency and solar collector efficiency must be found. In a first approximation this task can be solved by given mathematical pattern with conditioning of efficiency and cost. Herewith the domain of solar collector's application can be much wider.

This collector can be used for "solar houses". With it is usage the house can be supplied with heat and electricity completely or partially, depending on season. With the solar illumination on a unit the current coming from photoconverters will charge big electrochemical batteries [3]. At the same time, the heat of solar radiation will proceed from photoconvetrers to aluminum sheets and further to the heat carrier, flowing in tubes. Hot water accumulates in the reservoir – large water tank, which is located under the house's floor. For unit is operation during the cold weather seasons the antifreeze must circulate into the solar collector's profile. For this purpose the heat-exchange unit must be installed in the accumulator tank, and the profile itself must be pressurized with increased heat carrier's pressure, which assures solar collector's operation under higher temperature circumstances and protects user from antifreeze toxic influence. The main task during the solar radiation collector's combination is the preservation of high intake specifications under minimal expenses.

Therefore, for creation of combined solar collector's design procedure we have to bring in mathematical description of solar radiation concentrator and solve the concentrator-photovoltaic cell-absorber equation system in order to define the optimal temperature for maximal energy efficiency of photovoltaic cell and solar collector alike with qualifications by concentration level.

References

[1] Holographic solar batteries are dissecting the light before consumption 2006 [Electronic asset] – access mode:http://www.membrana.ru/articles/technic/2006/05/03/222900.html

[2] Products-Space-Cells-Single Junction Solar Cells-GaAs/Ge Single Junction Solar Cells 2009 [Electronic asset] – access mode:http://www.spectrolab.com.

[3] Bezruchko K V 2007 *Autonomous ground electric power installations on renewable sources* (schoolbook / K V Bezruchko and S V Gubin – Kharkov: National Aerospace University «Kharkov aviation institute») p 300

[4] Bordina N M 1986 *The simulation of volt-amps diagrams of edge-illuminated solar cells and solar batteries* (Bordina N M and Letin VA – M: Informelectro) p 62

[5] Andreeva O V 2008 *Applicative holography* (Schoolbook / Andreeva O V SPb: SPbUITMO) p 184

[6] Iurevych O and Gubin S 2011 Modélisation mathématique des installations solaires combinées avec concentrateurs du rayonnement solaire, *Congrès Français de Thermique* May 24-27 (Perpignan France)

IOP Conf. Series: Materials Science and Engineering **29** (2012) 012017 doi:10.1088/1757-899X/29/1/012017

Optoelectronic performance of gallium nitride devices: the role of tilt grain boundaries and point defects

A Béré[1], P Ruterana[2] and J Koulidiati[1]

[1] Laboratoire de Physique et de Chimie de l'Environnement, Université de Ouagadougou, 03 BP 7021 Ouagadougou 03, Burkina Faso
[2] Centre de Recherche sur les Ions, les Matériaux et la Photonique UMR6252, ENSICAEN, 6, Bd du Maréchal Juin, 14050 Caen Cedex France

E-mail: abere@univ-ouaga.bf, berebiya@yahoo.fr

Abstract. The electronic structures of perfect tilt grain boundaries in the gallium nitride (GaN) and the interaction of point defects with their boundaries have been investigated. Our calculations suggest that among the three possible atomic configurations describing the perfect tilt grain boundaries, namely, the 5/7-interface, the 4-interface and the 8-interface, it was found that the 8-interface introduces deep states in the very center and the upper half of the band gap whereas the 5/7-interface possess only states close to the conduction band. The N-vacancy in the 8-interface introduces filled or empty electronic states inside the band gap. Such states may give rise to transition towards the valence band, which could be a candidate to account for the yellow luminescence (YL) in GaN. For the gallium or nitrogen interstitial in interaction with the GaN tilt grain boundaries, it was shown that while the Ga-interstitial introduces empty deep-states in the very center and filled shallow-states in the half bottom of the band gap, the N-interstitial gives only rise to shallow-states at the band edges. Given that for an electrically neutral boundary the deep states are unoccupied (deep acceptors), such defects may contribute to the YL.

1. Introduction

Optoelectronic devices from III-nitride thin films exhibit high performances despite many remaining problems, in particular, the crystalline quality and the parasitic yellow luminescence (YL). The knowledge of the YL origin might assist further improvement of the III-nitrides, in particular, gallium nitride (GaN) at the basis of light-emitting devices. While it is now well known that the YL around 2.2 – 2.3 eV observed in undoped GaN layers results from a transition between band states and deep levels, the origin of the defects that are associated with this YL is still under debate. Some reports attributed it to impurities, native point defects [1], extended defects as low-angle grain boundaries [2] or dislocations [3]. Furthermore, the electronic states involved and the underlying mechanism are still under controversy. Experimental work has confirmed models involving transitions from a shallow donor to a deep acceptor [4-6], from a shallow donor to a deep acceptor [7] and from a deep donor to a shallow acceptor [8].

Gallium nitride (GaN) is a semiconductor with numerous applications in opto and electronic devices which is still mainly artificially produced in thin epitaxial layers on different substrates, such as sapphire [9] and SiC [10] which exhibit large differences in lattice parameters and lead to formation of defects during the growth [11, 12]. During the last years, GaN thin films have been extensively investigated, the extended defects that form upon growth have been identified [13]; they are threading dislocations [14], inversion domains [15], basal [16] and prismatic stacking faults [17-19]. Mechanisms of their formations [12, 20, 21] and possible ways of their reduction [22] have been proposed. Among the extended defect present in GaN layers and which have come out to be most difficult to eliminate are the threading dislocations [13, 14]. They can be isolated inside the epitaxial layers or be in form of grain boundaries [23], as has been clearly demonstrated: they exhibit three atomic configurations [23-26]. Apart from the extended defects, the nitride thin films also contain point defects, such as vacancies and foreign atomic impurities, which may associate with the extended defects and lead to a strong degradation of the performance of devices, this is mostly due to electronic states inside the band gap [27], or leakage channels [28]. Indeed, dislocation cores, which may not only exhibit dangling bonds [29], but also change the local strain configuration [26], can introduce deep levels in the band gap. Another origin for band gap states could be the impurities which may be incorporated present inside the layers or their segregation at interfaces during the growth [30]. This extensive work has been always a close combination of experimental investigations and extensive theoretical modelling. In this work, authors discuss on the electronic properties of the threading dislocations in GaN as carried out by theoretical modelling of the grain boundaries inside the layers grown along the conventional [0001] growth direction, which has been the basis of almost all the nitride devices such as the electroluminescent diodes [31] or high electron mobility transistors [28].

Atomic computer simulation of a variety of [0001] perfect tilt grain boundaries with misorientation angles from $\theta=9.3^o$ (corresponding to $\Sigma37$) to $\theta=44.8^o$ (corresponding to $\Sigma43$) shows that the structural units associated to all the boundaries could be explained in terms of combination of the three stable cores of the $1/3[2\bar{1}\bar{1}0]$ edge dislocation [32,33]. Structural modification, relative stability and electronic properties associated with such tilt grain boundaries in interaction with native point defects (vacancy and interstitial) have been investigated in the framework of Self-Consistent-Charge Density-Functional based Tight-Binding (SCC-DFTB) calculations [34-36].

As described in our previous study [38], the structures of the perfect tilt grain boundaries are obtained on the basis of the dichromatic complex and Coincidence Site Lattices (CSL) approach. In the atomic structure of lowest energy, the vacancy (respectively the interstitial) is created by adding (respectively by removing) an atom of gallium or nitrogen in a particular site, giving rise to a non stoichiometric Ga- or N-vacancy (respectively Ga- or N-interstitial) tilt boundary.

Total energy calculations and geometry optimizations were performed using the Self-Consistent-Charge Density-Functional based Tight-Binding (SCC-DFTB) method [39-42]. They offer the great advantage of giving an accurate evaluation of charge transfer which plays an important role in systems containing heterogeneous compound semiconductors. According to conjugate gradient algorithm, all atoms were allowed to relax to reach the minimum energies until the Hellmann-Feynman atomic forces became less than 0.0001 u.a. (≈ 0.02 eV/Å).

In the following, we present the main results which have been obtained on the electronic properties associated to these defects.

2. Results and discussion

2.1. *Electronic states associated with perfect tilt boundaries*

Figure 1 shows the three core structures of the $(3\bar{7}40)$ $\Sigma=37$ ($\theta = 9.3^o$), $(3\bar{5}20)$ $\Sigma=19$ ($\theta = 13.4^o$) and $(1\bar{3}20)$ $\Sigma=7$ ($\theta = 21.6^o$) perfect tilt grain boundaries in gallium nitride. These atomic configurations are formed of a mixture of six and five/seven atom rings (5/7-interface), a mixture of six and four atom rings (4-interface) or a mixture of six and eight atom rings (8-interface). These core structures are comparable to those obtained for the $1/3[2\bar{1}\bar{1}0]$ edge dislocation [32].

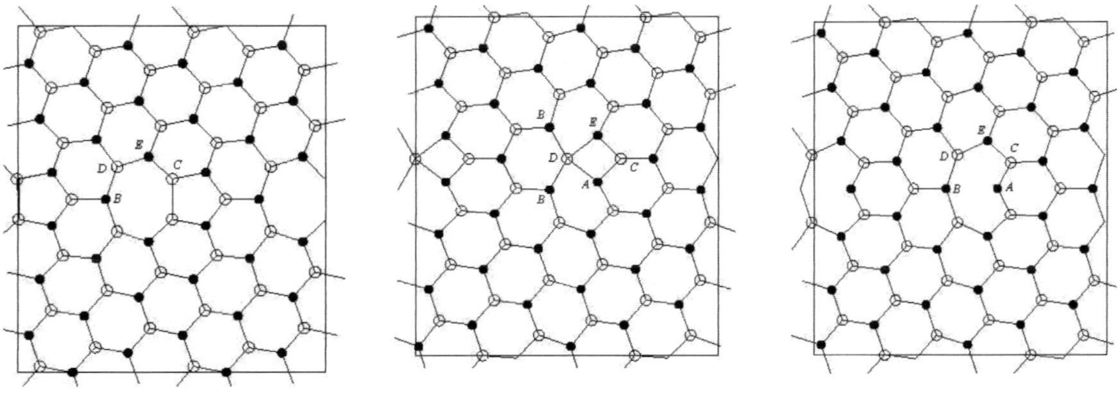

a) 5/7-Interface b) 4-Interface c) 8-Interface

Figure 1. Atomic structures of the [0001]/($1\bar{3}20$) Σ=7 tilt boundary. For all the figures, the full circles represent N atoms and the open circles the Ga atoms. Atomic structures of Σ=37 and those of Σ=19 in [33].

The electronic structure of these interfaces has been investigated [34] and typical behaviours have been pointed out as follows: the 8-interface introduces deep states inside the gap whereas for the 5/7-interface, only states close to the conduction band are predicted. Given that for an electrically neutral boundary the deep states are unoccupied (deep acceptors), the 8-interface may contribute to Yellow Luminescence in GaN layers. In any case, empty gap states will localize the carriers needed for the recombination. The analysis of density of states shows that each structural defect introduces electronic states at the edges of the valence and conduction bands. In contrast to the 5/7-core, the 8-core induces deep-states in the very center of the gap: one unoccupied state, 2-fold degenerate, centered at $E_{1a} = E_{VB} + 3.4\,eV$ for the Σ=37 and Σ=19 tilt boundaries and at $E_{1b} = E_{VB} + 3.7\,eV$ for the Σ=7 tilt boundary. Another unoccupied state, 2-fold degenerate, slightly above is obvious: $\mathrm{E_{2a} = E_{VB} + 4.7\,eV}$ for the Σ=37 and Σ=19 tilt boundaries and $\mathrm{E_{2b} = E_{VB} + 5.2\,eV}$ for the Σ=7 tilt boundary. Some states close to the conduction band are predicted for the 5/7-interface of each tilt boundary: $E_{3a} = E_{VB} + 5.3\,eV$ for the Σ=37 and Σ=19 tilt boundaries and $E_{3b} = E_{VB} + 5.8\,eV$ for the Σ=7 tilt boundary. The small difference for a given position between the Σ=37 and Σ=19 tilt boundaries on one side and the Σ=7 tilt boundary on the other side is in agreement with the fact that the former are considered as low-angle tilt boundaries whereas the latter is viewed as high-angle tilt boundary.
The presence of gap states is in good agreement with theoretical investigations of electronic sates associated with the threading dislocation in GaN [3, 42].

Density functional theory (DFT) usually underestimates the band gap. However, in the specific case of the minimal basis Density-Functional based Tight-Binding (DFTB) approach, the gap is overestimated [38]. Then, SCC-DFTB calculations give 6.52 eV as opposed to the experimental value of 3.4 eV. To correct deviations from the experimental gap value, in most cases, an ad-hoc correction is made, such as, for example by rigidly shifting all conduction bands upwards and shifting the gap levels according to their position by a scaling factor. A schematic diagram of the corrected positions of the deep-states for each configuration is shown in figure 2. The energy levels are in reasonable agreement with theoretical calculations that report deep states (unoccupied) at $E = E_{VB} + 2.0\,eV$ for the 8-core and $E = E_{VB} + 1.4\,eV$ for the 5/7-core of the edge dislocation [3, 42]. Hence, the 8-interface can act as a trap for electrons and might contribute to the YL as mentioned by Lee and al [42].

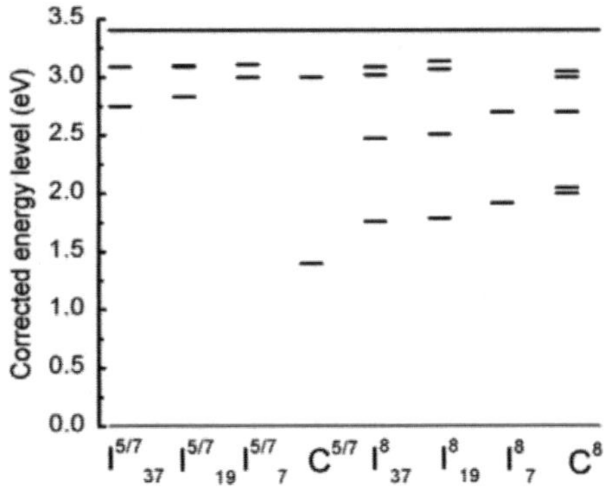

Figure 2. Schematic diagram of energy levels in the band gap for the studied interfaces. In the notation I^{β}_{α}, the symbol α refers to the tilt grain boundary and the symbol β to the interface type. The notation C^{β} designs the core type of the edge dislocation. The values correspond to those obtained after scaling all gap states and conduction band level energies, which give rise to calculated gap of the bulk GaN, equal to the experimental one (3.4 eV). Only values superior to 0.2 eV and inferior or equal to 3.1 eV are mentioned.

2.2. Electronic states associated with vacancy in [0001] tilt boundaries

Using a SCC-DFTB calculations, Lee and al [42] have reported that gallium and nitrogen vacancies in the core of threading edge dislocations can contribute to yellow luminescence even in the absence of impurities. A similar study has been done for the $(3\bar{7}40)$ $\Sigma = 37$ ($\theta = 9.3°$) tilt grain boundary [35]. In this study, the nitrogen vacancy is only created in the 8-interface by removing one N atom at the site indicated by A in figure 3(b). Given that we have two boundaries due to the periodic conditions, two A-sites are present and then, two atoms are removed.

The relaxed structure of the N-vacancy tilt grain boundary shows a small relaxation of neighbouring atoms localised around the vacancy. The N- vacancy is characterised by a strong interaction of the second nearest neighbours, the Ga atoms surrounding the vacancy. They form metallic like bonds similar to those in bulk Ga. The two Ga-N bonds at the vacancy are shorter (1.806 Å) compared to the ideal bond length of 1.95 Å, while the Ga-Ga bonds (2.84 Å) are 29% larger than in the bulk (2.2 Å). This is consistent with other calculations of bond length distortions describing the N-vacancy dislocation [42].

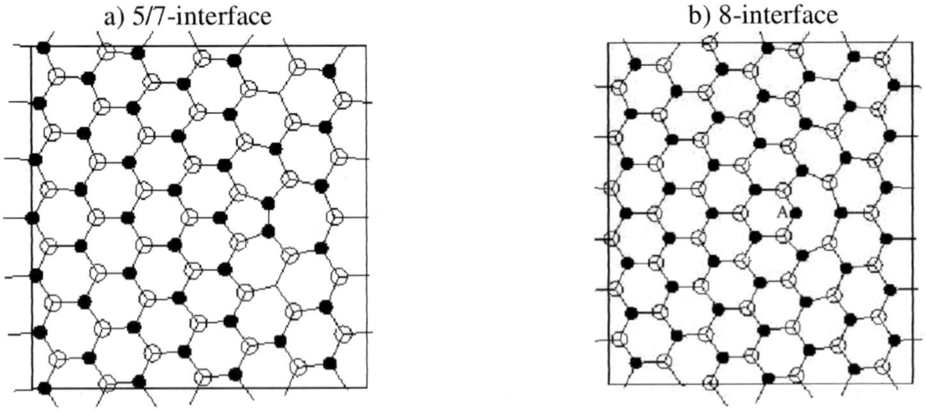

Figure 3. Atomic structures of the [0001]/$(3\bar{7}40)$ $\Sigma = 37$ tilt boundary. For all the figures, the full circles represent N atoms and the open circles the Ga atoms. The N-vacancy is created by removing a nitrogen atom in the A site.

The spectrum of the tight-binding eigenvalues shows that the N-vacancy introduces states in the band gap (figure 4(c)): two shallow and four deep states. This partially agrees with the previous study of the N-vacancy dislocation in which several gap states were shown to be spread over the entire gap [42]. As mentioned previously, the SCC-DFTB approach overestimates the band gap. In order to correct deviation from the experimental gap, we have used a linear scaling method, which consists in shifting the gap levels and conduction band. Then, the corrected positions of the N-vacancy are $E_1 = E_{VB} + 0.72eV$, $E_2 = E_{VB} + 2.20eV$ and $E_3 = E_{VB} + 2.43eV$. The energy level E_1 is fully occupied in reasonable agreement with previous calculations on the N-vacancy dislocation at $E = E_{VB} + 0.6\,eV$. The energy levels E_2 and E_3 are unoccupied, each 2-fold degenerate, and constitute deep levels. Hence, the N-vacancy tilt grain boundary can act as a source or a trap for electrons in a $E_2 - E_{VB}$ transition and might contribute to the yellow luminescence [42].

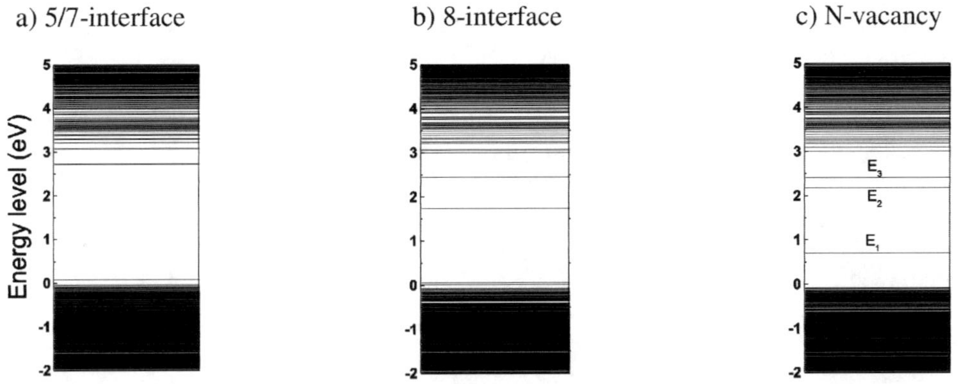

Figure 4. Spectra of the tight-binding eigenvalues

2.3. *Electronic states associated with interstitial in [0001] tilt boundaries*

Numerous works have been reported that Ga-interstitial (Ga_i) is among the dominant defects in epitaxial GaN films when they are exposed to high-energy electron radiation after growth [44, 45]. For our study, the higher energy 8-interface is chosen for the study of interstitials instead of the 5/7-interface due to the fact that the former is characterised by an atomic column exhibiting dangling bonds (position indicated by A in figure 5a). Then, a gallium (respectively nitrogen) interstitial is only created in the 8-interface by adding one gallium (respectively nitrogen) atom at position around the 8-coordinated channel indicated by P or Q in figure 5a. It should be noted that atoms at column A are three-fold coordinated. Likewise, atoms at columns E and D have strained with bond lengths ranging from 2.0 to 2.2 Å and thus give rise to a strain field which could act as a trap for intrinsic defects and impurities [41].

Figure 5b shows the relaxed structure of the Ga-interstitial boundary (Ga_i-interface) obtained when the Ga interstitial is initially placed in the position P or Q. We find that the stable position of the Ga-interstitial (I_{Ga}) is close to the position Q. Then, the structure of initial interstitial in the position Q is characterised by a small relaxation of neighbouring atoms localised around the interstitial. However, the interstitial initially located at the position P moves to the column A and pushes the Ga atom from this column to the I_{Ga} position. The stable positions of the Ga-interstitial in bulk GaN obtained with different potentials have been discussed [46]. However, a direct comparison with the present work is not possible due to geometry of the dislocation core at the interface.

From a structural point of view, the presence of the Ga interstitial at the position I_{Ga} leads to breaking of the Ga-N bonds between columns D and E, allowing the neighbouring atomic columns to relax further outwards. As a result, the relaxed structure presents 5- and 7-coordinated channel (Ga_i-interface) which is similar to the 5/7-interface of the perfect tilt boundary [34]. The main difference between these two interfaces is the presence of only Ga atom at column I_{Ga} of the Ga_i-interface and its consequences: the gallium atoms at columns E and A are three-fold coordinated as nitrogen atoms at column D.

IOP Conf. Series: Materials Science and Engineering **29** (2012) 012017 doi:10.1088/1757-899X/29/1/012017

Up to now, two methods have been used to obtain the 5/7-core also known as an "open" core. The first method based on the elastic theory, where it has been shown that when the origin of the dislocation line to obtain the field of atomic displacements is located in the glide set then, the 5/7-core structure is generated [32]. In the second method, the 5/7-core is obtained by removing the threefold coordinated atoms in column A from the 8-core [3, 41, 42]. The present result shows that another way to obtain the 5/7-core is to insert interstitials near the position Q of the 8-core (figure 5a).

Similar results are obtained when a N-interstitial atom is inserted at the position P or Q. The relaxed structure exhibits a 5/7-coordinated channel (figure 5c) with wrong (homo) bonds between N atoms located at I_{Ga}, A and E giving rise to an asymmetric interface in contrast to the Ga_i-interface. Gallium atoms at columns E and A and, nitrogen atom at column D exhibit default coordination in comparison to the fourfold coordination in bulk GaN.

The formation energy versus the chemical potential shows the Ga_i-interface is energetically more favourable than the stoichiometric 5/7-interface, which suggests that both configurations may exist [36]. This result stems from better relaxation of the Ga_i-interface than the 5/7-interface. It then raises the following question: What kind of 5/7-core? Stoichiometric or not? The results suggest that the core structure can fluctuate along the dislocation line and both Ga_i-core and pure 5/7-core can be present in the investigated samples. Further experimental work based on high resolution Z-contrast imaging is needed to clarify this issue.

a) 8-interface b) Ga_i-interface c) N_i-interface

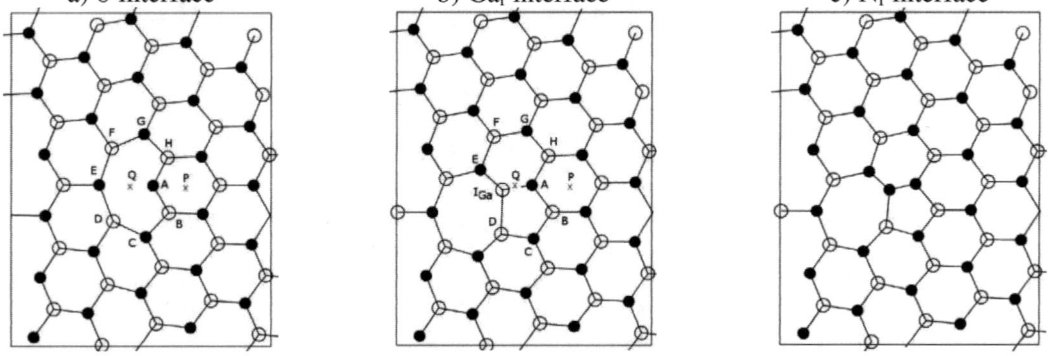

Figure 5. Atomic structures of the [0001]/($2\bar{5}30$) $\Sigma = 19$ grain boundary. For all the figures, the full circles represent N atoms and the open circles the Ga atoms. The Ga- or N-interstitial is created by adding a Ga or N atom in the position P or Q. Notice that one open circle superimposed one full circle and vice versa except at column I_{Ga}

A local density functional code, AIMPRO (Ab Initio Modelling PROgram), with localized and contracted basis sets of Gaussian orbitals [47] has been used to study the electronic structures $(2\bar{5}30)$ Σ=19 (θ=13.2°) tilt grain boundaries in interaction with gallium or nitrogen interstitial. Hartwigsen-Goedecker-Hutter pseudopotentials [48] with non-linear core corrections for Ga atoms are also used to eliminate core electrons. These pseudopotentials were optimised for Gaussian basis sets. The cutoff for the charge density Fourier expansion needs to be set to 350 Ry. A Monkhorst-Pack sampling scheme [49] was used and we checked the convergence of the calculations as a function of number of k-points in the Brillouin zone. Two k-points along the [0001] axis were found necessary to ensure the convergence of the total energy.

Figure 6 shows the band structures of the different configurations. The band gap of the GaN bulk is calculated to be 2.35 eV compared to the experimental value of 3.5 eV. This is not surprising due to the fact that density functional theory calculations usually underestimate the band gap. As previously mentioned, an ad-hoc correction is made to obtain the experimental band gap value, such as, for example by rigidly shifting all conduction bands upwards and shifting the gap levels according to their position by a scaling factor as previously mentioned.

The band structures reveal that each structural defect introduces electronic states at the edges of the valence and conduction bands. For the perfect tilt boundary, the 8-interface induces deep-states in the upper half of the band gap, $E_V + 1.12\,eV$ (figure 6b) in contrast to the 5/7-interface (figure 6a). This result is in good agreement with our previous results based on SCC-DFTB calculations and with theoretical investigations of electronic sates associated with the threading dislocation in GaN [3, 34, 42]. Concerning the interstitial tilt boundary, the Ga_i-interface induces deep-states in the center of the gap (figure 6c).

According to the Fermi levels shown as a dashed line, defect states at the edge of the valence band are filled for the 5/7- and 8-interfaces whereas it is not necessary the case for the interstitial tilt boundaries. Indeed, for the Ga_i-interface, states around $E_V + 0{,}46\,eV$ and $E_V + 1.34\,eV$ are filled and unfilled respectively. Concerning the N_i-interface, state around $E_V + 0.45\,eV$ is unfilled.

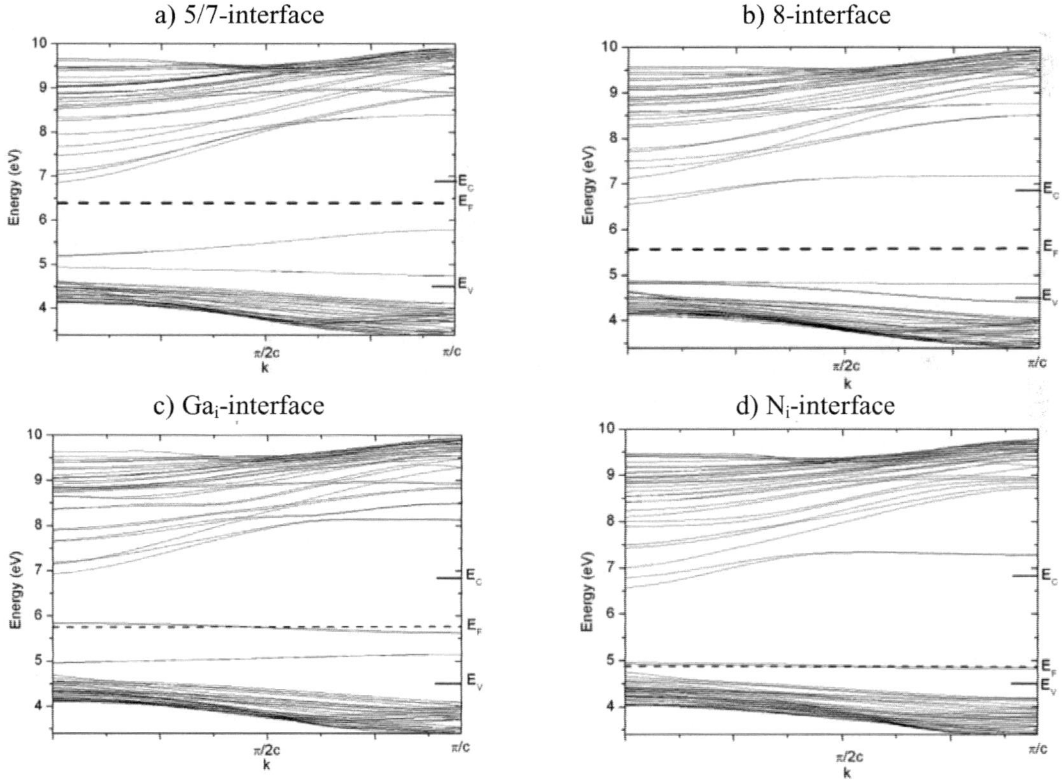

Figure 6. The band structure of atomic structures of the [0001]/($2\bar{5}30$) Σ = 19 grain boundary projected into the k-space axis. The Fermi level, up to which the states (bands) are occupied, is shown as a dashed line.

3. Conclusion

The Self-Consistent-Charge Density-Functional based Tight-Binding (SCC-DFTB) approach has been used to study the electronic structures of perfect tilt grain boundaries in gallium nitride and point defects associated to these boundaries. Concerning the perfect tilt boundaries, our calculations suggest that the interface formed by a mixture of six and eight atom rings (8-interface) introduces deep states inside the gap whereas for the interface constituted by a mixture of six and five/seven atom rings (5/7-interface), only states close to the conduction band are predicted.

The nitrogen vacancy located at a specific atomic site of the 8-interface introduces filled or empty electronic states inside the band gap. The gallium interstitial created in the 5/7-interface introduces also unoccupied states around the very center of the band gap. Some of the states may give rise to a transition with the valence band; this transition could be a candidate for the yellow luminescence in GaN.

Acknowledgement

Dr Nouet G is gratefully acknowledged for his contribution to this work.

References

[1] Van de Walle C G and Neugebauer J 2004 *Appl. Phys. Rev.* **95** 3851
[2] Ponce F A, Bour D P, Götz W and Wrigth P J 1996 *Appl. Phys. Lett.* **68** 57
[3] Blumenau A T, Fall C J, Elsner J, Jones R, Heggie M I and Frauenheim T 2003 Phys. *Stat. Sol. (c)* ***0** 1684*
[4] Ogino T, Aoki M 1980 *Jpn. J. Appl. Phys.* **19** 2395
[5] Shalish I, Kronik L, Segal G, Rosenwaks Y, Shapira Y, Tisch U and Salzman J 1999 *Phys. Rev. B* **59** 9748
[6] Morkoç H 2001 *Mater. Sci. Eng.* **R 33** 135
[7] Hoffmann D M, Kovalev D, Stende G, Meyer B K, Hoffmann A, Eckey L, Heitz R, Detchprom T, Amano A and Akasaki I 1995 *Phys. Rev. B* **52** 13326
[8] Glaser E R, Kennedy T A, Doverspike K, Rowland L B, Gaskill D K, Freitas Jr J A, Asif Khan M, Olson D T, Kuzina J N and Wickenden D K 1995 *Phys. Rev. B* **51** 13326
[9] Nakamura S, Senoh M, Iwasa N and Nagahama S 1995 *Jpn. J. Appl. Phys. (*Part 1) **134** p L797
[10] Tanaka S, Kern R S and Davis R F 1995 *Appl. Phys. Lett.* **66** 37
[11] Vermaut P, Ruterana P, Nouet G and Morkoç H 1997 P*hil. Mag. A* **75** 239
[12] Ruterana P, Potin V, Barbaray B and Nouet G 2000 *Phil. Mag. A* **80** 937
[13] Potin V, Vermaut P, Ruterana P and Nouet G 1998 *J. Electronic Materials* **27** 266
[14] Ning X J, Chien F R, Pirouz P, Yang J W and Asif Khan M 1996 *J. Mater. Res.* **11** 580
[15] Potin V, Ruterana P and Nouet G 1997 *J. Appl. Phys.* **82** 2176.
[16] Potin V, Ruterana P and Nouet G 2000 *J. Phys. Condensed Matter* **12** 10301
[17] Northrup E J 1998 *Appl. Phys. Lett.* **72** 2316
[18] Vermaut P, Nouet G and Ruterana P 1999 *Appl. Phys. Lett.* **74** 694
[19] Ruterana P, Barbaray B, Béré A, Vermaut P, Hairie A, Paumier E, Nouet G, Salvador A, Botchkarev A and Morkoc H 1999 *Phys. Rev. B* **59** 15917
[20] Ponce F A 1997 *MRS Bull.* **22** 51
[21] Ruterana P, Béré A and Nouet G 2000 *MRS Internet J. Nitride Semicond. Res.* **5S1** W3.55
[22] Ruterana P, Chen J, Potin V and Nouet G 2000 *J. Phys. Condensed Matter* **12** 10185
[23] Potin V, Ruterana P, Nouet G, Pond R C and Morkoç H 2000 *Phys. Rev. B* **61** 5587
[24] Ruterana P and Nouet G 2001 *Phys. Stat. Sol.(b)* **227** 177
[25] Chen J, Ruterana P and Nouet G 2001 *Mat. Sci and Eng.* **B82** 117
[26] Lymperakis L, Neugebauer J, Albrecht M, Remmele T and Strunk H P 2004 *Phys. Rev. Lett.* **93** 196401
[27] Look D C, Reynolds D C, Hemsky J W, Sizelove J R, Jones R L and Molnar R.J 1997 *Phys. Rev. Lett.* **79** 2273
[28] Hsu J W P, Manfra M J, Molnar R J, Heying B and Speck J S 2002 *Appl. Phys. Lett.* **81** 79
[29] Xin Y, Pennycook S J, Browning N, Nellist P, Sivanathan S, Ommnés F, Neaumont B, Faurie J and Gibart P, 2000 *Appl. Phys. Lett.* **72** 2680
[30] Hsu J W P, Manfra M J, Lang D V, Richter S, Chu S N G, Sergent A M, Kleiman, R N Pfeiffer L N and Molnar R J 2001 *Appl. Phys. Lett.* **78** 1685
[31] Nakamura S, Mukai T, Senoh M, Nagahama S and Iwasa N 1993 *J. Appl. Phys.* **74** 3911
[32] Béré A and Serra A 2002 *Phys. Rev. B* **65** 205323

[33] Béré A and Serra A 2002 *Phys. Rev. B* **66** 085330 and Béré A and Serra A 2001 *Interface Sci.* **9** 149
[34] Béré A, Ruterana P, Nouet Blumenau G A T, Sanna S, Frauenheim Th, Chen J and Koulidiati J 2005 *Phys. Rev. B* **71** 125211
[35] Béré A, Chen J, Nouet G, Ruterana P and Koulidiati J 2005 *Phys. Stat. Sol. (c)* **2** 2496
[36] Béré A, Belabbas I, Petit S, Nouet G, Ruterana P, Chen J and Koulidiati J 2006 *Phys. Stat. Sol. (a)* **203** 2176
[37] Béré A, Belabbas I, Petit S, Nouet G, Ruterana P, Chen J, G. Segda and Koulidiati J 2007 *Phys. Stat. Sol. (c)* **4** 2593.
[38] Béré A and Serra A 2006 *Philos. Mag.* **15** 2159
[39] Frauenheim Th, Seifert G, Elstner M, Hajnal Z, Jungnickel G, Porezag D, Suhai S and Scholz R 2000 *Phys. Stat. Sol. (b)* **217/1** 41
[40] Haugk M, Elsner J, Frauenheim Th, Seifert G and Sternberg M 2000 *Phys. Stat. Sol. (b)* **217/1** 473
[41] Elsner M, Porezag D, Jungnickel G, Elsner J, Haugk M, Frauenheim Th, Suhai S and Seifert G 1998 *Phys. Rev. B* **58** 7260
[42] Lee S M, Belkhir M A, Zhu XY, Lee Y H, Hwang Y G and Frauenheim Th 2000 *Phys. Rev. B* **61** 16033
[43] Fall C J, Jones R, Briddon P R, Blumenau A T, Frauenheim Th and Heggie M I 2002 *Phys. Rev. B* **65** 245304
[44] Chow K H, Watkins G D, Usui A and Mizuta M 2000 *Phys. Rev: B* **85** 2761
[45] Bozdog C, Przybylinska H, Watkins G D, Härle V, Scholz F, Mayer M, Kamp M, Wickenden A E, Koleske D D and Henry R L 1999 *Phys. Rev. B* **59** 12479
[46] Gao F, Bylaska E J and Weber W J 2004 *Phys. Rev. B* **70** 245208
[47] Briddon P and Jones R 2000 P*hys. Stat. Sol. (b)* **217/1** 131
[48] Hartwigsen C, Goedecker S and Hutter J 1998 *Phys. Rev. B* **58** 3641
[49] Monkhorst J H and Pack J D1976 *Phys. Rev. B* **13** 5188

IOP Conf. Series: Materials Science and Engineering **29** (2012) 012018 doi:10.1088/1757-899X/29/1/012018

3D determination of the minority carrier lifetime and the p-n junction recombination velocity of a polycrystalline silicon solar cell

R Sam[2], B Zouma[1], F Zougmoré[1], Z Koalaga[1], M Zoungrana[1] and I Zerbo[1]

[1]Laboratoire de Matériaux et Environnement, Département de Physique, UFR/SEA, 03 bp 7021 Ouagadougou 03, Burkina Faso
[2]Departement de Physique, Institut des Sciences Exactes et Appliquées, 01bp 1091 Bobo-Dioulasso 01, Burkina Faso

E-mail: samraguil@univ-ouaga.bf

Abstract. This work presents a theoretical and experimental transient tri-dimensional study conducted for the determination of the bulk component of the minority carrier lifetime and the p-n junction recombination velocity of a bifacial polycrystalline silicon solar cell. The theoretical analysis is based on the columnar model of the grains in the polycrystalline silicon solar cell. The boundaries conditions are defined in order to use Green's function to solve the three-dimensional diffusion equation. This leads to a new expression of the transient photovoltage. The value of the constraint coefficients at interfaces of the grain are computed while those of the effective minority carrier lifetime τ_{eff} is extracted from the experimental curve of transient voltage. The bulk lifetime and the p-n junction recombination velocity are deduced and have been compared to those obtained from transient state by one-dimensional modelling of carrier's diffusion. This comparative study permitted us to show grain effects on the lifetime and consequently the inadequacy of one-dimensional modelling of carrier's diffusion in the polycrystalline silicon solar cells.

1. Introduction

The minority carrier's lifetime and the p-n junction recombination velocity are two important recombination parameters that affect strongly the efficiency of the solar cells. It is the reason why many techniques have been developed, especially dynamic methods, in order to determine these parameters [1, 2]. But these methods developed in different laboratories often lead to overestimated or underestimated values [3]. This is probably due to the fact that the one-dimensional model is not appropriate to describe the diffusion of carriers because it does not take into account the effect of grain boundaries. In the case of polycrystalline cell, the action of grain boundaries recombination generates a high diffusion of carriers from grain center to its boundaries [4]. In these conditions, the one-dimensional model is not appropriate to describe the diffusion of carriers. So the purpose of this work is to develop mathematical solutions of the time dependent and three-dimensional diffusion equation for minority carriers in the base of a polycrystalline p-n junction solar cell and to propose a new experimental procedure to measure minority carriers' lifetime and p-n junction recombination velocity by photovoltage decay analysis. In this study, mathematical approach in 3D description of carriers' diffusion is based on the following hypothesis [5]:

- The surfaces between two adjacent grains and perpendicular to the junction are characterized by the same carrier recombination processes evaluated by a grain boundary recombination velocity Sg
- The electric field of crystal lattice is negligible
- Only the contribution of the base in the processes of generation is considered

2. Theoretical analysis

Polycrystalline devices contain many grains randomly oriented or relatively ordered. In order to do the modeling of the processes of generation, diffusion and recombination, the polycrystalline solar cell will be considered as a regular array of many units cells connected in parallel with dimensions 2a, 2b and H (figure 1). This modeling of polycrystalline solar cell structure was described in detail by the authors of references [5, 6, and 7]. On these hypotheses, the properties of polycrystalline solar cell can be described by a study of generation, diffusion and recombination processes only in one grain.
Figure 1 shows the theoretical model for a grain taken as a unit cell inside the sample and the Cartesian coordinates used to solve the diffusion equation.

For simplicity it is considered a square section for the grain (i.e. grain size $Xg = 2a = 2b$).

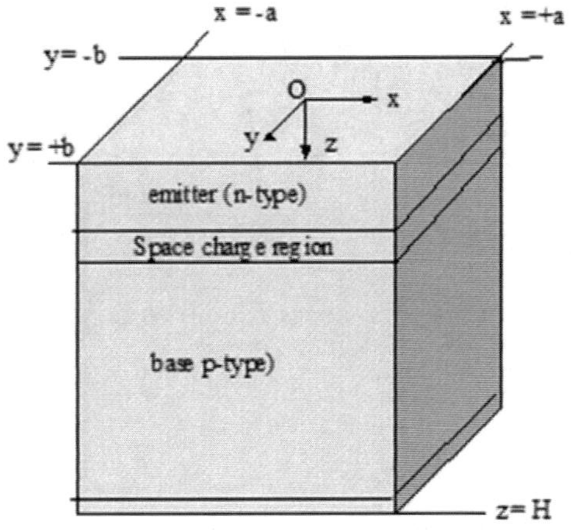

Figure 1. Theoretical model of grain

When the cell is submitted to a multispectral flash illumination on its surface defined by $z = 0$, the excess minority carrier density $\delta(x,y,z,t)$ in the base region is governed by the inhomogeneous partial diffusion equation of the form:

$$\frac{\partial\delta(x,y,z,t)}{\partial t} - D.\left[\vec{\nabla}^2\delta(x,y,z,t) - \frac{\delta(x,y,z,t}{L^2}\right] = g(x,y,z,t) \qquad (1)$$

The generation rate $g(x,y,z,t)$ is governed by equations:

$$g(x,y,z,t) = \begin{cases} \sum_{m=1}^{3} a_m.\exp(-b_{m.}.z) & \text{if } \; 0 \le t \le Te \\ 0 & \text{if } \; t > Te \end{cases} \qquad (2)$$

The coefficients a_m and b_m are the modeling coefficients of AM1.5 solar spectrum. These coefficients were evaluated by [8]; D and L are respectively the diffusion coefficient and the diffusion length of the carriers; Te is the duration of the flash illumination.

The differential equation defined at (1) is integrated with the following boundary conditions:

- at p-n junction defined by z = 0

$$\left.\frac{\partial\delta(x,y,z,t)}{\partial z}\right|_{z=0} = \frac{Sf}{D}\delta(x,y,0,t) \tag{3}$$

- at back surface defined by $z = H$

$$\left.\frac{\partial\delta(x,y,z,t)}{\partial z}\right|_{z=H} = -\frac{Sb}{D}.\delta(x,y,H,t) \tag{4}$$

- at surfaces limited by $x = \pm a$ and $y = \pm b$

$$\left.\frac{\partial\delta(x,y,z,t)}{\partial x}\right|_{x=\pm a} = \mp\frac{Sg}{D}.\delta(\mp a,y,z,t) \tag{5}$$

$$\left.\frac{\partial\delta(x,y,z,t)}{\partial y}\right|_{y=\pm b} = \mp\frac{Sg}{D}.\delta(x,\mp b,z,t) \tag{6}$$

Sf, *Sb* and *Sg* are the recombination velocity of minority carriers respectively at surfaces $z = 0$, $z = \mathrm{H}$ and $x = \pm a$ (or $y = \pm b$); *a*, *b* and *H* are the grain sizes as indicated on figure 1.

The transient voltage decay is defined by equation (7):

$$V(t) = V_T.\ln\left\{1 + Fc(k_1,l_1,\mu_1)q\exp\left[-\beta_{1,1,1}\left(t - Te\right)\right]\right\} \tag{7}$$

where $Fc(k_1,\ l_1,\mu_1)$ is the reduced magnitude of transient voltage defined by :

$$Fc(k_1,l_1,\mu_1) = \frac{XO_{1,1,1}}{XI_{1,1,1}} \tag{8}$$

The value of this function depends on the characteristics of the cell.

The quantities $XO_{1,1,1}$, $XI_{1,1,1}$, q and $\beta_{1,1,1}$ are giving by :

$$XO_{1,1,1} = \int_{-a}^{a}\int_{-b}^{b} Z_{1,1,1,}\left(x,y,0\right)\mathrm{d}x\mathrm{d}y \tag{9}$$

$$XI_{1,1,1} = \int_{-a}^{a}\int_{-b}^{b} \delta_{1,1,1}(x,y,0)\mathrm{d}x\mathrm{d}y \tag{10}$$

$$q = \exp(\Delta V) - 1 \tag{11}$$

$$\beta_{1,1,1} = D.\left(k_1^2 + l_1^2 + \mu_1^2 + \frac{1}{L^2}\right) \tag{12}$$

$Z_{1,1,1}\left(x,y,0\right)$ and $\delta_{1,1,1}(x,y,0)$ are respectively the spatial component of $\delta(x,y,z,t)$ and the minority carriers density during the phase of illumination. The parameters k_1, l_1 and μ_1 are the fundamental eigenvalues obtained from the boundary conditions.

These parameters depend only on the grain sizes and the grain boundary recombination activities. Expression (7) leads to predict two type of transient voltage decay: linear and exponential decay. To understand the role that the grain play in limiting the solar cell efficiency, the concept of effective minority carrier lifetime used in the case of polycrystalline silicon solar is introduced. The effective minority carrier lifetime is given by:

$$\frac{1}{\tau_{eff}} = \frac{1}{\tau_s} + \frac{1}{\tau_b} \tag{13}$$

where τ_b is the bulk lifetime and τ_s is the surface component which is controlled by the grain boundary recombination velocity and the grain size through k_1 and l_1, the base thickness, the back and p-n junction surfaces recombination velocity through μ_1. Then the surface lifetime can be expressed as:

$$\frac{1}{\tau_s} = D.\left(k_1^2 + l_1^2 + \mu_1^2\right) \qquad (14)$$

3. Experimental details

The experimental setup is presented on figure 2. It is composed of : a bifacial solar cell of MOTECH INDUSTRIES; pulse light source MINISTROB, PHYWE, model BOX-1203 BBE; a digital scope, TEKTRONIX, model TDS 210; a computer, Intel 586, 1GHz; a variable white light source and a variable resistance.

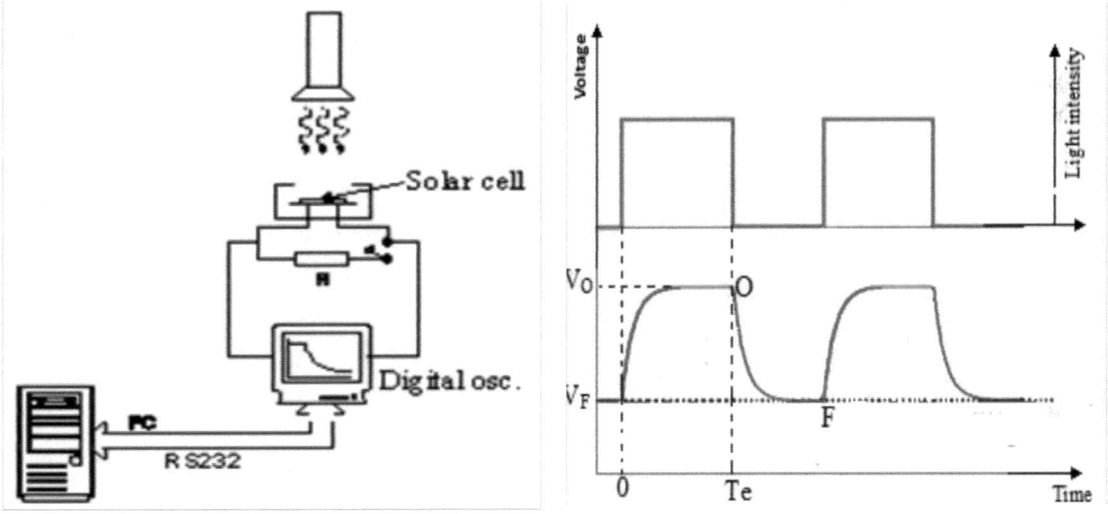

Figure 2. Experimental setup and operating mode

At time t = 0, the cell is illuminated with the multispectral flash which establishes a steady state characterized by the potential V^F corresponding to an operating point called *F* (figure2). At *t* = *Te* the flash is abruptly cut off. The voltage V^F drops from V^F to V^0 corresponding to a new operating point denoted O. The decay of voltage from V^F to V^o is recorded on a digital scope connected to a computer via a RS-232 cable. Experimental data are stored in the computer to be used later for the reconstruction of the signal response of the solar cell and analysis.

Let *t' = 0* be the time of beginning of the exponential part of *V(t)*. At this time corresponds the ordinate *V(t' = 0)*. The couple *(t' = 0, V (t' = 0))* is solution of the equation:

$$V(t') = V_T.Fc(k_1, l_1, \mu_1)q\exp\left(-\beta_{1,1,1}.t'\right) \qquad (15)$$

V (t') is given by the best fit of experimental points corresponding to an exponential decay from which τ_{eff} is deduced:

$$\left[V\left(t'\right) - V_{exp}\right] \to 0 \qquad (16)$$

The other experimental points not fulfilling the condition defined by (16) are not solutions of equation (15) and are therefore taken out of the fit. The error estimation is given by the correlation factor R^2 obtained from the fit.

The experimental value of $Fc(k_1, l_1, \mu_1)$, the reduced magnitude of transient voltage, is then obtained by:

$$Fc(k_1, l_1, \mu_1) = \frac{V(t'=0)}{q.V_T} \tag{17}$$

$V(t'=0)$ is as defined above; q is deduced from expression (11) using ΔV defined by: $\Delta V = V_O - V_F$. The values of $V(t'=0^+)$ and q lead to the experimental value of reduced magnitude of transient voltage: Fc_{exp}. The diffusion coefficient D for silicon is a function of dopant density N_B and expressed as:

$$D = V_T.\frac{1350}{\left[1+\frac{81.N_B}{N_B+3,2.10^{18}}\right]^{\frac{1}{2}}} \tag{18}$$

The diffusion length can be expressed as function of D, τ_{eff}, k_l, l_l and μ_l by

$$L = \left[\frac{D.\tau_{eff}}{1-D.\tau_{eff}.\left(k_{1,\alpha}^2 + l_{1,\alpha}^2 + \mu_{1,\alpha}^2\right)}\right]^{\frac{1}{2}} \tag{19}$$

The boundary conditions at p-n junction surface and back surface allow to express Sf as:

$$Sf = -\mu_l D \tan\left[\tan^{-1}\left(\frac{Sb}{\mu_l D}\right) - \mu_l H\right] \tag{20}$$

Likewise, with grain boundary conditions, the common value of the laterals size of grain can be expressed as function of *Sg*. The value of grain boundary recombination velocity Sg evaluated by optical scanning allows us to express *Sg* as:

$$Sg = Sg_m = m \times 10^4 \ cm/s \tag{21}$$

where *m* is a real >1.

These expressions allow to write $Fc(k_l, l_l, \mu_l)$ as a function of parameters N_B, *H*, *Sb*, *m* and τ_{eff}. As *H*, *D*, *Sb*, and τ_{eff} are known, therefore, the analytical expression of the function $Fc(k_l, l_l, \mu_l)$ is completely described by the parameter m.

The hypothesis of square section of grains leads to $k_1 = l_1$. In these conditions, reduced magnitude of transient voltage is expressed as only a function of k_l and μ_l. So these parameters can be obtained by solving equation (22)

$$Fc_m(k_1, \mu_1) = Fc_{exp} \tag{22}$$

A homemade software allows us to get the experimental values of constraints coefficients k_l and μ_l After calculation of the remarkable points, the values of k_l, μ_l and m are found by dichotomy. This dichotomic search is done using an iterative method of calculation based on NEWTON's algorithm. The incrementations are submitted to the inequality expressed by equation (23):

$$\left| Fc_m(k_1, l_1, \mu_1) - \frac{V(t'=0^+)}{q.V_T} \right| \leq 10^{-6} \tag{23}$$

The surface lifetime is calculated using expression (14) and the bulk lifetime is deduced from the equation (13). Likewise Sf is obtained by means of expression (20).

On figure 4 to figure 6 are presented experimental curves of transient voltage and their fit.

IOP Conf. Series: Materials Science and Engineering **29** (2012) 012018 doi:10.1088/1757-899X/29/1/012018

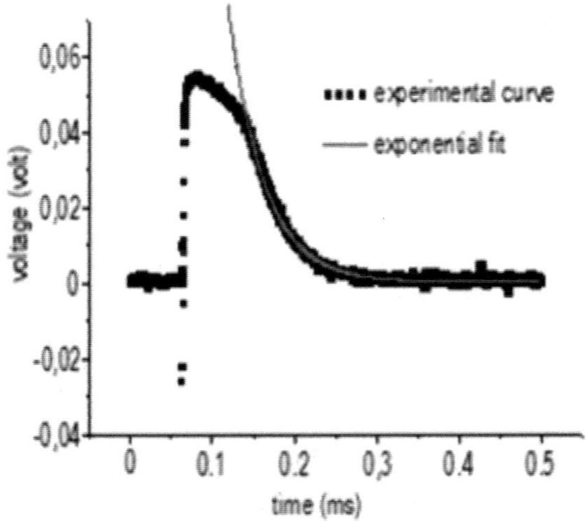

Figure 3. Transient voltage decay for load R = 0,7kΩ

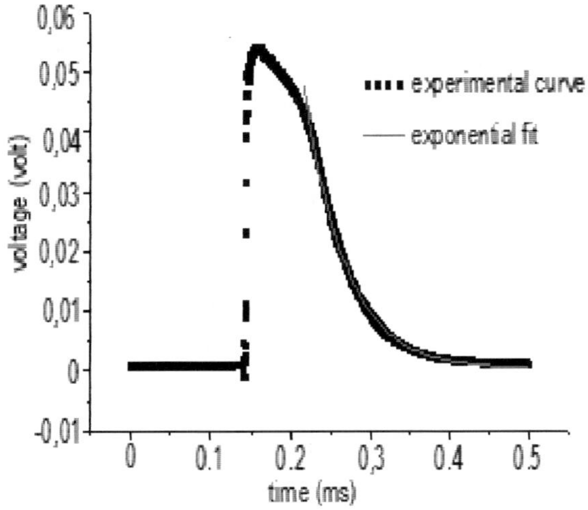

Figure 4. Transient voltage decay for load R=1kΩ

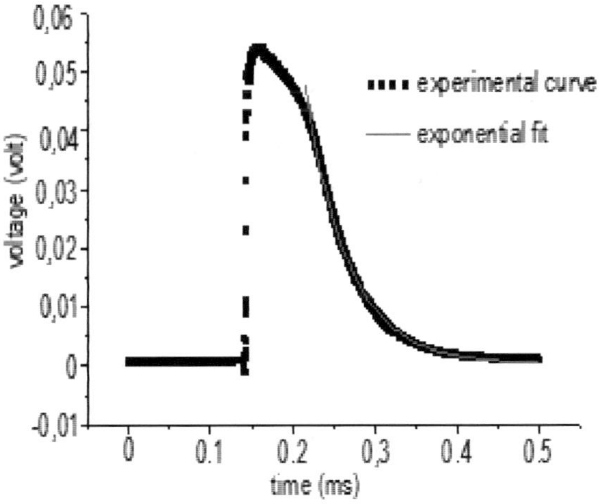

Figure 5. Transient voltage decay and corresponding exponential fit; R =1,5kΩ

Technique described above leads to the results presented on the table 1. The solar cell used in our experiment has the following characteristics: H = 0,02 cm ; D=26 cm^2/s.

Table 1. Lifetime and p-n junction surface velocity for different values of load

	$\tau_b(\mu s)$	$\tau_{eff}(\mu s)$	Sf(cm/s)	$R^{\wedge 2}$
R=0.7kΩ	10,9	9,3	$1.2.10^4$	0,98
R=1kΩ	4	40	1.10^3	0,95
R=1.5k Ω	4,3	43	900	0,91

A similar procedure as described in this paper based on one-dimensional model of carrier diffusion and developed by the author of the reference [9] leads to the same values of τ_{eff} . However, a profound disagreement is revealed on the value of bulk component of lifetime τ_b.

Even if the appearance of interference impedances during the transient state disrupts the measured values, this disagreement can't be justified only by the effects of interference impedances. This disagreement can be explained by the lateral diffusion of minority [10].

Indeed, by superimposing the expression of bulk lifetime τ obtained by one- dimensional modeling to those established in this study, we obtain through the three dimensional analysis of carrier diffusion, the expression:

$$\frac{1}{\tau_b}-\frac{1}{\tau}=\omega_0^2-D\left(k_1^2+l_1^2+\mu_1^2\right) \tag{24}$$

The constants ω_0 and μ_1 are controlled by the same parameters i.e. *Sf*, *Sb* and *H*. Then it occurs

$$\frac{1}{\tau_b}-\frac{1}{\tau}\simeq -2\mathrm{D}\mathrm{k}_1^2 \tag{25}$$

With the experimental value of, k_l the part of grains activity on the disagreement of experimental values of bulk lifetime obtained by one and three-dimensional analysis of carrier diffusion can be expressed. This contribution of grains can be estimated by the value of coefficient η defined:

$$\eta=\left|\frac{-2Dk_1^2}{\frac{1}{\tau_b}-\frac{1}{\tau}}\right| \tag{26}$$

The difference between the experimental value of bulk lifetime obtained in this study and those published by the authors of the reference [9] leads to a value of η greater than 0, 9.

4. Conclusion

In this 3 D study, we have presented a new mathematical description of carrier's diffusion as well as a new experimental procedure for minority carrier lifetime and p-n junction recombination velocity determination. This study led us to establish a new expression of transient voltage containing all the characteristic parameters of the geometrical grains and their interface states.

With a classic experimental setup allowing the observation of the transient state, we have reconstituted on computer and analyzed the response of cell submitted to a multispectral flash illumination. The treatment of transient voltage decay curve reconstituted from data acquisition allowed us to obtain the value of reduced magnitude of transient voltage as well as the effective minority carrier lifetime. The eigenvalues, solution of equations expressing the cell's boundary constraints are then calculated by software program based on NEWTON's algorithm. The values of bulk lifetime and p-n junction recombination velocity are then deduced.

The experimental results obtained reveal a profound disagreement with those obtained on the same cell by one-dimensional modeling. Using the experimental values of coefficients k_l and l_l, we have demonstrated that the disagreement observed is due to the fact that we take into account the lateral diffusion of minority carrier. This analysis highlights the part of grains effect in indetermination of the bulk lifetime of carriers and the shortcoming of one-dimensional model in the description of polycrystalline silicon wafers.

The experimental results also showed that the values of the effective lifetime increases with the value of load. This increase may be related to the capacitance of the PV modules which manifests itself at proximity of open circuit conditions. During the experimental operations, it appeared that the value of bulk lifetime can be stabilized by increasing the illumination level. This observation will be examined in details in the future by a global increase of injection level.

References

[1] Ogita Y-I 1996 *J. Appl. Phys.* **79** 6954-60
[2] Green M A 1984 *Solar cells* **11** 147-61
[3] Salach-Bielecki R, Pisarkiewicz T, Stapinski T and Wojcik P 2004 *Opto-Electronics Rev.* **12** 79-83
[4] Kirev P 1975 *La physique des semi-conducteurs* (Moscow : Editions Mir)
[5] Dugas J. and Oualid J 1987 *Revue Physique Appliquée* **22** 677-85
[6] Ba B and Kane M 1995 *Solar Energy Materials and Solar Cells*, **37** 259-71
[7] Ba B, Kane M, Fickou A and Sissoko G 1993 *Solar Energy Materials and Solar Cells* **31** 33-49
[8] Mohammad S N 1987 *J. Appl. Phys.* **61** 767-72
[9] Barro F I, Nanema E, Zougmoré F, Wereme A, Ndiaye A and Sissoko G 2003 *Journal des Sciences* **3** 10-14
[10] Dimitriadis C A 1985 *Solid State Commun.* **56** 925-27

Silicon solar cell under electromagnetic wave in steady state: effect of the telecommunication source's power of radiation

I Zerbo[1], M Zoungrana[1], A D Seré[2], F Zougmoré[1]

[1]Laboratoire de Matériaux et Environnement, Département de Physique, UFR-SEA, Université de Ouagadougou, BURKINA FASO.
[2]Département de Physique, Institut des Sciences Exactes et Appliquées, Université Polytechnique de Bobo Dioulasso, BURKINA FASO

E-mail : izerbo@univ-ouaga.bf

Abstract. The characterisation of minority carriers' recombination parameters is important to design any optoelectronic component, since they play an important role in the device performance. Many diagnostic techniques, steady state, quasi-steady state, transient and modulation methods for both the experimental and theoretical measurement have been developed for the determination of single or several recombination parameters. Many studies on recombination and electrical parameters of solar cells under constant magnetic field and / or electric field in the bulk due to carrier concentration have been proposed with one or three dimensional approach. In previous works, we study the effects of electromagnetic waves on silicon solar cell recombination parameters by changing the distance between the 2 MW power of radiation AM antenna and the solar cell. This work presents a one dimensional study of the influence of telecommunication source's power of radiation on a silicon solar cell illuminated by a white light in steady state. The telecommunication sources which can be an AM radio antenna or a FM radio antenna radiate electromagnetic waves. Electric field intensity depends on the power of radiation and the distance between the electric field intensity measurement point and the antenna. The magneto transport equations of excess minority carriers are solved with new boundary conditions. New analytical expressions of minority carriers' density, photo-current density, back surface recombination velocity, intrinsic junction recombination velocity and photo voltage depending on electromagnetic field intensity are established. The effects of electromagnetic field on photo-current density, back surface recombination velocity, intrinsic junction recombination velocity and photo- voltage by making vary the antenna power of radiation are then analysed. We analyze also the effects of electromagnetic field on the electric power delivered by the solar cell to an external charge. Thus, using the maximum of electric power delivered by the solar cell to an external charge, we deduce the solar cell conversion efficiency. The analysis of electromagnetic field effects on the silicon solar cell put out the influence of AM radio antenna or FM radio antenna on recombination parameters.

1. Introduction

In previous work we have studied the influence of electromagnetic waves, produced by a radio antenna functioning in amplitude modulation AM, on silicon solar cell recombination parameters [1]. In the present article we study the influence of a telecommunication antenna's power of radiation on the same parameter. This telecommunication antenna can be either a radio antenna functioning in frequency modulation FM or a radio antenna functioning in amplitude modulation AM. This radio station antenna is the source of production of progressive monochromatic plane electromagnetic waves linearly polarized whose electric field intensity measured at a distance r from the source is dependant on the power radiated by the source and the distance r. Therefore we will study the influence of electromagnetic field on photo-current density, back surface recombination velocity, intrinsic junction recombination velocity and photo- voltage by varying the antenna power of radiation. We will study the influence of the antenna power of radiation on the electric power delivered by the solar cell to an external load circuit and the solar cell conversion efficiency. Finally we will discuss the influence of each type of antenna on the solar cell recombination parameters .

2. Theory

2.1. Excess minority carriers' density

We consider a polycrystalline back surface field silicon solar cell with n^+-p-p^+ structure [2]. Since the base has a greater contribution to photo conversion than the emitter [3-5], the one-dimensional analysis will be developed only on this region using the Quasi-Neutral Base assumption. Moreover we are located in the quasi neutral basis hypothesis (Q.N.B) while neglecting the crystalline field that exists within the solar cell.

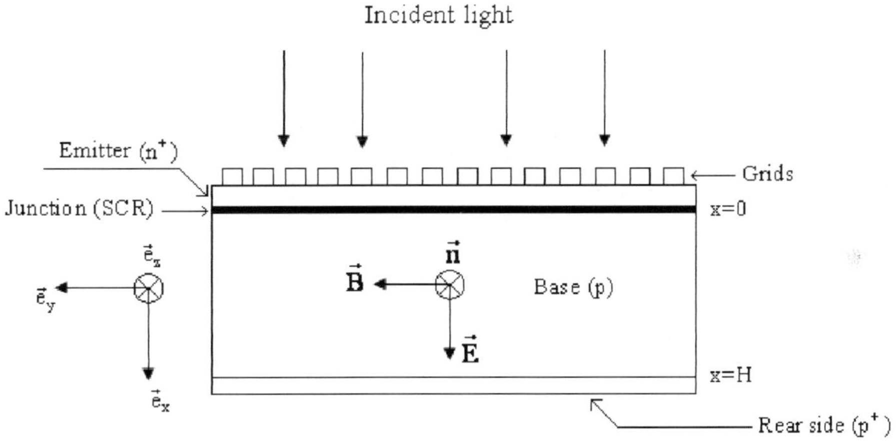

Figure 1. Silicon solar cell illuminated by white light and under electromagnetic waves influence

We suppose that the solar cell represented on figure 1 above is submitted to the action of a progressive monochromatic plane wave linearly polarized, polarized in Ox direction and propagating in the sense of the increasing z.

$\vec{n}$ is the unit vector of the electromagnetic wave direction of propagation as the trihedral $(\vec{E}, \vec{B}, \vec{n})$ is direct.

The complex expression of the electric field is:

$$\vec{E} = E_0 \cdot \vec{e}_x \cdot \exp j(\omega \cdot t - k \cdot z) \qquad (1.a)$$

either in real notation

$$\vec{E} = E_0 \cdot \vec{e}_x \cdot \cos(\omega \cdot t - k \cdot z) \qquad (1.b)$$

The magnetic field of a progressive plane wave is obtained by the relation: (2)

$$\vec{B} = \frac{\vec{e}_z \times \vec{E}}{c}$$

$$\vec{B} = B_0 \cdot \vec{e}_y \cdot \exp j(\omega \cdot t - k \cdot z) \tag{2.a}$$

either in real notation

$$\vec{B} = B_0 \cdot \vec{e}_y \cdot \cos(\omega \cdot t - k \cdot z) \tag{2.b}$$

We choose as origin of space and time the moment when the electromagnetic wave meets the solar cell. At t = 0, z = 0 so $\vec{E} = E_0 \cdot \vec{e}_x$ and $\vec{B} = B_0 \cdot \vec{e}_y$

E_0 and B_0 are respectively the amplitudes of electric and magnetic field.

In telecommunication, the level of reception in a given place of the signal radiated by a transmitter is measured through the intensity of electric field E_0 of this wave.
For an isotropic antenna radiating a power P_r (W) in free space, the electric field intensity E_0 (V /m) depending on the distance r (m) is given by the formula [6]:

$$E_0 = \frac{1}{2 \cdot r} \cdot \sqrt{\frac{P_r \cdot Z_0}{\pi}} \tag{3}$$

r being the distance that separates the source of radiation at the measurement point of electric field intensity E_0 and Z_0 is the characteristic impedance in free space.

When the solar cell is illuminated with a multi-spectral white light and submitted to the action of an electromagnetic field, the continuity equation relative to excess minority carriers (electrons) density photo generated in the base region can be written as [2]:

$$\frac{\partial^2 \delta(x)}{\partial x^2} + \frac{L_E}{L_n^2} \cdot \frac{\partial \delta(x)}{\partial x} - \frac{\delta(x)}{L_n^{*2}} + \frac{G(x)}{D_n^*} = 0 \tag{4}$$

In this expression $L_E = \frac{\mu_n \cdot E_0 \cdot L_n^2}{D_n}$ is a coefficient that accounts for the migration phenomena within the solar cell base, μ_n is the electrons mobility, E_0 is the electric field intensity, L_n the electrons diffusion length, L_n^* the electrons diffusion length in presence of magnetic field. δ(x) and G(x) represent respectively carriers' density and the optic generation rate.

The term $\frac{L_E}{L_n^2} \cdot \frac{\partial \delta(x)}{\partial x}$ of the previous equation being a first derivative in relation with position x, is similar to a term of damping.

The electrons-holes pair's optic generation rate for a multi-spectral incident light is given by [7]

$$G(x) = \sum_{i=1}^{3} a_i \cdot e^{-b_i \cdot x} \tag{5}$$

Parameters a_i and b_i are coefficients deduced from modelling of the generation rate considered for over all the solar radiation spectrum under Air Mass 1, 5 standard conditions [8].

The excess minority carriers (electrons) density, solution of equation (9) is given by expression:

$$\delta(x) = e^{\alpha \cdot x} \cdot [A \cdot ch(\beta \cdot x) + B \cdot sh(\beta \cdot x)] - \sum_{i=1}^{3} K_i \cdot e^{-b_i \cdot x} \tag{6}$$

with $K_i = \dfrac{-a_i}{D_n^* \cdot \left[b_i^2 - \dfrac{1}{L_n^{*2}} - \dfrac{L_E \cdot b_i}{L_n^2}\right]}$, $\alpha = -\dfrac{L_E}{2 \cdot L_n^2}$ et $\beta = \dfrac{\left(L_E^2 + 4 \cdot L_n^2 \cdot \left[1 + (\mu_n \cdot B_0)^2\right]\right)^{1/2}}{2 \cdot L_n^2}$

Constants A and B are determined using the two boundary conditions below:

- At the junction emitter-base (x = 0)

$$\frac{\partial \delta(x)}{\partial x}\Big)_{x=0} = Sf \cdot \frac{\delta(0)}{D_n} \quad (7.a)$$

- At the rear side of the solar cell (x = H)

$$\frac{\partial \delta(x)}{\partial x}\Big)_{x=H} = -Sb \cdot \frac{\delta(H)}{D_n} \quad (7.b)$$

Sf is the sum of two contributions: Sf_0 which is the intrinsic junction recombination velocity induced by the shunt resistance and Sf_j which translates the current flow imposed by an external load and defines the operating point of the cell [9]:

$$Sf = Sf_0 + Sf_j \quad (8)$$

Sb is the effective back surface recombination velocity.

Table 1 gives information concerning the power of radiation of AM and FM radio antennas [6].

Table 1. Power of radiation of AM and FM radio antennas

Type of antenna	Power of radiation
FM antenna	50 W – 5 kW
AM antenna	500 W – 2 MW

The results of the previous work [2] showed that the solar cell conversion efficiency is much better as the distance that separates the solar cell of the antenna is weak. Consequently we will carry out the simulation by supposing that the solar cell is placed at 10 m from the electromagnetic waves transmitting antenna.

The table 2 below gives information concerning the antenna power of radiation and the electric and magnetic fields intensity used in the numeric simulation for a distance solar cell-antenna r =10 m.

Table 2. Electric and magnetic fields intensity versus power of radiation

Alphabetic letter	Power of radiation (W)	E_0 (V/m)	B_0 (T)
a	0	0	0
b	5000	38.7	$1.29.10^{-7}$
c	50000	122.4	$4.081.10^{-7}$
d	500000	387.1	$1.29.10^{-6}$
e	2000000	774.3	2.58110^{-6}

2.2. Photo-current density

While applying Fick's law at the solar cell junction, we get photocurrent density expression in presence of electromagnetic field:

$$Jph = q \cdot \left[D_n \cdot \frac{\partial \delta(x)}{\partial x})_{x=0} + \frac{\mu_n \cdot E_0}{1 + (\mu_n \cdot B_0)^2} \cdot \delta(0) \right] \tag{9}$$

where q is the electronic charge.

The photo-current density curves versus junction recombination velocity are plotted on **figure 2** for different values of electromagnetic field intensity or for different values of the power radiated by the antenna.

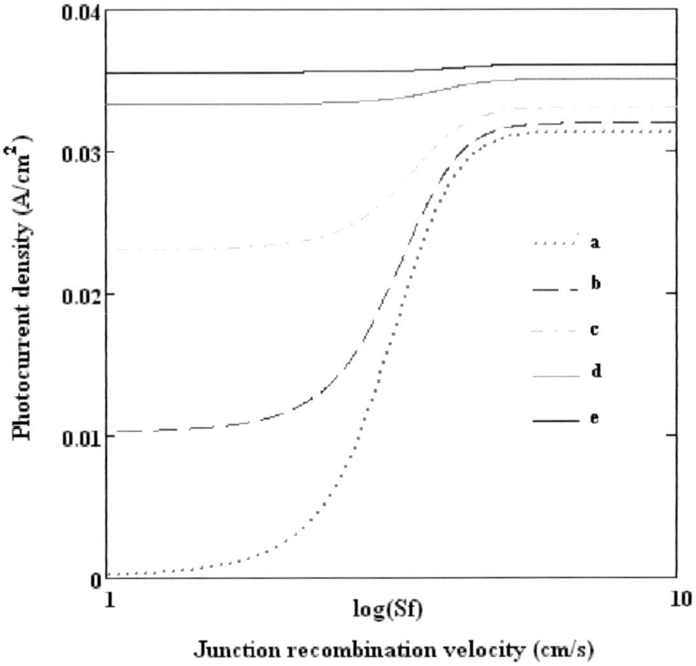

Figure 2. Photo-current density versus junction recombination velocity for different values of power of radiation (L=0.02 cm; H=0.03 cm; D=26 cm^2/s; μ_n=1350 $cm^2/V.s$)

Curves of figure 2 show that the short circuit photo-current and the leak current or diode current are increasing functions of the antenna power of radiation. One also notes that the open circuit photo-current or diode current is proportional to the antenna power of radiation and stretches toward the value of short circuit photo-current for large values of the antenna power of radiation.

Curves **(a)** and **(b)** are practically confused in short circuit situation while they are different one from the other in open circuit. Curve **(b)** represents the maximum power radiated by a FM radio antenna. We deduce that photo-current produced by the solar under the influence of AM radio waves is more important than the one produced by the FM radio waves.

On figure 3 below we plot photo-current density curves versus solar cell back surface recombination velocity for different values of electromagnetic field intensity or for different values of the power radiated by the antenna. We notice on figure 3, that for low values of back surface recombination velocity (Sb $\leq 10^2$ $cm.s^{-1}$), the photo-current density is maximal whatever the power of radiation. Then, it decreases with increase of back surface recombination velocity to reach an asymptotic minimal value for Sb $\geq 10^4$ $cm.s^{-1}$. Whatever is the part of the curves; photo-current density is an increasing function of the power of radiation.

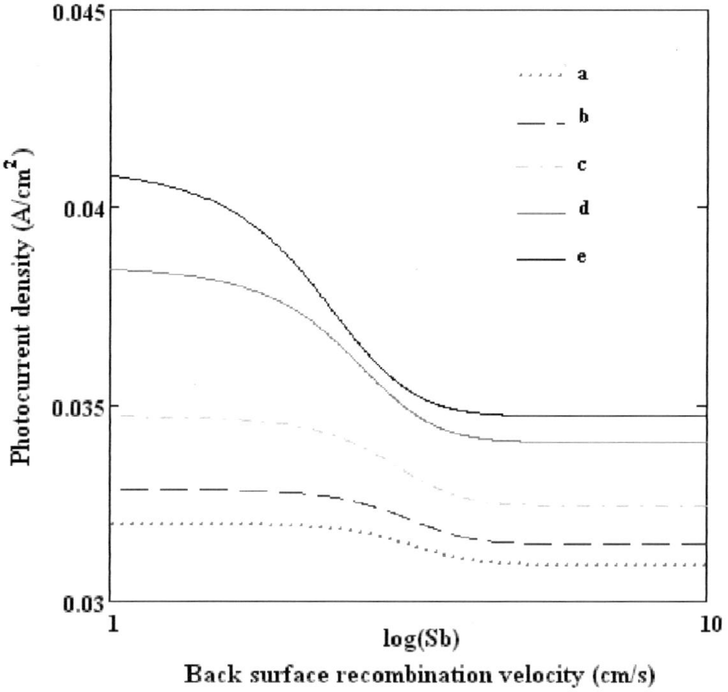

Figure 3. Photo-current density versus back surface recombination velocity for different radiation powers (L=0.02 cm; H=0.03 cm; D=26 cm^2/s; μ_n=1350 $cm^2/V.s$)

2.3. Recombination velocities

Photo-current density curves versus junction recombination velocity (**figure 2**) are asymptotic for large values of Sf ($Sf \geq 10^6$ $cm.s^{-1}$); their derivative with respect of Sf equals zero and leads to the back surface recombination velocity [3-5] from equation (10) below:

$$\frac{\partial Jph}{\partial Sf} = 0 \qquad (10)$$

Solving the equation (10) enables us to obtain back surface recombination velocity expression:

$$Sb = D_n \cdot \sum_{i=1}^{3} \frac{\beta \cdot b_i \cdot \left[ch(\beta \cdot H) - e^{-(b_i+\alpha)\cdot H} \right] + \left[\alpha \cdot (b_i + \alpha) - \beta^2 \right] sh(\beta \cdot H)}{\beta \cdot \left[ch(\beta \cdot H) - e^{-(b_i+\alpha)\cdot H} \right] - (\alpha + b_i) sh(\beta \cdot H)} \qquad (11)$$

Photo-current density curves versus back surface recombination velocity (figure 3) are also asymptotic for large values of Sb ($Sb \geq 10^4$ $cm.s^{-1}$); their derivative with respect of Sb equals zero and leads to the intrinsic junction recombination velocity [3-5] from equation (12) below:

$$\frac{\partial Jph}{\partial Sb} = 0 \qquad (12)$$

Resolution of equation (12) permits to obtain also intrinsic junction recombination velocity expression:

$$Sf_0 = D_n \cdot \sum_{i=1}^{3} \frac{e^{-(b_i+\alpha)\cdot H} \cdot \left[(\beta^2 - \alpha^2) sh(\beta \cdot H) + b_i (\beta \cdot ch(\beta \cdot H) - \alpha \cdot sh(\beta \cdot H)) \right] - \beta \cdot b_i}{\beta - e^{-(b_i+\alpha)\cdot H} \left[\beta \cdot ch(\beta \cdot H) + (b_i + \alpha) \cdot sh(\beta \cdot H) \right]} \qquad (13)$$

On figure 4 below, we plot back surface recombination velocity curves versus carriers' diffusion length for different values of the antenna power of radiation.

We notice that solar cell back surface recombination velocity is a decreasing function of the diffusion length and the antenna power of radiation.

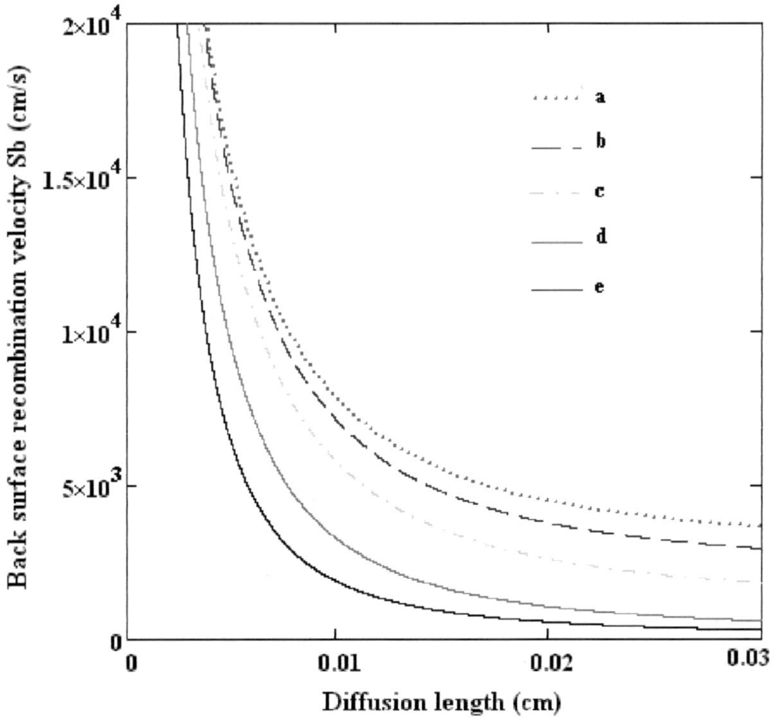

Figure 4. Back surface recombination velocity versus diffusion length for different values of power of radiation (H=0.03 cm; D=26 cm^2/s; μ_n=1350 $cm^2/V.s$)

Electric field intensity being also an increasing function of the antenna power of radiation and considering the previous work results [2], the electric field intensifies carriers' migration toward the solar cell junction. It has for consequence the decrease of back surface recombination velocity so a reduction of the losses of carriers in this solar cell zone and an improvement of the Back Surface Field effect.

The intensification of carriers' migration toward the solar cell junction is also at the origin of short circuit current increase whereas in open circuit situation it is at the origin of the open circuit current observed on the curves of figure 2.

Thus, solar cell back surface recombination velocity is weaker when the incident wave is an AM radio wave.

We plot on figure 5 below, solar cell intrinsic junction recombination velocity curves versus carriers' diffusion length for different values of the antenna power of radiation.

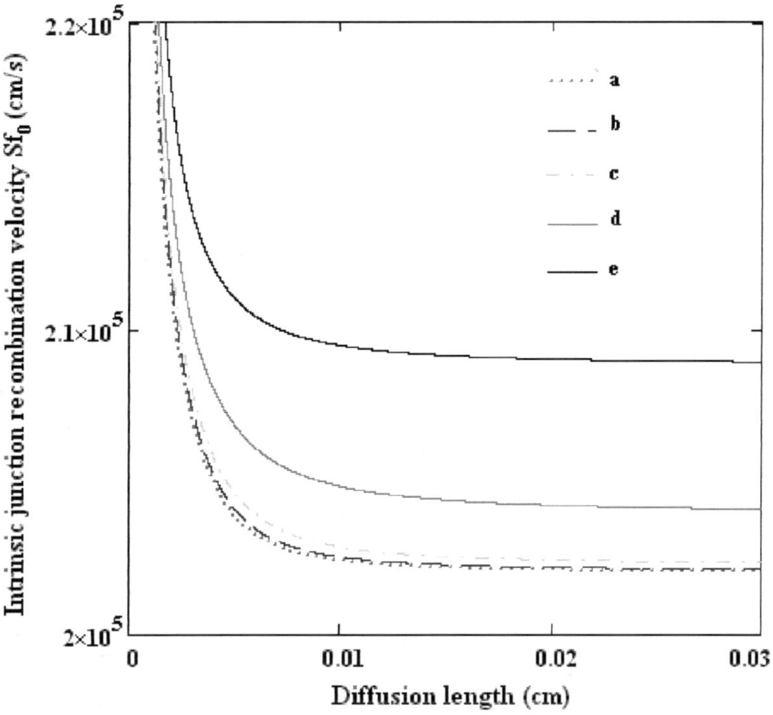

Figure 5. Intrinsic junction recombination velocity versus diffusion length for different values of power of radiation (H=0.03 cm; D=26 cm^2/s; μ_n=1350 $cm^2/V.s$)

The shape of the different curves shows that solar cell intrinsic junction recombination velocity is a decreasing function of diffusion length and an increasing function of the antenna power of radiation.

Indeed, the intrinsic junction recombination velocity is a microscopic parameter that accounts for the losses of carriers at the solar cell junction.

These losses of carriers constitute the leak current at the solar cell junction. Thus, the higher the intrinsic junction recombination velocity, the more the losses of carriers or even the leak current at the solar cell junction is enormous. This is in agreement with the curves of figure 2 where we observed a leak current in open circuit, which current is strongly tributary of the antenna power of radiation.

It is finally evident from the analysis of the different curves that the intrinsic junction recombination velocity induced by the AM radio waves is more important than the one induced by the FM radio waves.

2.4. Junction photo-voltage

The photo-voltage across the junction is expressed by means of excess minority carriers' density at the junction (x = 0) according to Boltzmann's relation:

$$Vph = \frac{k_B \cdot T}{q} \cdot \ln\left(\frac{\delta(0)}{n_0} + 1\right) \quad (14)$$

With $n_0 = n_i^{-2} \cdot N_B$ and $V_T = k_B \cdot T / q$ where n_i is the intrinsic carrier's density at thermal equilibrium, N_B the doping density of the base and V_T the thermal voltage.

We plot on figure 6 photo-voltage curves versus junction recombination velocity for different values of the antenna power of radiation

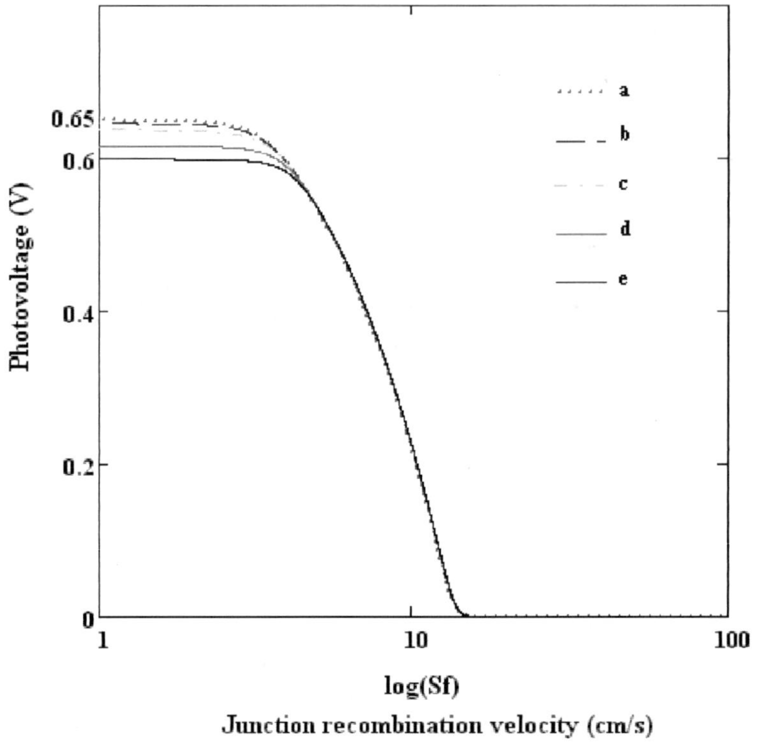

Figure 6. Photo-voltage versus junction recombination velocity for different values of power of radiation (L=0.02 cm; H=0.03 cm; D=26 cm^2/s; μ_n=1350 cm^2/V.s)

Curves of figure 6 show that open circuit voltage is a decreasing function of the power of radiation. Indeed, we have observed on the curves of figure 2 the existence of a current in open circuit named leak current due to the passage of some carriers through the solar cell junction

Thus, a decrease of the open circuit voltage translated a decrease of the carriers stocked at the junction. In fact, some carriers having crossed the junction produce a leak current that is quantified, from a microscopic view point, by the intrinsic junction recombination velocity.

A decrease of the open circuit voltage translates so an increase of the losses of carriers at the junction or even an increase of the intrinsic junction recombination velocity at the solar cell junction. The analysis of the different curves permits to say that the theoretical open circuit voltage produced by the solar cell under the influence of FM radio waves is more important than the one produced by the AM radio waves.

2.5. Solar cell electric power and conversion efficiency

The electric power delivered by the solar cell base to an external load circuit is expressed using the equation (14) below:

$$P(Sf) = Vph(Sf) \cdot Jph(Sf)_T \qquad (15)$$

With $Jph(Sf)_T = q \cdot Sf_j \cdot \delta(0)$

$Jph(Sf)_T$ is the photo-current that crosses the external load resistance.

Electric power variations versus junction recombination velocity for different values of the electromagnetic waves transmitting antenna power of radiation are plotted on figure 7.

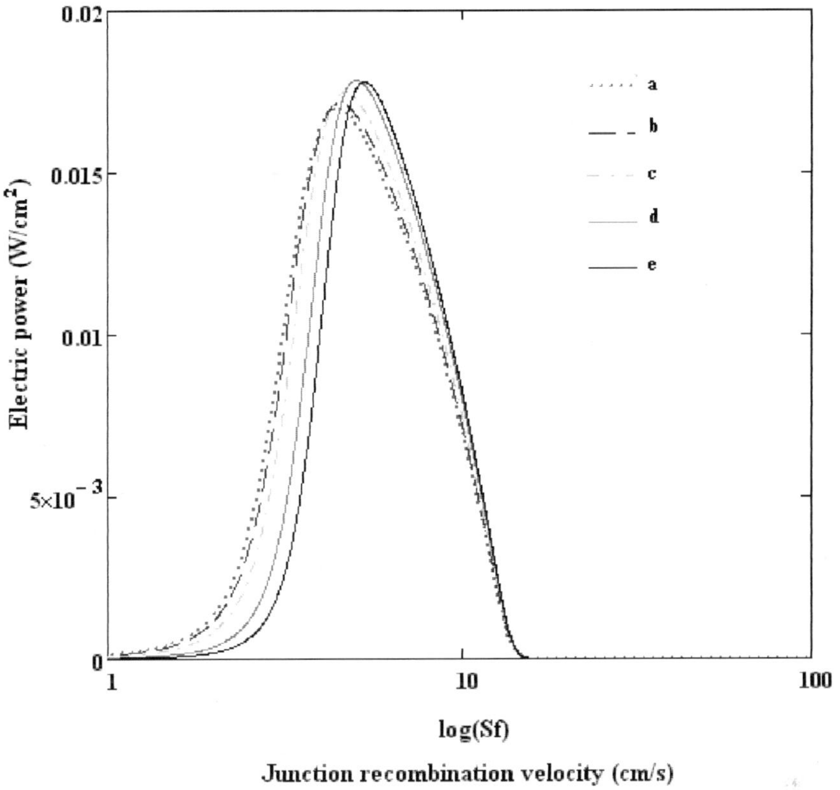

Figure 7. Electric power delivered by the solar cell versus junction recombination velocity for different values of power of radiation (L=0.02 cm; H=0.03 cm; D=26 cm^2/s; μ_n=1350 $cm^2/V.s$)

The analysis of the different curves shows that the electric power delivered by the solar cell base is null near the open circuit and the short circuit. It passes by a maximum that is located in an intermediate operating point. This operating point for which the electric power is maximal moves toward the short circuit (large values of Sf) when the electromagnetic field intensity increases or when the antenna power of radiation increases. The maximum electric power delivered by the solar cell to an external circuit permits to calculate the solar cell conversion efficiency.
Thus, this conversion efficiency is the report of the maximum electric power delivered by the solar cell to an external circuit and the power of incident light flux received by the solar cell.

$$\eta = \frac{P_{el}(Sf)_{max}}{P_{inc}} \quad (16)$$

P_{inc} is the power of the incident light flux and P_{inc} = 100 mW/cm^2 in Air Mass 1.5 standard conditions.

We determined the values of maximum electric power delivered by the solar cell to an external circuit and the values of junction recombination velocity corresponding, and then we calculated the solar cell conversion efficiency.

These results are consigned in the table 3. The maximum electric power delivered by the solar cell base to an external circuit increases with electromagnetic field intensity that is to say with the antenna power of radiation. The analysis of the results consigned in table 3 permits to say that the FM radio antennas that produce weak values of electromagnetic field have less influence on the solar cell than AM radio antennas. Besides, the solar cell conversion efficiency and the maximum electric power delivered by the solar cell to an external circuit evolve in the same sense.

Table 3. Maximum electric power delivered by the solar cell, corresponding junction recombination velocity and solar cell conversion efficiency for different power of radiation

Alphabetic letter	Power of radiation (W)	E_0 (V/m)	B_0 (T)	Sf (cm/s)	Pel_{max} (W/cm^2)	Efficiency η (%)
a	0 (absence antenna)	0	0	$2{,}884.10^4$	0,017000	17,000
b	5000	38,7	$1{,}29.10^{-7}$	$3{,}236.10^4$	0,017167	17,167
c	50000	122,4	$4{,}081.10^{-7}$	$5{,}129.10^4$	0,017486	17,486
d	500000	387,1	$1{,}29.10^{-6}$	$1{,}047.10^5$	0,017854	17,854
e	2000000	774,3	$2{,}581.10^{-6}$	$1{,}95.10^5$	0,017798	17,798

3. Conclusion

A theoretical study of electromagnetic waves power of radiation influence on silicon solar cell recombination parameters is presented. The study of photo current density shows that the short circuit photo-current and the open circuit photo-current (leak current or diode current) are increasing functions of the antenna power of radiation. The study of back surface recombination velocity shows that the electric field which is an increasing function of the antenna power of radiation intensifies carriers' migration toward the solar cell junction. The consequence of this migration phenomenon is the decrease of back surface recombination velocity, which means a reduction of the losses of carriers at the solar cell junction and an improvement of the Back Surface Field effect. The same migration phenomenon explains the existence of the open circuit current. As for the intrinsic junction recombination velocity which is a microscopic parameter that accounts for the losses of carriers at the solar cell junction, it is an increasing function of the antenna power of radiation. The decrease of the open circuit voltage, decreasing function of the power of radiation, translates an increase of the losses of carriers at the solar cell junction or even an increase of the intrinsic junction recombination velocity at the same junction. According to the analysis above, the short circuit current, the open circuit current and the intrinsic junction recombination velocity of the solar cell under the influence of AM radio waves are more important than that of the same solar cell under the influence of FM radio waves. The same analysis also put in evidence a decrease of solar cell open circuit voltage and a back surface recombination velocity with the increase of the antenna power of radiation. The numeric calculations put in evidence an increase of the maximum electric power and the solar cell conversion efficiency with the increase of the antenna power of radiation. These results permit to conclude that the influence of FM radio waves on the solar cell is lesser than that of AM radio waves.

References

[1] Zerbo I, Zoungrana M and Zougmoré F 2011 *4th IUPAP Int. Conf. of Women In Physics ICWIP Stellenbosch South Africa*

[2] Le Quang N, Rodot M, Nijs J, Ghannam M and Coppye J 1992 *Journal de Physique III* **2 (7)**1305-16

[3] Barro F I, Nanema E, Wereme A, Zougmore F and Sissoko G 2001 *Journal des Sciences* **1 (1)** 76-80

[4] Barro F I, Nanema E, Zougmore F, Wereme A, Ndiaye A L and Sissoko G 2003 *Journal des Sciences* **3 (1)**10-14

[5] Zerbo I, Koalaga Z, Barro F I, Zougmore F, Ndiaye A L , Diao A and Sissoko G 2004 *Journal des Sciences* **4 (2)** 42-46

[6] Ulrich Freyer 1994 *Nachrichten-Uebertragungstechnik* (Munich Carl Hanser)

[7] Furlan J and Amon S 1985 *Solid State Electronics* **28(12)**1241-43

[8] Mohammad S N 1987 *J. Appl. Phys.* **61(2)** 767-772

[9] Diallo H L, Maiga A S, Wereme A. and Sissoko G 2008 *European Phy. J.: Appl. Phys.* **42** 203-211

IOP Conf. Series: Materials Science and Engineering **29** (2012) 012020 doi:10.1088/1757-899X/29/1/012020

3D modelling of magnetic field and light concentration effects on a bifacial silicon solar cell illuminated by its rear side

M Zoungrana[1], I Zerbo[1], F Ouédraogo[2], B Zouma[1], F Zougmoré[1]

[1]Laboratoire des Matériaux et Environnement, Département de Physique, UFR-SEA, Université de Ouagadougou, BURKINA FASO.
[2]Institut Universitaire de Technologie (IUT), Université Polytechnique de Bobo Dioulasso, BURKINA FASO

E-mail: zmartial@univ-ouaga.bf, zmartial2003@yahoo.fr

Abstract. According to their installation places, the solar cells are often under the effect of the electro-magnetic field coming from outside sources as telecommunication antennas, radio antennas... These magnetic fields have a big influence on the distribution and mobility of charge carriers in the base of the solar cell and therefore on its efficiency. This work presents a tree-dimensional study of bifacial silicon solar cell under intense light concentration (more than 50 suns) and under a constant magnetic field. This approach is based on the resolution of the minority continuity equation, taking into account the distribution of the electric field in the bulk evaluated as a function of both majority and minority carrier densities. In this approach new analytical expression of carrier's density is established for rear side illumination. The effects of light concentration and magnetic field on excess minority carriers distribution, photocurrent and photovoltage in solar cell base are analysed.

1. Introduction

The solar cells are submitted to the influences of magnetic fields of various origins and intensities: terrestrial magnetic field (5.10^{-5} T), magnetic component of the electromagnetic wave coming from radios emitters, televisions emitters and telecommunications emitters [1]: (*AM antenna, power of radiation: 50 W - 5 kW,* $B \leq 1{,}29.10^{-7}$ *T; FM antenna, power of radiation: 500 W – 2 MW,* $4.10^{-8}\ T \leq B \leq 2{,}58.10^{-6}\ T$). However, many experimental studies proved magnetic field influences on the solar cells properties: magnetic field influence on carrier's mobility and photocurrent [2, 3], magnetic field influence on photovoltage and electron effective mass [4].

The impact of the magnetic field on the solar cells is the consequence of its influences on electrons trajectories, solar cells electronic parameters (Dn, Ln, Sb...), solar cells electric parameters (Jph, Vph) and carriers distribution in the volume of the base.

In this work, we propose a three dimensional study of bifacial silicon solar cell under intense light concentration (more than 50 suns) and under external variable magnetic field. The calculations were carried out in the case of a polycrystalline solar cell with columnar grain orientation under rear side illumination. This approach is based on the resolution of the continuity equation, taking into account the distribution of the electric field in the bulk due to the carrier gradient concentration. This electric field is evaluated as a function of both majority and minority carrier densities. In this approach, new analytical expressions of recombination and electric parameters, excess minority carrier density were established for rear side illumination through continuity equation resolution. The effects of magnetic

IOP Conf. Series: Materials Science and Engineering **29** (2012) 012020 doi:10.1088/1757-899X/29/1/012020

field and light intensity on diffusion length, diffusion coefficient, excess minority carrier generation and profile, photocurrent and photovoltage are then analysed.

The expressions of diffusion parameters, carriers density, back surface recombination velocity, photocurrent photovoltage are gotten by analytic resolution of the continuity equation. The numerical calculations and the curves drawn are gotten from programming on Mathcad software.

2. Model and assumption

2.1. Analytical formulation

We consider a bifacial polycrystalline Si-solar cell operating under a concentrated multispectral light. This kind of solar cell, because of its manufacture technique is made up of several grains of different sizes and forms separated by grain boundaries. These grain boundaries are recombination centers [5, 6, 7]. Macroscopic and microscopic parameters of the solar cell are deeply dependent on grain boundaries. These grains boundaries are either grains junctions or either local distortions caused by solar cell manufacturing foulness.

Each grain has a square section ($g_x = g_y$) and a thickness H as shown in Figure 1 below. The recombination planes are assumed to be thin surfaces between two adjacent grains and are located at $x = \pm\frac{g_x}{2}$ and $y = \pm\frac{g_y}{2}$, respectively, perpendicular to the x- and y-axis of cartesian coordinates (O;x; y; z) [8]. The grain boundaries are perpendicular to the junction and their effective recombination velocity S_{gb} is constant.

The cell is excited uniformly on the rear surface by an intense light concentration (more than 50 suns). The light penetrates the junction at the plane $z = H$. Because of the light intensity, carrier concentration in the base is not uniform. So, we take into account the electric field $E(z)$ due to the difference of carrier concentration on z axis [9]. To investigate the magnetic field influence on carrier behaviour and photocurrent, we applied a variable external magnetic field with the induction ***B*** parallel to the surface of the p-p$^+$ junction [2, 10, 11, 12].

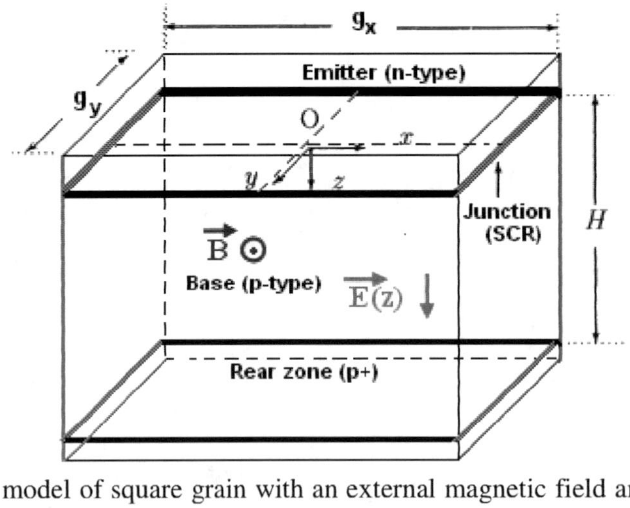

Figure 1. Theoretical model of square grain with an external magnetic field and the internal electric field.

To characterize the electric field analytical expression on every axis, we considered a silicon solar cell under intense light concentration [9]. We take into account the electric field $E(z)$ due to electrons concentration gradient in the base, and its expression is given by equation 1 [9] :

$$E(z)=\frac{D_p-D_n}{\mu_p+\mu_n}\cdot\frac{1}{\delta(x,y,z)}\frac{\partial\delta(x,y,z)}{\partial z} \qquad (1)$$

Introducing this expression of the electric field in the model of figure 1 and taking into account our assumptions, the expression of excess minority carrier distribution in the solar cell is:

$$\frac{\partial\delta(x,y,z)}{\partial t}=\frac{1}{e}\vec{\nabla}.\overrightarrow{J_n}+G(z)-R(z) \qquad (2)$$

$G(z)$ is the excess minority carrier generation at position z , and $R(z)$ the recombination rate at this position: for a rear side illumination [13, 14] $G(z)=C\cdot\sum_{i=1}^{3}a_i\cdot e^{-b_i(H-z)}$, a_i and b_i are deduced from modelling of the generation rate considered for over all the solar radiation spectrum [10, 13, 14]. Recombination rate expression is $R(z)=\frac{\delta(x,y,z)}{\tau_n}$ where τ_n is electron lifetime in the base, and $\delta(x,y,z)$ the excess minority carrier density in the base of our illuminated solar cell.

The expression of $\overrightarrow{J_n}$ is given by the equation of transportation phenomenon [2]:

$$\overrightarrow{J_n}=\overrightarrow{J_d}+\overrightarrow{J_{In}}+\overrightarrow{J_c}=e.D_n.\vec{\nabla}\delta(x,y,z)-\mu_n.\overrightarrow{J_n}\Lambda\vec{B}+e.\mu_n.\delta.(x,y,z).\vec{E} \qquad (3)$$

In this expression:

$\overrightarrow{J_d}$ represents carriers diffusion current on *x,y,z* axis, $\overrightarrow{J_{In}}$ represents the *y* axis magnetic field induced current and $\overrightarrow{J_c}$ represents conduction current on z axis, with $\vec{B}=B\vec{j}$, $B(x)=B(z)=0$ $\vec{E}=E(z)\vec{k}$.

When the solar cell rear side is illuminated, we suppose a uniform carrier distribution along *x and y* axis at any base depth. Therefore, conduction current doesn't exist according to these axes, so that:

$$\frac{\partial J_c}{\partial x}=\frac{\partial J_c}{\partial y}=0$$

On the basis of these assumptions, we deduce the expressions of electron density current $\overrightarrow{J_n}$ and the components of $\vec{\nabla}.\overrightarrow{J_n}$.

Taking into account the coordinates of $\overrightarrow{J_n}$ equation 2 can be written in permanent regime $(\frac{\partial\delta(x,y,z)}{\partial t}=0)$ as:

$$C_x\cdot\frac{\partial^2\delta(x,y,z)}{\partial x^2}+C_y\cdot\frac{\partial^2\delta(x,y,z)}{\partial y^2}+\frac{\partial^2\delta(x,y,z)}{\partial z^2}+\frac{G(z)}{D^*}-\frac{\delta(x,y,z)}{L^{*2}}=0 \qquad (4)$$

with: $C_x=\frac{D_n(\mu_p+\mu_n)}{D_n.\mu_p+D_p.\mu_n}$ and $C_y=\frac{D_n(\mu_p+\mu_n).(1+\mu_n^2B^2)}{D_n.\mu_p+D_p.\mu_n}$

$D^*=\frac{(D_n+\mu_n.A)}{1+\mu_n^2\cdot B^2}$, $L^{*2}=\tau_n\cdot D^*$ with $A=\frac{D_p-D_n}{\mu_p+\mu_n}$ and $R(z)=\frac{\delta(x,y,z)}{\tau_n}$

Equation 4 is the continuity equation of a solar cell under constant multispectral rear side illumination and with magnetic field. Our model lead to the news expressions of carrier diffusion length (L^*) and carrier diffusion coefficient (D^*), which depend on magnetic field, electrons and holes mobility, and diffusion coefficients (D_p, D_n).

We present on the following table, the values of diffusion parameters in our model for null value of magnetic field, comparatively to the literature values [15, 16].

Table 1. Comparative table values of electron diffusion parameters

	Diffusion length	*Diffusion coefficient*
Literature values	$L_n = 0.02$ cm	$D_n = 26$ cm².s⁻¹
Rear side illumination	$L^* = 0.0098$ cm	$D^* = 6.2$ cm².s⁻¹

We note that for a rear side illumination, diffusion parameters values decrease when taking into account excess minority carrier's gradient concentration electric field.

This behaviour implies an increase of the recombination in volume with taking into account excess minority carrier's gradient concentration electric field. It is therefore evident through these results that excess minority carrier's gradient concentration electric field has a very large influence on solar cell optoelectronic parameters.

Figures 2 and 3 give the carriers diffusion length (L^*) and carriers diffusion coefficient (D^*) profiles versus magnetic field.

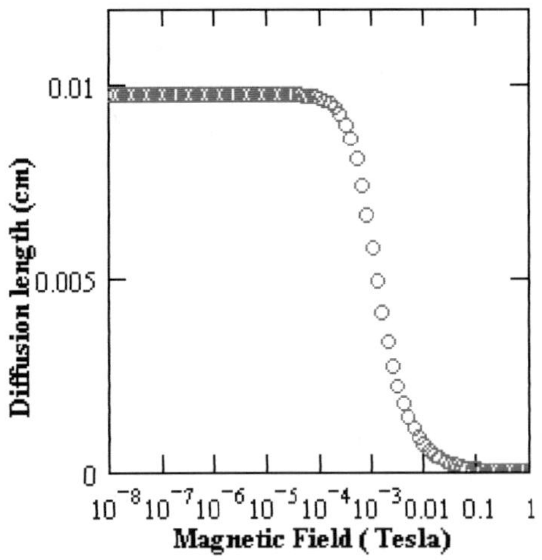

Figure 2. Effect of magnetic field on Diffusion length

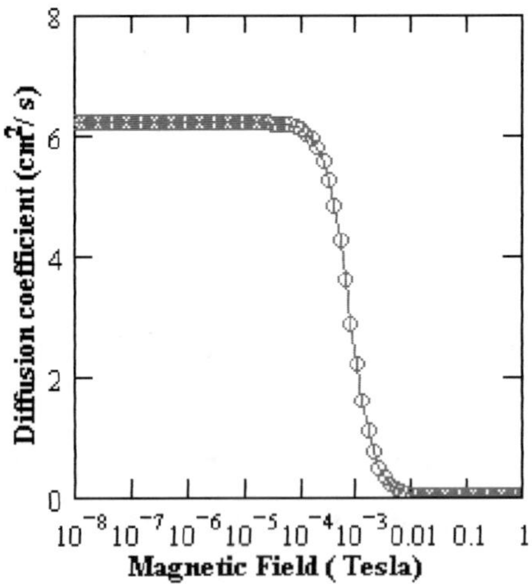

Figure 3. Effect of magnetic field on Diffusion coefficient.

In figures 2 and 3, we observe that diffusion length and diffusion coefficient are both decreasing with magnetic field increase. These figures also show that for very low magnetic field ($B<7.10^{-5}$ T), diffusion length and carrier diffusion coefficient are not magnetic field dependant. On the other hand, for values of magnetic field between 7.10^{-5} T and 0.12 T, carrier diffusion length and carrier diffusion coefficient values decrease before annulling themselves around 0.12 T.

2.2. Solution of the continuity equation

Equations 4 is a partial derivative differential equation for which the general shape of the solution can be written [8, 10, 12] as:

$$\delta(x,y,z)=\sum_j\sum_k Z_{j,k}(z)\cdot\cos(C_{xj}x)\cdot\cos(C_{yk}y),\ \text{with}\ C_{xj}=\frac{C_j}{C_x}\ \text{and}\ C_{yk}=\frac{C_k}{C_y} \quad (5)$$

The coefficients C_j and C_k are determinated by the boundary conditions on the grain boundaries for $x=\pm\frac{g_x}{2}$ *and* $y=\pm\frac{g_y}{2}$ respectively:

$$\left[\frac{\partial\delta(x,y,z)}{\partial x}\right]_{x=\pm\frac{g_x}{2}}=\pm\frac{Sgb}{D^*}\cdot\delta(\pm\frac{g_x}{2},y,z)\quad and\quad \left[\frac{\partial\delta(x,y,z)}{\partial y}\right]_{y=\pm\frac{g_y}{2}}=\pm\frac{Sgb}{D^*}\cdot\delta(x,\pm\frac{g_y}{2},z) \quad (6)$$

The coefficients C_{xj} and C_{yk} are given by the transcendental equations:

$$C_{xj}\tan(C_{xj}\cdot\frac{g_x}{2})=\frac{Sgb}{D^*}\quad and\quad C_{yk}\tan(C_{yk}\cdot\frac{g_y}{2})=\frac{Sgb}{D^*} \quad (7)$$

While taking into account that $\cos(C_{xj}x)$ (and $\cos(C_{yk}y)$) are orthogonal functions, the z dependence of $\delta(x,y,z)$ can be expressed for a rear side illumination as [8]

$$Z_{jk}(z)=A_{jk}\cosh\left(\frac{z}{L_{j,k}}\right)+B_{jk}\sinh\left(\frac{z}{L_{j,k}}\right)-\sum_{i=1}^{3}K_i e^{-bi\cdot(H-z)}\quad with\quad K_i=C\cdot\frac{a_i\cdot L_{j,k}^2}{D_{j,k}\cdot\left[b_i^2\cdot L_{j,k}^2-1\right]} \quad (8)$$

$$\frac{1}{L_{j,k}}=\left[C_j^2+C_k^2+\frac{1}{L_n^2}\right]\quad and\quad \frac{1}{D_{j,k}}=\frac{16\cdot\sin(C_{xj}\cdot\frac{g_x}{2})\cdot\sin(C_{yk}\cdot\frac{g_y}{2})}{D^*\cdot\left[\sin(C_{xj}\cdot g_x)+C_{xj}\cdot g_x\right]\cdot\left[\sin(C_{yk}\cdot g_y)+C_{yk}\cdot g_y\right]}$$

The expressions of the coefficients A_{jk} and B_{jk} are obtained by solving equation 8, with the following boundary conditions (equation 9) at the two edges of the base region [5,8]

- the junction interface at $z=0$:

$$S_f=\frac{D^*}{\delta(x,y,0)}\cdot\left[\frac{\partial\delta(x,y,z)}{\partial z}\right]_{z=0} \quad (9)$$

- the rear side at $z=H$

$$S_b=-\frac{D^*}{\delta(x,y,H)}\cdot\left[\frac{\partial\delta(x,y,z)}{\partial z}\right]_{z=H}$$

This lead to:

$$A_{jk}=\sum_{i=1}^{3}K_i\cdot\frac{\frac{1}{L_{2j,k}}\cdot\left(\frac{S_b}{D^*}+bi\right)-e^{-b_iH}\cdot\beta_{j,k}\cdot\left(bi-\frac{S_f}{D^*}\right)}{\frac{S_f}{D^*}\cdot\beta_{j,k}+\frac{1}{L_{j,k}}\cdot\alpha_{j,k}}\quad \text{and}$$

$$(10)$$

$$B_{jk}\sum_{i=1}^{3}K_i\cdot\frac{\alpha_{j,k}\cdot e^{-b_iH}\cdot\left(bi-\frac{S_f}{D^*}\right)+\frac{S_f}{D^*}\cdot\left(\frac{S_b}{D^*}+bi\right)}{\frac{S_f}{D^*}\cdot\beta_{j,k}+\frac{1}{L_{j,k}}\cdot\alpha_{j,k}}\quad with:$$

$$\alpha_{j,k} = \frac{1}{L_{j,k}} \cdot \sinh\left(\frac{H}{L_{j,k}}\right) + \frac{S_b}{D^*} \cdot \cosh\left(\frac{H}{L_{j,k}}\right) \quad \text{and} \quad \beta_{j,k} = \frac{1}{L_{j,k}} \cdot \cosh\left(\frac{H}{L_{j,k}}\right) + \frac{S_b}{D^*} \cdot \sinh\left(\frac{H}{L_{j,k}}\right)$$

3. Effects of magnetic field and light concentration on back surface recombination velocity S_b

We can get the expression of back surface recombination velocity from the resolution of the equation (11) below [17, 18]:

$$\left[\frac{\partial J_{ph}}{\partial S_f}\right]_{S_f \to \infty} = 0 \qquad (11)$$

Its expression is given by the equation (12) below:

$$S_b = D^* \frac{\sum_j \sum_k \frac{R_{jk}}{L_{j,k}^2} \sum_{i=1}^{3} K_i \left[b_i e^{-b_i H} \cosh\left(\frac{H}{L_{j,k}}\right) - b_i - \frac{1}{L_{j,k}} e^{-b_i H} \sinh\left(\frac{H}{L_{j,k}}\right)\right]}{\sum_j \sum_k \frac{R_{jk}}{L_{j,k}^2} \sum_{i=1}^{3} K_i \left[e^{-b_i H} \cosh\left(\frac{H}{L_{j,k}}\right) - b_i L_{jk} e^{-b_i H} \sinh\left(\frac{H}{L_{j,k}}\right) + 1\right]} \qquad (12)$$

The figures 4 and 5 below illustrate the influence of the magnetic field and light intensity on back surface recombination velocity. These curves have been drawn with the software Mathcad in semi-logarithmic scale.

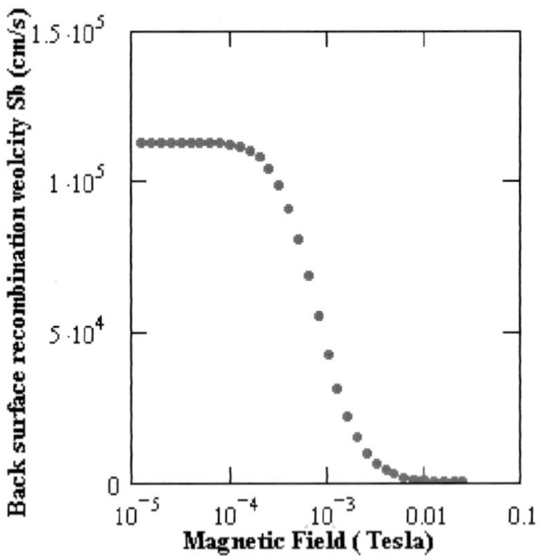

Figure 4. Effect of magnetic field on back surface recombination velocity: $g_x=g_y=3.10^{-3}cm$, $Sgb=10^2 cm.s^{-1}$, $C=200$ suns, $H = 0.03cm$.

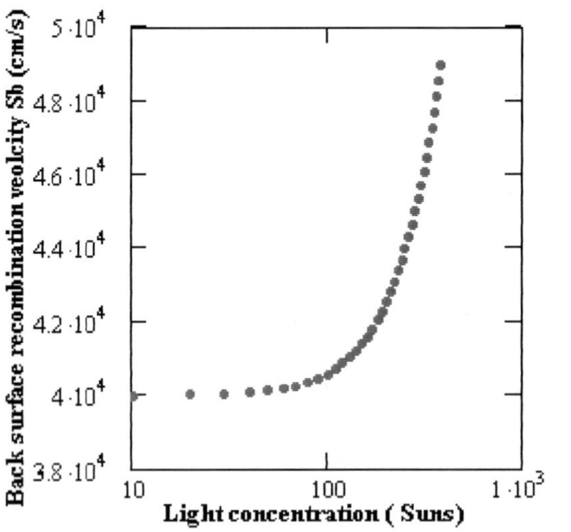

Figure 5. Effect of light concentration on back surface recombination velocity: $g_x=g_y=3.10^{-3}cm$, $Sgb=10^2 cm.s^{-1}$, $C=200$ suns, $H = 0.03cm$.

Under the effect of magnetic field (figure 4), the curve of back surface recombination velocity presents three parts:

- A first part where the magnetic field doesn't have any effect on back surface recombination velocity ($B < 9.10^{-5}T$): This result characterizes the weak influence of magnetic field on carrier's mobility for its weak values. It also justifies the results of the paragraph 2.1 where, it appears that the weak values of magnetic field don't have any influence on diffusion parameters.
- A second part where the back surface recombination velocity decreases with the increase of

magnetic field: In this part, the electron rays curvature that are inversely proportional to the magnetic field cut down very appreciably with the increase of magnetic field. The strength of Lorentz helping, a part of the electrons photogenerated in the rear zone base is deviated towards the lateral surfaces where they undergo some grain boundary recombination velocity. This phenomenon reduces the possibilities of electrons recombination in the rear zone of the base and is especially intense at the big values of magnetic field.

- A third part where the back surface recombination velocity cancels itself out ($B > 8.10^{-3}$ T): This part characterizes critical values of magnetic field for which, almost all the electrons photogenerated in the rear zone are deviated toward the lateral surfaces for recombination.

In the same way we observe on figure 5 that the back surface recombination velocity presents a null gradient for illumination of a few score of suns (< 50 suns). However, for more intense illumination, back surface recombination velocity increases with the increase of light intensity.

For a rear side illumination, the maximum of light absorption is made in the rear zone of the base. Therefore, the number of carriers photogenerated in this zone increases with light intensity. However, since the density of the traps doesn't vary in this zone, the increase of the number of carriers photogenerated comes ineluctably with an increase of the number of carriers likely to be trapped, that explains the increase of back surface recombination velocity for important illumination level.

4. Effects of magnetic field and light concentration on carrier density

4.1. Effect of magnetic field and light concentration on electron distribution close to the junction

The figures 6 and 7 illustrate the influence of the magnetic field and light intensity on the quantity of carriers photogenerated in the volume of the base for a rear side illumination. These figures illustrate the distribution of the carriers on a plan (2x, 2y) at the height z of base volume. The depth of the base where we study the photogeneration is the region close to the junction (z =10^{-4}cm), and the three applied values of the magnetic field are: $B= 5.10^{-5}\,T$, $B= 20.10^{-5}\,T$, $B= 50.10^{-5}\,T$.

One observes on the figure 6 bellow that whatever the value of the applied magnetic field, the concentration of the carriers to the surface (2x, 2y) increases from grain sides toward the center of the surface. This phenomenon is the consequence of grain boundary recombination, whose effects are more intense on grains junction surfaces.

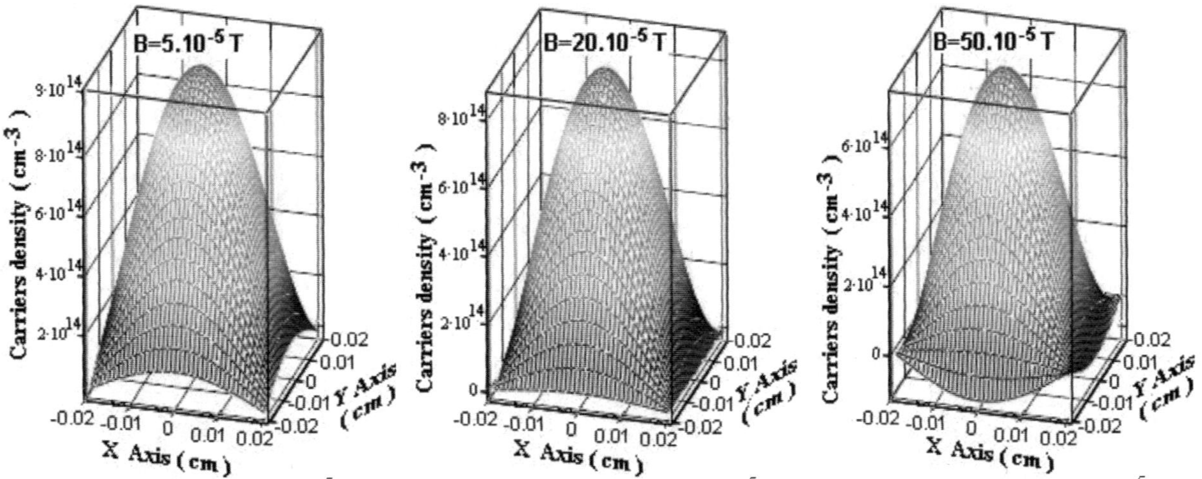

Magnetic field: $B = 5.10^{-5}$ T *Magnetic field: $B = 20.10^{-5}$ T* *Magnetic field: $B = 50.10^{-5}$ T*

Figure 6. Effect of magnetic field on carrier distribution in the region close to the junction: $S_f=10^1cm.s^{-1}$, $S_b=10^4cm.s^{-1}$, $Sgb=10^2cm.s^{-1}$, $C=200$ suns, $z=10^{-4}cm$.

We also observe on this figure 6 that when the solar cell is illuminated by the rear side the concentration of the carriers in the region close to the junction decreases with magnetic field increase. This phenomenon is explained not only by the decrease of the diffusion parameters with the magnetic field increase observed above, but also by the consequence of Hall effect which drags carriers deviation towards grain lateral surfaces to boundaries for recombination.

On the other hand, we observe on figure 7 below that for the solar cell rear side illumination, carriers presence in the region close to the junction increases with the increase of the light concentration. This phenomenon is the consequence of the increase not only of the depth of light penetration but also of the proportion of light absorbed in base volume with the increase of light intensity. These two phenomena contribute to the increase of the number of carriers photogenerated in base volume and beyond the number of carriers capable to reach the junction.

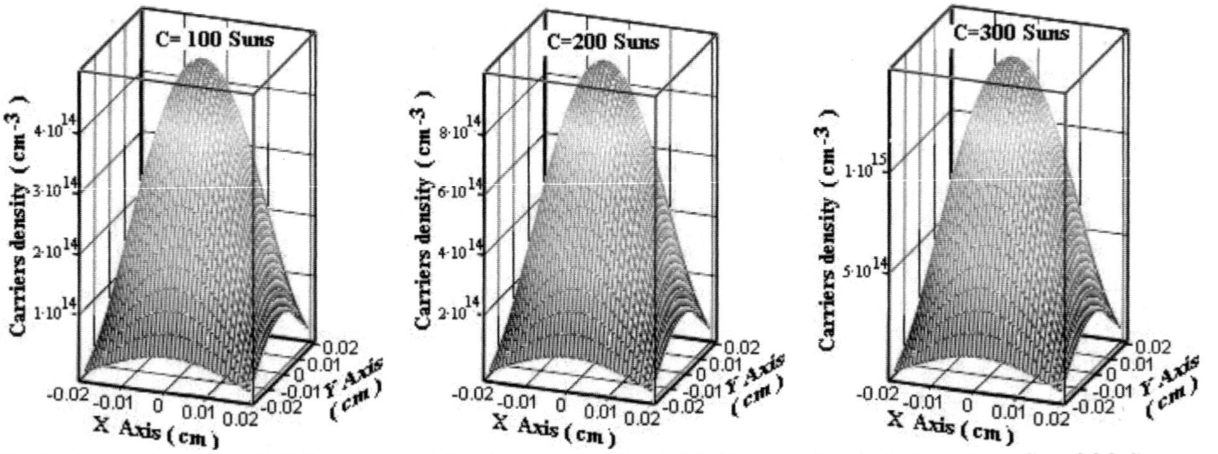

Light intensity : C =100 Suns *Light intensity : C = 200 Suns* *Light intensity : C = 300 Suns*

Figure 7. Effect of light concentration on carrier distribution in the region close to the junction: $S_f=10^1 cm.s^{-1}$, $S_b=10^4 cm.s^{-1}$, $Sgb=10^2 cm.s^{-1}$, $B=5.10^{-5}T$, $z=10^{-4}cm$.

4.2. Effect of magnetic field and light concentration on the profile of carrier density

We study in this paragraph, the influences of magnetic field and light concentration on the profile of carriers' density for the solar cell rear side illumination.

We observe on the figures 8 and 9 below that with respect of base depth, the gradient of carrier density is positive. We also observe a very strong generation of carriers in the rear zone of the base. The positive nature of carriers gradient means that for a rear side illumination, all the carriers photogenerated in the base, if they are not recombined, can flow into the junction and participate to the photocurrent. The strong photogeneration in rear zone is the consequence of the fact that for a rear side illumination, the maximum of energy is accumulated in rear zone of the base.

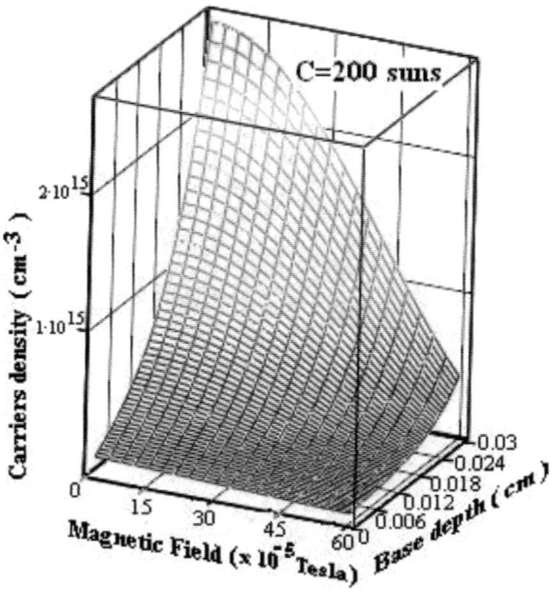

Figure 8. Effect of magnetic field on excess minority carriers profile: $g_x=g_y=3.10^{-3}cm$, $Sgb=10^2cm.s^{-1}$, $S_f=10^1cm.s^{-1}$, $S_b=10^4cm.s^{-1}$, $H=0.03cm$.

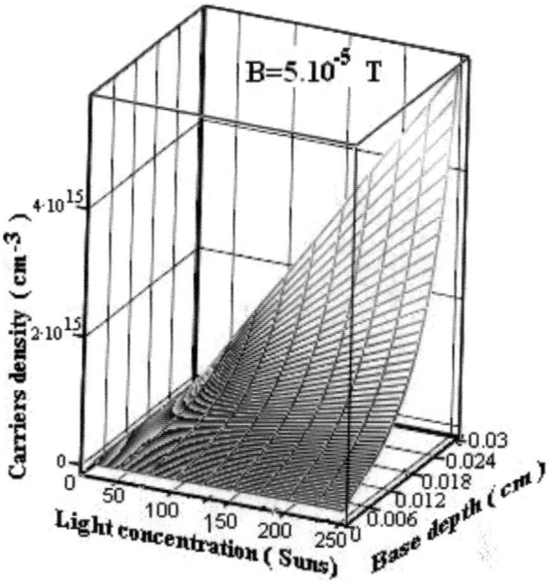

Figure 9. Effect of light concentration on excess minority carriers profile: $g_x=g_y=3.10^{-3}cm$, $Sgb=10^2cm.s^{-1}$, $S_f=10^1cm.s^{-1}$, $S_b=10^4cm.s^{-1}$, $H=0.03cm$.

We observe on figure 8 above that the curves of carrier density decrease strongly with the increase of magnetic field. This phenomenon is more sensitive in rear zone of the base. This phenomenon is again the consequence of Hall effect that causes carriers deviation towards grain lateral surfaces that are big recombination centers.

We observe on the other hand on figure 9 that for a rear side illumination, the density of carrier photogenerated increases very quickly with the increase of light concentration.

For this illumination mode, the maximum of energy being absorbed in rear zone of the base, the increase of illumination intensity provokes more photogeneration in rear zone. Therefore, the number of carriers that escape the recombination in rear zone increases and this fact increases the presence of carriers in base volume.

5. Effects of magnetic field and light concentration on photocurrent and photovoltage

5.1. Effect of magnetic field and light concentration on photocurrent

The general expression of photocurrent density is given by J Dugas [8] according to the following relation

$$J_{ph} = \frac{q \cdot D^*}{g_x \cdot g_y} \cdot \int_{\frac{-g_x}{2}}^{\frac{g_x}{2}} \int_{\frac{-g_y}{2}}^{\frac{g_y}{2}} \left[\frac{\partial \delta(x,y,z)}{\partial z} \right]_{z=0} dxdy \qquad (13)$$

with

$$\delta(x,y,z) = \sum_j \sum_k Z_{j,k}(z) \cdot \cos(C_{xj}x) \cdot \cos(C_{yk}y)$$

If we replace $\delta(x,y,z)$ and the equations (8) and (10) in the equation (13) and we integrate, we get the expression of the photocurrent:

$$J_{ph} = q \cdot D * \cdot \sum_j \sum_k R_{jk} \cdot \left[\frac{B_{jk}}{L_{jk}} + \sum_{i=1}^{3} K_i b_i \cdot e^{-b_i H} \right] \quad (14)$$

with

$$R_{jk} = \frac{\left(1 + \mu_n^2 \cdot B^2\right) \cdot 4\sin\left(C_{xj} \cdot \frac{g_x}{2}\right) \cdot \sin\left(C_{yk} \cdot \frac{g_y}{2}\right)}{g_y \cdot g_x \cdot C_j \cdot C_k}$$

In figures 10 and 11 below are illustrated the influences of magnetic effect and light intensity on electrons photocurrent plotted versus variation junction recombination velocity.

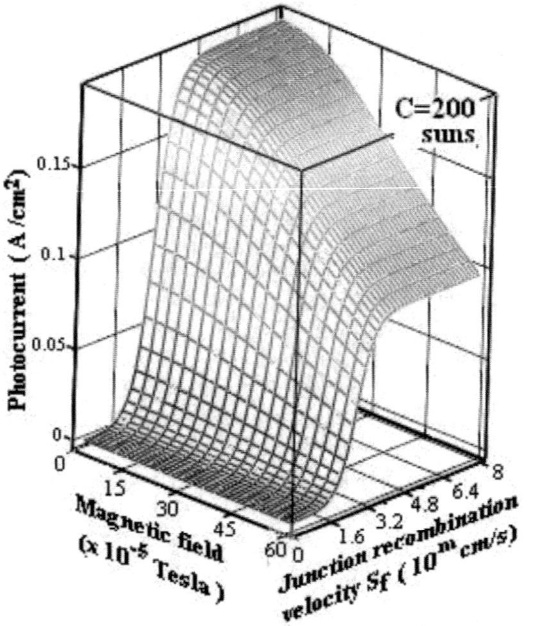

Figure 10. Effect of magnetic field and junction recombination velocity on photocurrent: $g_x=g_y=3.10^{-3}cm$, $Sgb=10^2 cm.s^{-1}$, $S_b=10^4 cm.s^{-1}$.

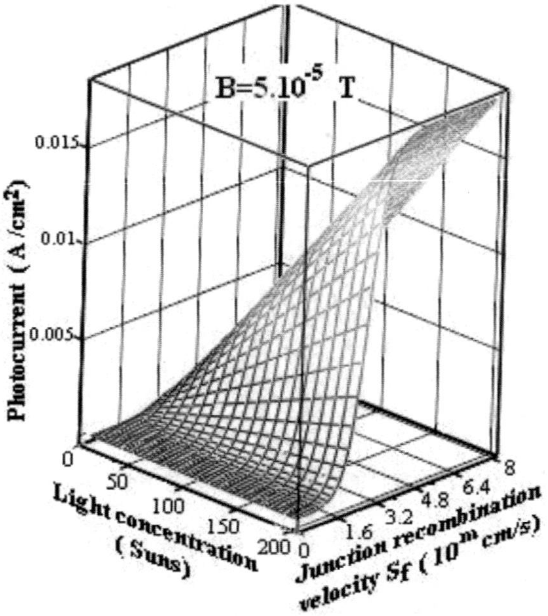

Figure 11. Effect of light concentration and junction recombination velocity on photocurrent: $g_x=g_y=3.10^{-3}cm$, $Sgb=10^2 cm.s^{-1}$, $S_b=10^4 cm.s^{-1}$.

We observe on these figures that photocurrent density decreases with the increase of the magnetic field and it increases with the intensification of light concentration. One also notices that the influence of magnetic field and light intensity on the photocurrent is more pronounced close to short circuit ($S_f>10^6$ cm.s^{-1}).

The decrease of the photocurrent with magnetic field increase characterizes a decrease of the number of carrier that crosses the junction with the increase of magnetic field. This phenomenon confirms the fall of carrier density profile with the increase of magnetic field observed previously.
In the same way, the increase of photocurrent density with light intensification confirms also the above results in paragraph 4 that highlights an increase of the photogeneration in the volume of the base with the increase of light concentration.

5.2. Effect of magnetic field and light concentration on photovoltage

The general expression of the photovoltage is given by Boltzmann law in three dimensional approximation hypothesis by [8]:

$$V_{ph} = V_T . \ln\left[1 + \frac{N_B}{n_i^2}\int_{-\frac{g_x}{2}}^{\frac{g_x}{2}}\int_{-\frac{g_y}{2}}^{\frac{g_y}{2}} \delta(x,y,0)dxdy\right] \quad (15)$$

In this expression V_T represents the thermal voltage: $V_T = k_B.T/q$ ($V_T = 26\ mV$ à $T=300\ K$), and n_o the represents electron density in thermodynamic equilibrium: $n_o = n_i^2/N_B$.
n_i represents the electron intrinsic concentration ($n_i = 10^{10}\ cm^{-3}$) for silicon. N_B represents the base doping level ($N_B = 10^{16}\ cm^{-3}$) and k_B the constant of Boltzmann.

The resolution of equation 15 lead to the expression of phototension:

$$V_{ph} = V_T . \ln\left[1 + \frac{N_B}{n_i^2}\cdot\sum_j\sum_k R_{jk}\cdot\left[A_{jk} - \sum_{i=1}^{3} K_i \cdot \exp(-b_i H)\right]\right] \quad (16)$$

In figures 12 et 13 below are illustrated the influence of magnetic field effect and light intensity on electrons photovoltage plotted versus variation junction recombination velocity.

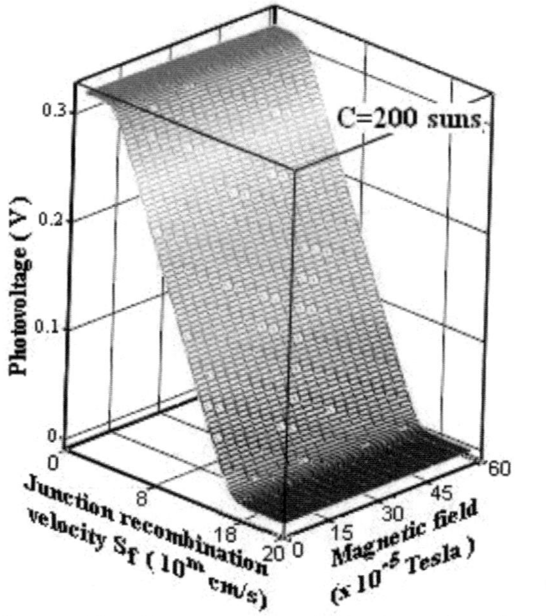

Figure 12. Effect of magnetic field and junction recombination velocity on photovoltage: $g_x=g_y=3.10^{-3}cm$, $Sgb=10^2 cm.s^{-1}$, $S_b=10^4 cm.s^{-1}$.

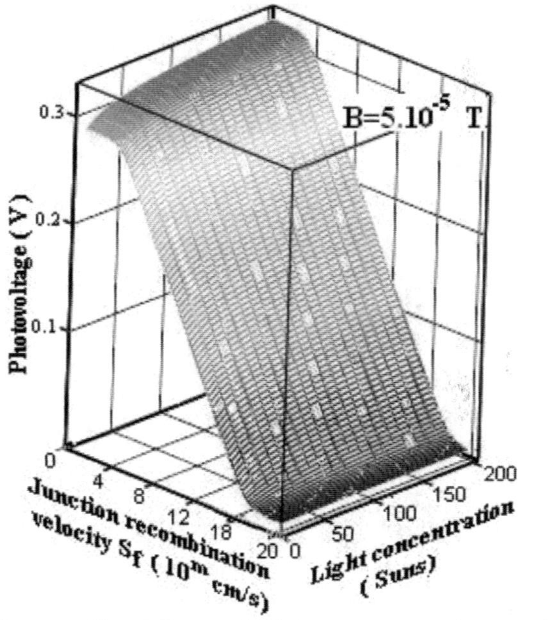

Figure 13. Effect of light concentration and junction recombination velocity on photovoltage: $g_x=g_y=3.10^{-3}cm$, $Sgb=10^2 cm.s^{-1}$, $S_b=10^4 cm.s^{-1}$.

It appears on these figures that the magnetic field and the light intensity don't have an influence on the short circuit voltage. On the other hand, to the close of the open circuit, photovoltage increases with the increase of light concentration and decreases with the increase of magnetic field.

These results also confirm the above ones (paragraph 4 and 5.1) those highlight the influences of magnetic field and the light intensity on the photogeneration in the volume of the base and on photocurrent.

6. Conclusion

This study in modelling of magnetic effect, electric effect and light intensity effect on the silicon solar cell leads us to new expressions of continuity equation, diffusion length, diffusion coefficient, back surface recombination velocity. The curves of carrier diffusion length and carrier diffusion coefficient decrease with magnetic field increase. The effects of magnetic field on carrier diffusion length (L^*), carrier diffusion coefficient (D^*) are more sensitive for magnetic field values greater than 7.10^{-5}T ($B>7.10^{-5}$ T). We also observed that the fact of taking into account carrier gradient concentration electric field lowers diffusion coefficient values under the non approximated electric field ones.

It appears that, magnetic field has a big influence on carrier generation and distribution in the bulk of the base. Back surface recombination velocity, carrier distribution, carrier density profile, photocurrent and photovoltage decrease with the increase of magnetic effect.
In the same way, we have shown that these parameters increase with the increase of light concentration.

References

[1] Freyer U 1994 *Nachrichten-Uebertragungstechnik, Edition Carl Hanser (Munich)*
[2] Betser Y, Ritter D, Bahir G, Cohen S and Sperling J 1995 *App. Phys. Lett.* **67(13)** 1883
[3] Vardanyan R R 1998 *2nd World Conference and Exhibition Photovoltaic Solar Energy Conversion* 191
[4] Erel S 2002 *Solar Energy Materials and Solar cells* **71** 273
[5] Ba B, Kane M and Sarr J 2003 *Solar Energy Materials & Solar Cells* **80** 143
[6] Rocher A 1987 *Revue de Physique Appliqué* **22** 591
[7] Priester L 1980 *Revue de Physique Appliqué* **15** 789
[8] Dugas J 1994 *Solar Energy Materials and Solar cells* **32** 71
[9] Pelanchon F, Sudre C and Moreau Y 1992 *11th European Photovoltaic Solar Energy Conference Montreux* **11** 265
[10] Samb M L, Zoungrana M, Sam R, Dione M M, Deme M M and Sissoko G 2010 *Journal des Sciences* **10(4)** 23
[11] Touré F, Zoungrana M, Sam R, Diop M T D, Barro FI and Sissoko G 2010 *Journal des Sciences* **10(4)** 16
[12] Zouma B, Maiga A S, Dieng M, Zougmore F and Sissoko G 2009 Global *Journal of Pure and Applied Science* **15(1)** 117
[13] Barro F I, Zerbo I, Lemrabott O H, Zougmore F, and Sissoko G 2001 17th *European Photovoltaic Solar Energy Conf. p* 368
[14] Mohammad S N 1987 *Journal of Applied Physics* **61(2)** 767
[15] Equer B 1993 *Energie Solaire Photovoltaique*, **Vol 1** Collection *Ellipses*
[16] Diallo H L , Maiga A S, Wereme A and SissokoG 2008 *European Physical Journal Applied Physics* **42** 203.
[17] Barro F I, Nanema E, Wereme A, Zougmoré F and Sissoko G 2001 *Journal des Sciences* **1(1)** 76
[18] Barro F I, Mbodji S, Ndiaye A L, Zerbo I, Madougou S, Zougmore F, and Sissoko G 2004 19th *European Photovoltaic Solar Energy Conf. and Exhibition*

Institute of Physics
Dirac House, Temple Back
Bristol BS1 6BE UK

CURRAN ASSOCIATES INC
proceedings.com